Genetic Diversity Assessment and Marker-Assisted Selection in Crops

Genetic Diversity Assessment and Marker-Assisted Selection in Crops

Editors

Francesco Mercati
Francesco Sunseri

MDPI • Basel • Beijing • Wuhan • Barcelona • Belgrade • Manchester • Tokyo • Cluj • Tianjin

Editors
Francesco Mercati
National Research Council
Italy

Francesco Sunseri
University Mediterranea of
Reggio Calabria
Italy

Editorial Office
MDPI
St. Alban-Anlage 66
4052 Basel, Switzerland

This is a reprint of articles from the Special Issue published online in the open access journal *Genes* (ISSN 2073-4425) (available at: https://www.mdpi.com/journal/genes/special_issues/marker-assisted_selection).

For citation purposes, cite each article independently as indicated on the article page online and as indicated below:

LastName, A.A.; LastName, B.B.; LastName, C.C. Article Title. *Journal Name* **Year**, *Volume Number*, Page Range.

ISBN 978-3-0365-0854-2 (Hbk)
ISBN 978-3-0365-0855-9 (PDF)

Contents

About the Editors

Francesco Mercati was born in Milan, received a degree in Biological Science in 2003, and is a researcher at the Institute of Biosciences and BioResources (IBBR) UOS Palermo, CNR—National Research Council of Italy (e-mail: francesco.mercati@ibbr.cnr.it). He obtained his PhD in Biology Applied to Agri-Food and Forest systems at the University Mediterranea of Reggio Calabria. He has also been a visiting scientist at the Department of Plant Biology, University of Georgia (USA). The main research activities of Dr. Mercati are focused on biodiversity, marker assisted selection, and the study of genes associated to trait of interest. He is the author of approximately 80 scientific publications, the most important published in ISIS and Scopus indexed journals, available at https: //www.cnr.it/people/francesco.mercati.

Research Interest:

Genetic structure of plant biodiversity. Next-generation sequencing (NGS) technologies and their implications for plant genetics, DNA barcode, and structure population. Genotyping-by-Sequencing (GBS). Molecular Assisted Selection (MAS). Olive and grape cultivars (cultivated and wild populations) from Sicily genotyping by using different molecular markers (ISSR, AFLP, SSR). Molecular markers investigation for somatic protoplast fusion hybrids and cybrids.

Francesco Sunseri was born in Palermo, received a degree in Agricultural Science in 1983, and received a permanent position at the University Mediterranea as Associate Professor on Plant Genetics and Breeding in 2008. He is the leader of many projects focused on plant biodiversity conservation and exploitation, as well as molecular assisted selection on horticultural crops such as asparagus, tomato, and eggplant. A member of the Italian Society of Plant Genetics (S.I.G.A.) and the International Society Horticultural Science (I.S.H.S.). Associate Editor of BMC Plant Biology and Genes (MDPI) and occasional reviewer for several scientific journals devoted to Plant Genetics and Horticultural Science. Author of approximately 80 scientific publications indexed in ISI and Scopus databases.

Research Interest:

Genetic structure of plant biodiversity. Genomics and Transcriptomics. Genomic Wide Association Selection and Mapping (GWAS). Genotyping-by-Sequencing (GBS). Molecular Assisted Selection (MAS). QTL analysis and cloning. Plant abiotic stress with particular interest in Nitrogen Use Efficiency (NUE).

 *genes*

Editorial

Genetic Diversity Assessment and Marker-Assisted Selection in Crops

Francesco Mercati [1,*] **and Francesco Sunseri** [2,*]

[1]   CNR—Consiglio Nazionale delle Ricerche, Istituto di Bioscienze e Biorisorse (IBBR), Corso Calatafimi 414, 90129 Palermo, Italy

[2]   Dipartimento di Agraria, Università Mediterranea degli Studi di Reggio Calabria, Loc. Feo di Vito, 89124 Reggio Calabria, Italy

*   Correspondence: francesco.mercati@ibbr.cnr.it (F.M.); francesco.sunseri@unirc.it (F.S.)

Received: 16 November 2020; Accepted: 7 December 2020; Published: 9 December 2020

Global warming is negatively impacting on crop yield and Earth's climate changes can bring possible negative effects on the growth and reproductive success of crops. Therefore, the exploitation of biodiversity is essential to select more resilient genotypes employable in more sustainable cropping systems.

The assessment of genetic diversity from the major crops and their wild relatives together with its exploitation have been always among the main challenges for plant breeding, as recently highlighted [1,2]. The wide utilization of molecular markers for mapping traits of agronomic interest in specific genomic regions appears to return back another pivotal effort for the future development of novel cultivars [3]. Indeed, the improvement of plant breeding efficacy has always gone through the construction of exotic genetic libraries, exploiting the genetic resources [4].

Nowadays, there is evidence that MAGIC and other exotic populations will play a major role in the coming years in allowing for impressive gains in plant breeding for developing new generations of improved cultivars [5].

This Special Issue focused on the application of such advanced technologies devoted to crop improvement, exploiting the available biodiversity in crops. In detail, next-generation sequencing (NGS) technologies supported the development of high-density genotyping arrays for different plants included in this issue.

By using a high throughput approach, here we report a new high-resolution eggplant (*Solanum melongena* L.) genetic map based on a RIL population and Genotyping by Sequencing (GBS) analysis by which 7249 SNPs were assigned to the 12 chromosomes spanning 2169.23 cM [6]. Afterwards, the phenotyping of the RIL population at three locations allowed us to elucidate the genetic bases of seven traits related to anthocyanin content in eggplant as well as seed vigor. Overall, between 7 and 17 QTLs (at least one major QTL) were identified for each trait [6]. Otherwise, a genome-wide association scan (GWAS) using 121 accessions and a 9K single nucleotide polymorphisms (SNPs) chip were also reported to clarify the genetic determinants underlying drought tolerance in barley (*Hordeum vulgare* L.) [7]. Overall, a total number of 101 significant SNPs, distributed over all seven barley chromosomes, were found to be highly associated with the studied traits, of which five genomic regions were associated with candidate genes at chromosomes 2 and 3 [7].

The limited availability of simple sequence repeats (SSR) in *Paeonia lactiflora*, a flowering crop with great economic value, triggered a study to develop a novel SSR panel with Illumina RNA sequencing for also assessing the role of these variants in gene regulation. The results showed that dinucleotides with AG/CT repeats were the most abundant type of repeat motif in *P. lactiflora* and were preferentially distributed in untranslated regions. Significant differences in SSR size were observed among motif types and locations [8]. This new set of SSRs will aid programs for accession identification, marker-trait association and molecular assisted breeding in *P. lactiflora* [8].

QTL-related Lethal Necrosis (LN) tolerance/resistance in maize (*Zea mays* L.) has been studied by using five hundred selected kompetitive allele specific PCR (KASP) SNPs and multiple mapping populations [9]. To understand the status of previously identified quantitative trait loci (QTL) in diverse genetic backgrounds, F_3 progenies derived from seven bi-parental populations were genotyped and phenotyped under artificial LN inoculation for three seasons. Joint linkage association mapping revealed at least seven major QTL spread across the 7-biparetal populations, for resistance to LN infections potentially useful for marker-assisted breeding [9].

A particular resequencing approach was utilized for exploring the natural variation and the domestication selection of *ZmPGP1*, involved in the polar auxin transport and associated to plant height, leaf angle, yield traits, and root development in maize (*Z. mays* L.) [10]. Li et al. (2019) [10] re-sequenced this gene in 349 inbred lines, 68 landraces, and 32 teosintes. Sequence polymorphisms, nucleotide diversity, and neutral tests revealed that *ZmPGP1* might be selected during domestication and improvement processes. Marker–trait association analysis identified 11 variants significantly associated with 4 plant architecture and 5 ear traits, revealing that significant variants in *ZmPGP1* can be used to develop functional markers to improve plant architecture and ear traits in maize [10].

Another particular approach was reported for two popular fruit crops such as strawberry (*Fragaria* × *ananassa* Duchesne) and raspberry (*Rubus idaeus* L.). Here, Lebedev et al. (2020) [11] reported the potential transferability between the species of a large SSR panel for their employment in breeding programs assisted by functional DNA markers [11]. One hundred eighteen (118) microsatellite loci in the flavonoid biosynthesis were developed to assess the genetic diversity of 48 *Fragaria* and *Rubus* accessions, including wild species and rare cultivars, which differ in berry color, ploidy, and origin. SSR panel may be a useful molecular tool in strawberry and raspberry breeding programs for improvement anthocyanin related traits [11].

Informative molecular markers such as SSR were adopted also for detecting hybridity and homozygosity in breeding segregant populations in lettuce (*Lactuca sativa* L.). In this study, a panel of 16 SSR was used to genotype 71 putative parental lines and to plan 89 controlled crosses designed to maximize the genetic recombination [12]. Unexpected genotypes were detected (5%), consistent with this species' spontaneous out-pollination rate. Overall, the synergistic advantages of conventional and molecular selection applied in different steps of a breeding programs aimed at developing new varieties were demonstrated [12].

Two other manuscripts reported the usefulness of molecular analysis for analyzing germplasm collections [13,14]. In the first, an SSR panel was adopted to analyze the official Algerian olive (*Olea europea* L.) collection highlighting a biodiversity hotspot in the Mediterranean Basin [13]. The olive germplasm was characterized using 16 nuclear (nuSSR) and six chloroplast (cpSSR) microsatellites, useful to underline the presence of an exclusive genetic core represented by 13 cultivars located in a mountainous area in the North-East of Algeria, named Little Kabylie. The genetic relationship of Algerian and Mediterranean olive germplasm was assessed, suggesting possible events of secondary domestication and/or crossing and hybridization across the Mediterranean area [13]. The second manuscript described the genetic diversity in sweet potato (*Ipomea batatas* L.) genetic resources by morphological and molecular markers [14]. The EU market of this orphan crop has recently increased by 100%, and its cultivation in southern European countries is a new opportunity for the EU to exploit and introduce new genotypes. In this view, the origins of the principal Italian sweet potato clones, compared with a core collection of genotypes from Central and Southern America, were investigated by combining genetic analysis with morphological and chemical measurements [14]. Overall, these markers combination resulted as being effective to cluster the sweet potato clones in agreement with their geographical origin [14]. The last report included in this Special Issue focused on molecular markers supporting the in situ conservation of faba bean (*Vicia faba* L.) landraces in Tunisia [15]. The seed phenotypic features of the collected samples were analyzed, together with the genetic diversity and population structure, by using simple sequence repeat markers, highlighting the genetic stability of the population under study. These findings suggested that farmers applied international best practices for the in situ conservation of agricultural crops [15].

In conclusion, the Special Issue focused on the development and application of such technologies associated with adaptation and functional crop improvement, exploiting the available biodiversity in very different crops, from vegetables and legumes (eggplant, lettuce, sweet potato and faba bean), through important cereals (barley and corn) to very important Mediterranean trees (olive). This issue has allowed a scientific journey through the use of consolidated molecular markers, such as SSR, as well as novel classes of molecular markers obtainable by the new technologies (NGS). These were applied at the genetic analysis of germplasm collections, but also to the findings of new markers and QTL for assisted breeding programs.

Conflicts of Interest: The authors declare no conflict of interest.

References

1. Dzyubenko, N.I. Vavilov's Collection of Worldwide Crop Genetic Resources in the 21st Century. *Biopreserv. Biobank.* **2018**, *16*, 377–383. [CrossRef] [PubMed]
2. Nguyen, G.N.; Norton, S.L. Genebank Phenomics: A Strategic Approach to Enhance Value and Utilization of Crop Germplasm. *Plants* **2020**, *9*, 817. [CrossRef] [PubMed]
3. Cobb, J.N.; Biswas, P.S.; Platten, J.D. Back to the future: Revisiting MAS as a tool for modern plant breeding. *Theor. Appl. Genet.* **2019**, *132*, 647–667. [CrossRef] [PubMed]
4. Zamir, D. Improving plant breeding with exotic genetic libraries. *Nat. Rev. Genet.* **2001**, *2*, 983–989. [CrossRef] [PubMed]
5. Arrones, A.; Vilanova, S.; Plazas, M.; Mangino, G.; Pascual, L.; Díez, M.J.; Prohens, J.; Gramazio, P. The Dawn of the Age of Multi-Parent MAGIC Populations in Plant Breeding: Novel Powerful Next-Generation Resources for Genetic Analysis and Selection of Recombinant Elite Material. *Biology* **2020**, *9*, 229. [CrossRef] [PubMed]
6. Toppino, L.; Barchi, L.; Mercati, F.; Acciarri, N.; Perrone, D.; Martina, M.; Gattolin, S.; Sala, T.; Fadda, S.; Mauceri, A.; et al. A New Intra-Specific and High-Resolution Genetic Map of Eggplant Based on a RIL Population, and Location of QTLs Related to Plant Anthocyanin Pigmentation and Seed Vigour. *Genes* **2020**, *11*, 745. [CrossRef] [PubMed]
7. Thabet, S.G.; Moursi, Y.S.; Karam, M.A.; Börner, A.; Alqudah, A.M. Natural Variation Uncovers Candidate Genes for Barley Spikelet Number and Grain Yield under Drought Stress. *Genes* **2020**, *11*, 533. [CrossRef] [PubMed]
8. Wan, Y.; Zhang, M.; Hong, A.; Zhang, Y.; Liu, Y. Characteristics of Microsatellites Mined from Transcriptome Data and the Development of Novel Markers in *Paeonia lactiflora*. *Genes* **2020**, *11*, 214. [CrossRef] [PubMed]
9. Awata, L.A.O.; Beyene, Y.; Gowda, M.; Suresh, L.M.; Jumbo, M.B.; Tongoona, P.; Danquah, E.; Ifie, B.E.; Marchelo-Dragga, P.W.; Olsen, M.; et al. Genetic Analysis of QTL for Resistance to Maize Lethal Necrosis in Multiple Mapping Populations. *Genes* **2020**, *11*, 32. [CrossRef] [PubMed]
10. Li, P.; Wei, J.; Wang, H.; Fang, Y.; Yin, S.; Xu, Y.; Liu, J.; Yang, Z.; Xu, C. Natural Variation and Domestication Selection of *ZmPGP1* Affects Plant Architecture and Yield-Related Traits in Maize. *Genes* **2019**, *10*, 664. [CrossRef] [PubMed]
11. Lebedev, V.G.; Subbotina, N.M.; Maluchenko, O.P.; Lebedeva, T.N.; Krutovsky, K.V.; Shestibratov, K.A. Transferability and Polymorphism of SSR Markers Located in Flavonoid Pathway Genes in *Fragaria* and *Rubus* Species. *Genes* **2020**, *11*, 11. [CrossRef] [PubMed]
12. Patella, A.; Palumbo, F.; Galla, G.; Barcaccia, G. The Molecular Determination of Hybridity and Homozygosity Estimates in Breeding Populations of Lettuce (*Lactuca sativa* L.). *Genes* **2019**, *10*, 916. [CrossRef] [PubMed]
13. Haddad, B.; Gristina, A.S.; Mercati, F.; Saadi, A.E.; Aiter, N.; Martorana, A.; Sharaf, A.; Carimi, F. Molecular Analysis of the Official Algerian Olive Collection Highlighted a Hotspot of Biodiversity in the Central Mediterranean Basin. *Genes* **2020**, *11*, 303. [CrossRef] [PubMed]
14. Palumbo, F.; Galvao, A.C.; Nicoletto, C.; Sambo, P.; Barcaccia, G. Diversity Analysis of Sweet Potato Genetic Resources Using Morphological and Qualitative Traits and Molecular Markers. *Genes* **2019**, *10*, 840. [CrossRef] [PubMed]

15. Babay, E.; Khamassi, K.; Sabetta, W.; Miazzi, M.M.; Montemurro, C.; Pignone, D.; Danzi, D.; Finetti-Sialer, M.M.; Mangini, G. Serendipitous In Situ Conservation of Faba Bean Landraces in Tunisia: A Case Study. *Genes* **2020**, *11*, 236. [CrossRef] [PubMed]

Publisher's Note: MDPI stays neutral with regard to jurisdictional claims in published maps and institutional affiliations.

Article

A New Intra-Specific and High-Resolution Genetic Map of Eggplant Based on a RIL Population, and Location of QTLs Related to Plant Anthocyanin Pigmentation and Seed Vigour

Laura Toppino [1], Lorenzo Barchi [2,*], Francesco Mercati [3], Nazzareno Acciarri [4],
Domenico Perrone [5,6], Matteo Martina [2], Stefano Gattolin [7], Tea Sala [1], Stefano Fadda [1],
Antonio Mauceri [8], Tommaso Ciriaci [4], Francesco Carimi [3], Ezio Portis [2], Francesco Sunseri [8],
Sergio Lanteri [2] and Giuseppe Leonardo Rotino [1]

[1] CREA, Research Centre for Genomics and Bioinformatics, 26836 Montanaso Lombardo (LO), Italy; laura.toppino@crea.gov.it (L.T.); tea.sala@crea.gov.it (T.S.); stefano.fadda@crea.gov.it (S.F.); giuseppeleonardo.rotino@crea.gov.it (G.L.R.)
[2] DISAFA–Plant genetics and breeding—University of Turin, 10095 Grugliasco (TO), Italy; matteo.martina@unito.it (M.M.); ezio.portis@unito.it (E.P.); sergio.lanteri@unito.it (S.L.)
[3] Institute of Biosciences and BioResources, Division of Palermo—National Research Council (CNR), 90129 Palermo, Italy; francesco.mercati@ibbr.cnr.it (F.M.); francesco.carimi@ibbr.cnr.it (F.C.)
[4] CREA, Research Centre for Vegetable and Ornamental Crops, 63030 Monsampolo del Tronto (AP), Italy; acciarri@libero.it (N.A.); t.ciriaci@libero.it (T.C.)
[5] CREA, Research Centre for Vegetable and Ornamental Crops, 84098 Pontecagnano (SA), Italy; domenico.perrone@crea.gov.it
[6] CREA, Research Centre for Plant Protection and Certification, 84091 Battipaglia, Italy
[7] Institute of Agricultural Biology and Biotechnology (IBBA)—National Research Council (CNR), 20133 Milano, Italy; stefano.gattolin@ibba.cnr.it
[8] Department AGRARIA, University Mediterranea of Reggio Calabria, 89124 Reggio Calabria, Italy; antonio.mauceri87@unirc.it (A.M.); francesco.sunseri@unirc.it (F.S.)
* Correspondence: lorenzo.barchi@unito.it

Received: 8 June 2020; Accepted: 2 July 2020; Published: 4 July 2020

Abstract: Eggplant is the second most important solanaceous berry-producing crop after tomato. Despite mapping studies based on bi-parental progenies and GWAS approaches having been performed, an eggplant intraspecific high-resolution map is still lacking. We developed a RIL population from the intraspecific cross '305E40', (androgenetic introgressed line carrying the locus *Rfo-Sa1* conferring *Fusarium* resistance) x '67/3' (breeding line whose genome sequence was recently released). One hundred and sixty-three RILs were genotyped by a genotype-by-sequencing (GBS) approach, which allowed us to identify 10,361 polymorphic sites. Overall, 267 Gb of sequencing data were generated and ~773 M Illumina paired end (PE) reads were mapped against the reference sequence. A new linkage map was developed, including 7249 SNPs assigned to the 12 chromosomes and spanning 2169.23 cM, with iaci@liberoan average distance of 0.4 cM between adjacent markers. This was used to elucidate the genetic bases of seven traits related to anthocyanin content in different organs recorded in three locations as well as seed vigor. Overall, from 7 to 17 QTLs (at least one major QTL) were identified for each trait. These results demonstrate that our newly developed map supplies valuable information for QTL fine mapping, candidate gene identification, and the development of molecular markers for marker assisted selection (MAS) of favorable alleles.

Keywords: linkage map; RAD; QTL; *Solanum melongena*

---

1. Introduction

Eggplant (*Solanum melongena* L., 2n = 2x = 24) is a member of the Solanaceae, a large plant family comprising over 3000 species and including important crops such as tomato, potato, pepper and tobacco. Unlike most of the other solanaceous crops, which are native to the New World [1–3], eggplant has a phylogenetic uniqueness due to its exclusive Asian origin [4]. It has been reported that the species resulted from two or three independent domestication events [5,6], although a recent study suggested a single domestication event [7]. Eggplant worldwide production is estimated as about 54 Mt, with China, India and Indonesia being the major producing countries, while Egypt, Turkey and Italy represent the main producers in the Mediterranean region (FAO 2018; [8]). Breeding efforts in eggplant, like in most crops, have been focused on increasing yield, resistance/tolerance to biotic and abiotic stress, and fruit shelf-life, but also on improving some plant morphological distinguishing traits (reduced prickliness and leaf hairiness) as well as raising the content of health-promoting metabolites (e.g., anthocyanins and chlorogenic acid) or reducing the anti-nutritional content (e.g., steroidal glycoalkaloid, saponins) in the berries. Furthermore, studies have been carried out with the goal to improve seed germination and seedling emergence [9–11], which affect the crop performance.

In eggplant, several inter-specific genetic maps were developed by applying pre-next generation sequencing (NGS) techniques (RFLP, AFLP, RAPD, SSR, etc.). They were based on inter-specific crosses between cultivated *S. melongena* and *S. linnaeanum* (=*S. sodomaeum*) or *S. incanum* and used for Quantitative Trait Loci (QTL) analyses of domestication and morphological traits [12–16], as well as to locate genes involved in polyphenol biosynthesis [17] and resistance to *Verticillium* spp. [18]. Intra-specific maps were also built [19–22] and, more recently, Fukuoka et al. [23] generated two intra-specific genetic maps based on F_2 populations, which were then combined into one on the basis of common markers for studying macro-syntenic relationships between eggplant and tomato, as well as for QTL analysis of parthenocarpy [24] and resistance to *Fusarium oxysporum* [25].

The advent of NGS-based marker technologies, by increasing the speed, throughput, and cost effectiveness of genotyping and providing genome-wide marker coverage, has allowed the development of the so-called 'second generation' maps. Barchi et al. [26], by applying the RAD-seq protocol from Baird et al. [27] on an intra-specific F_2 population, identified 10,000 single nucleotide polymorphisms (SNPs) as well as nearly 1000 polymorphic indels, and more than 2000 SNPs were found of potential use for genotyping on the basis of a GoldenGate© assay. Afterwards, the first 'second generation genetic map' was developed [28], which included 415 markers assigned to the 12 chromosomes. The latter was used to identify the genetic bases of traits associated with anthocyanin content [28] and, more recently, for detecting QTL affecting key horticultural traits [29], fruit metabolic content [30] and resistance to soil-borne diseases [31]. Furthermore, the previously identified loci were validated, and new linked marker/trait associations were detected, through a genome-wide association (GWA) mapping approach [32,33]. Second generation intra-specific genetic maps were also generated [2] for anchoring the first draft genome sequence of eggplant and for mapping resistance QTLs to *Ralstonia* strains by SNPs developed through Illumina sequencing of the parents of a Recombinant inbreed line (RIL) mapping population as well as AFLP, SSR and SRAP markers [34].

Despite recent efforts, the linkage maps used for identifying the genetic basis of traits of breeding interest are still not saturated, hampering the fine mapping of QTL regions and the identification of candidate genes associated with the phenotypic traits. Up to now, the only available high-resolution SNP-based linkage map was developed on a F_2 population from the inter-specific cross (*S. melongena* × *S. linneanum*) and was employed to highlight QTLs affecting stem height and fruit and leaf morphology [35].

We previously developed a RIL mapping population of 170 F_6-F_7 lines from the intra-specific cross between the breeding lines '305E40' (female parent) and '67/3' (male parent). Furthermore, the first high quality eggplant genome sequence of the breeding line '67/3' was released [36] and, through the resequencing of the female parent ('305E40') and a low coverage Illumina sequencing of each RIL, we constructed a first linkage map aimed at anchoring the scaffolds to the 12 chromosomes.

The map also demonstrated efficient mapping metabolomic traits of interest related to the metabolomics composition of fruit flesh and peel [37].

The genetic basis of anthocyanin synthesis and accumulation has been widely studied in the Solanaceae [38–43]. In the last decade, QTL-related studies using family-based or GWA mapping approaches allowed us to shed light on the genetic bases of anthocyanin distribution in eggplant as well as to identify its syntenic relationships with tomato [28,30,32,44]. By contrast, no information is available on QTLs controlling seed vigor in term of speed of seedling emergence, which diversifies the parents of our RIL mapping population. Here, we propose a more breeder-friendly map developed through a genotype-by-sequencing (GBS) approach with our RIL mapping population, whose reliability for mapping studies has been proved by identifying QTLs related to plant anthocyanin pigmentation and seed vigor.

2. Materials and Methods

2.1. Plant Material

A population of 163 F_7 plants, previously obtained by the single seed descent approach from a cross between eggplant lines '305E40' and '67/3' [28,36], was employed. The two parental lines were contrasting for a wide number of key agronomic and metabolic traits [28–31], as well as for their seed vigor. The '305E40' line (female parent) is a double haploid derived from the inter-specific somatic hybrid [*Solanum aethiopicum* gr. gilo(+)*S. melongena* cv. Dourga], which was repeatedly backcrossed with the recurrent eggplant genotypes (lines DR2 and Tal1/1) prior to selfing and anther culture. This line carries the locus *Rfo-sa1* from *S. aethiopicum*, which confers complete resistance to the soil-borne fungus *Fusarium oxysporum f. sp. melongenae (Fom)* [45] and is partially resistant to *Verticillium dahliae* [31]. Plants of '305E40' display a slight anthocyanin overall pigmentation, produces pink flowers and long, highly pigmented dark purple fruits characterized by the presence of the anthocyanin delphinidin-3-rutinoside (D3R) as well as a higher glycoalkaloids and organic acid content than the ones of '67/3' [28–30]. The '67/3' line is an F8 selection from the intra-specific cross cv. 'Purpura' × cv. 'CIN2', which lacks the *Rfo-sa1* locus and is fully susceptible to *Verticillium* [31]. Its plants display higher anthocyanin pigmentation than '305E40' in leaves and stems and produce violet flowers and round, violet colored fruits with white peel colour both under and next to the calyx. The fruits are characterized by the presence of the anthocyanin nasunin in the peel, higher soluble solids, sugars and chlorogenic acid content in the flesh compared to 305E40.

The mapping population was sown, along with both parents and the F_1 hybrid, in glasshouses at Montanaso Lombardo in 2012. The seeds were sown in plastic trays consisting of 104 holes (8 rows of 13 holes each) filled with peat and placed over an electric warmed carpet at 24 °C. For each RIL, we sowed 52 seeds split in two replicates of 26 seeds. Each replicate was sown in two adjacent randomly chosen rows containing 13 holes of the replicate-specific tray, using a single seed per hole; each replicate was kept under the same conditions but in a different glasshouse at Montanaso Lombardo. All plantlets were grown in heated glasshouses (minimum temperature of 15 °C ensured) until the 3rd–4th leaf, and then were transplanted in three field trials in northern (Montanaso Lombardo, ML, 45°20′ N, 9°26′ E), central (Monsampolo del Tronto, MT, 42°53′ N,13°47′ E) and southern (Battipaglia, BP, 40°36′ N, 14°59′ E) Italy. Mulched twin rows of 1.1 m width were arranged using plastic black PE (0.05 mm), and plantlets were transplanted at 45 cm between each other along the rows. A drip irrigation system was employed for watering and fertilizing, and local standard horticultural practices were applied. In each site, the material was transplanted in the field according to a randomized block design (3 replicate blocks, 4 plants per block) to score the phenotypic traits.

2.2. Library Construction and Sequencing

DNA from the RIL population, parental lines and F_1 hybrid was extracted following a modified CTAB method [46] as indicated elsewhere [47]. Library construction was performed as proposed by Acquadro et al. [48] by using a HindIII-MseI enzyme combination and adding a final biotin/streptavidin-coated beads-based purification step. Quality, quantity and reproducibility of libraries were assessed with a Bioanalyzer instrument (DNA High Sensitivity chip) as well as qPCR using KAPA SYBR FAST Universal 2X qPCR Master Mix (Kapa Biosystems, Boston, MA, USA). On the basis of the quantitation, DNA libraries were pooled and sequenced on Illumina HiSeq 2500 platform (Illumina Inc., San Diego, CA, USA), following the manufacturer protocol using 150 PE chemistry at Biodiversa srl (Rovereto (TN), Italy).

2.3. Sequence Analysis and Map Construction

Raw reads were analyzed with Scythe (https://github.com/vsbuffalo/scythe) for filtering out contaminant substrings, and Sickle (https://github.com/najoshi/sickle), for removing reads with poor quality ends (Q < 30). Illumina reads were de-multiplexed on the basis of the Illumina TruSeq index using Stacks process rad tags. Alignment to the reference eggplant genome [36] was carried out using the Burrows-Wheeler Aligner (BWA) aligner [49] (i.e., mem command) with default parameters and avoiding multiple-mapping reads. BAM files were processed and use for the SNP calling using bcftools mpileup/call/norm utilities [50] with default parameters, except for the use of multiallelic calling model (-m option), minimum mapping quality (Q = 20) and filtering out multimapping events ($-q > 1$). Only SNPs with at most 20% of missing data and a mean-minDP of 20 were retained for linkage analyses. Polymorphic markers were grouped in linkage groups with "R/qtl" package [51], with minimum LOD = 8, rec ≤ 0.15. For each linkage group identified, identical loci were removed with the Exclude identical function and the remaining loci were ordered with Joinmap software (version 4, [52]), using a LOD of 8 and the Kosambi function to estimate distance and the Maximum-likelihood function to infer correct order. Markers exhibiting segregation distortion were identified applying the chi-square (X^2)-goodness-of-fit test ($p < 0.001$) and also integrated into the map. The ordering step was iterated several times, each time by correcting genotype calls with the "SMOOTH.pl" script, which is a Perl implementation of the SMOOTH software, as developed by van Os et al. [53]. Finally, visual inspection of genotypes was applied to identify and correct the remaining genotype errors. Linkage groups were visualized in MapChart (version 2.32, [54]).

2.4. Phenotypic Traits Evaluation

The speed of emergence and hypocotyl anthocyanin distribution traits were assessed on 56 plantlets of each RIL, of the two parental lines and their F_1 hybrids, which were obtained from as many seeds sown as previously described in Plant material.

The speed of emergence index (*sei*) was evaluated as the time, expressed in days, needed for the emergence of 50% + 1 plantlets from the soil. Hypocotyl anthocyanin (*hyan*) trait was assessed on plantlets at the second–third leaf stage (Figure 1d) according to a 0–5 scale, with "0" representing no visible anthocyanin coloration (i.e., completely green tissues) and "5" representing complete dark violet coloration.

Figure 1. Phenotypic trait evaluation. (**a**) *adlan* (range from 0 to 5); (**b**) *lvean* (0–5); (**c**) *stean* (1,3,5); (**d**) *hyan* (0, 5); (**e**) *corcol* (clockwise from top-right: light violet, light pink, dark pink, dark violet); (**f**) *flian* (1,3,5); (**g**) *toan* (violet-reddish); (**h**) *sei* (picture at 10 days after sowing).

In all the three field trials (ML, MT and BT), the material was arranged as a set of three randomized complete blocks with 4 replicate plants per entry per block. Phenotyping was based on the European Cooperative program for Plant Genetic Resource descriptors panel for Solanaceae (ECPGR, 2008) and the International Board for Plant Genetic Resource descriptors for eggplant (IBPGR, 1990). The traits assayed, reported in Figure 1 and detailed in Table 1, were: adaxial leaf lamina anthocyanin (*adlan*), corolla colour (*corcol*), flower anthocyanin intensity (*flian*), hypocotyl anthocyanin (*hyan*), leaf venation anthocyanin (*lvean*), stem anthocyanin (*stean*), anthocyanin tonality (*toan*).

The anthocyanin content of stems, leaves and flower calyxes was scored on a 0 (no visible coloration) to 5 (complete dark violet coloration) scale. Anthocyanin content in leaves and leaf venation was evaluated on 4 leaves per RIL in each block, chosen in the upper middle part of the plant. Stem anthocyanin content was measured as an average value based on 3 stems per block. Flower anthocyanin intensity resulted from averaging 5 flowers per block. Anthocyanin tonality was scored as "1" reddish, "3" intermediate, or "5" violet. Finally, for the corolla colour, the trait was coded as "1", pink; "2", dark pink; "3", light violet; "4", violet-pink and "5", violet.

Table 1. List of the traits analyzed and their code, means, standard deviations (SD), coefficients of variation (cv), broad sense heritability and transgressive genotypes for the traits in study.

Trait	Trait Code	305E40 Mean	±SD	67/3 Mean	±SD	F1 Mean	±SD	cv	RIL pop Means	±SD	cv	Shapiro-Wilks	Skewness	SE	Kurtosis	SE	Heritability	Transgressive Respect 305E40	Transgressive Respect 67/3
Adaxial Leaf Lamina Anthocyanin	adlan_BT	0.42	0.20	5.00	0.00	1.67	1.17	0.70	1.76	1.58	0.89	0.87	0.72	0.08	−0.75	0.16	0.97	2	0
	adlan_ML	0.42	0.20	4.67	0.41	2.25	0.42	0.19	2.04	1.58	0.77	0.90	0.43	0.08	−1.15	0.16	0.94	0	0
	adlan_MT	0.00	0.00	2.67	0.87	1.11	0.60	0.54	0.69	1.03	1.49	0.71	1.63	0.07	2.14	0.13	0.92	0	0
Stem Anthocyanin	stean_BT	2.17	0.25	5.00	0.00	4.83	0.25	0.05	3.80	1.02	0.27	0.92	−0.62	0.07	−0.14	0.13	0.96	3	0
	stean_ML	2.33	0.43	5.00	0.00	4.78	0.36	0.08	3.80	0.98	0.26	0.92	−0.35	0.07	−0.80	0.13	0.96	0	0
	stean_MT	1.11	0.33	4.89	0.33	4.44	0.53	0.12	3.39	1.08	0.32	0.91	−0.22	0.07	−0.53	0.13	0.95	0	0
Leaf Venation Anthocyanin	lvean_BT	1.67	0.26	5.00	0.00	4.33	0.26	0.06	3.61	1.33	0.37	0.76	4.26	0.08	71.86	0.16	0.86	1	0
	lvean_ML	2.75	0.42	4.92	0.20	4.75	0.27	0.06	3.74	0.99	0.27	0.92	−0.54	0.08	−0.24	0.16	0.87	2	0
	lvean_MT	1.00	0.71	4.22	0.97	4.44	1.01	0.23	3.06	1.53	0.50	0.91	−0.38	0.07	−0.95	0.13	0.87	0	0
Flower Anthocyanin Intensity	flian_BT	2.28	1.12	5.00	0.00	4.89	0.22	0.05	4.27	0.89	0.21	0.79	−1.05	0.07	0.40	0.13	0.94	0	0
	flian_ML	2.89	0.89	4.89	0.33	4.67	0.43	0.09	3.95	1.06	0.27	0.83	−0.75	0.07	−0.06	0.13	0.95	0	0
	flian_MT	3.22	1.48	4.89	0.33	5.00	0.00	0.00	4.12	0.98	0.24	0.84	−0.79	0.07	−0.43	0.13	0.95	0	0
Anthocianin Tonality	toan_BT	1.00	0.00	5.00	0.00	3.00	0.00	0.00	3.02	1.86	0.62	0.72	−0.02	0.11	−1.85	0.23	0.94	0	0
	toan_ML	1.00	0.00	5.00	0.00	3.00	0.00	0.00	3.21	1.86	0.58	0.71	−0.21	0.11	−1.80	0.22	0.89	0	0
Corolla Colour	corcol_BT	1.00	0.00	5.00	0.00	4.11	0.33	0.08	3.00	1.78	0.59	0.76	−0.01	0.07	−1.78	0.13	0.97	0	0
	corcol_ML	1.00	0.00	5.00	0.00	3.33	1.32	0.40	3.23	1.78	0.55	0.76	−0.20	0.07	−1.75	0.13	0.97	0	0
	corcol_MT	4.33	1.00	5.00	0.00	4.67	0.50	0.11	4.49	1.10	0.24	0.51	−2.08	0.07	3.26	0.13	0.87	0	0
Hypocotyl Anthocyanin	hyan	0.67	0.29	4.83	1.26	3.00	0.50	0.17	2.88	1.34	0.47	0.94	−0.18	0.12	−0.98	0.23	0.98	1	0
Speed of emergence index	sei	11.67	1.53	5.67	1.53	5.67	1.15	0.20	13.65	3.00	0.22	0.90	1.27	0.12	4.80	0.23	0.98	44	0

2.5. Statistical Analyses and QTL Detection

Statistical analyses were performed using R software [55]. A conventional analysis of variance was applied to estimate genotype and environment effects based on the linear model $Y_{ij} = \mu + g_i + b_j + e_{ij}$, where μ, g, b and e represent, respectively, the overall mean, the genotypic effect, the block effect and the error. Broad-sense heritability values were given by $\sigma^2_G/([\sigma^2_G + \sigma^2_E]/n)$, where "σ2G" represents the genetic variance, "σ2E" the residual variance and "n" the number of blocks. Correlations between traits were estimated using the Spearman coefficient, and normality, kurtosis and skewness were assessed with the Shapiro–Wilks test ($\alpha = 0.05$). Segregation was considered as transgressive when at least one individual RIL recorded a trait value higher or lower by at least two standard deviations than the higher or lower scoring parental line. QTL detection was performed considering each location independently and was based on the newly developed map using MQM [56,57] mapping, as implemented in MapQTL v4 software [58]. QTLs were initially identified using interval mapping, after which one linked marker per putative QTL was treated as a co-factor in the approximate multiple QTL model. Co-factor selection and MQM analysis were repeated until no new QTL could be identified. LOD thresholds for declaring a QTL to be significant at the 5% genome-wide probability level were established empirically by applying 1000 permutations per trait [59]. Additive and dominance genetic effects, as well as the percentage of the phenotypic variation (PVE) explained by each QTL, were obtained from the final multiple QTL model. Individual QTLs were prefixed by a trait abbreviation, followed by the relevant chromosome designation—BT, ML or MT—which was added as a suffix when a QTL was expressed in a site-specific manner. Confidence interval of the QTL was calculated at a LODmax−1 interval or at least by considering 0.5Mb upstream and downstream (if not explicitly reported in the text) the marker identified at the QTL. CMplot was used for drawing QTL results [60]. No site-specific suffix was added to the *hyan* and *sei* QTLs, as these two traits were assessed in a single environment.

3. Results

3.1. Sequencing and Linkage Map Construction

A total of 855 million paired-ends (PE) reads were produced, corresponding to about 257 Gb of data. After demultiplexing, cleaning and trimming, a total of 745 M Illumina PE reads were retained, corresponding to an average number of PE reads per sample of 4.48 M, with a standard deviation of 2.87 M (Figure S1). The sequence data were deposited into NCBI Short Read Archive under the Bioproject PRJNA635547.

Reads were then aligned to the reference eggplant genome [36]; close to 100% of reads were successfully mapped to single regions (no multiple mapping was permitted). A total of 10,316 polymorphic sites (i.e., markers) were identified after SNP calling using conservative filtering parameters, and were used for mapping purposes by applying a combination of R/qtl and Joinmap. At first, all markers were fed to R/qtl for linkage group identification and eventually ordered with Joinmap. Overall, 7249 markers were successfully retained and assigned to the 12 linkage groups (LG) corresponding to the haploid chromosome number of the species (Figure 2 and Figure S2). A total of 1744 markers showed segregation distortion with $p < 0.001$, covering about 24% of the total mapped markers. Chromosome E02 showed the largest segregation distortion for 1423 markers, followed by E05, with 72 markers (Table 2).

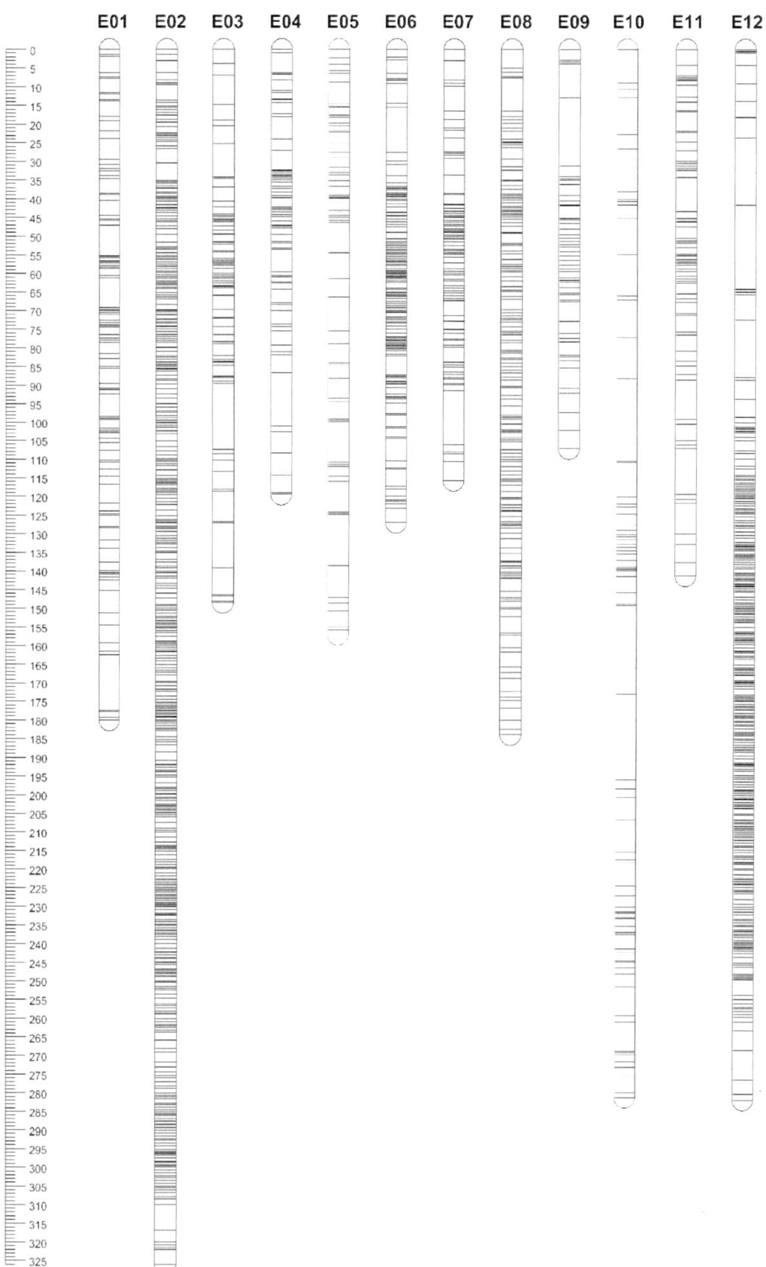

Figure 2. Eggplant linkage map depicting the size of the chromosomes and markers distribution. Marker names and map distances (in cM) are detailed in Figure S2.

The marker names, chromosome, and genetic position of all markers on the map and RILs haplotypes are included in Table S1. The linkage map spans 2169.23 cM (Table 2), with E02 being the longest (326 cM) chromosome and containing the highest number of markers (i.e., 1454), while E09 was the shortest (107 cM) and contained 230 markers. Some markers belonging to contigs previously assigned to CH0 [36] were mapped to the 12 LGs, of which the majority were mapped on E06, while just 47 were mapped to other chromosomes. The genome-wide mean inter-locus separation was reduced (mean of 0.4 cM), with the highest value (0.7 cM) for E05 and E10. The mapping of the SNP markers on the eggplant genome sequence revealed a total coverage of 95.88% (1095.72 Mb) of the diploid genome (1142.80 Mb). The genome-wide mean inter-locus separation was 216.5 kb, with the highest value (411.2 kb) in E04. The largest gap was observed on E12 (22.45 cM), while almost all gaps were less than 5 cM. The recombination rate of different chromosomes was estimated as the quotient between the genetic distance (cM) covered by the corresponding LG and the size in Mb of the chromosome fragment covered with markers. This value ranged from 0.81 cM/Mb on E07 to 3.91 cM/Mb on E02.

To evaluate the quality of the map obtained, we used heat maps of recombinant frequency, which highlighted that the mapped markers were ordered correctly, as the pair-wise recombination rates were noticeably low between adjacent markers (diagonal distribution of the yellow color indicates the lowest recombination rate) in the heat map for each chromosome, except E02, E06, E08 and E12 (Figure S3). Similar results were obtained when collinearity between the genetic distances of mapped SNP on each linkage group and their corresponding physical position on the eggplant chromosomes was spotted (Figure S4).

3.2. Phenotypic Variation and Inter-Trait Correlations

A summary of the phenotypic performance for each trait in the parental lines, hybrid F_1 and RILs, together with the skewness, kurtosis, broad sense heritability (h_{2BS}) values and presence of transgressive genotypes for each trait, are listed in Table 2. As expected, the parental lines contrasted for each trait. The female line '305E40' had a delayed emergence from the soil compared to '67/3', as evidenced by its higher *sei* (6 days). The '305E40' line showed lower anthocyanin content in leaves, leaf venations, flower calyxes and stems; its hypocotyl was characterized by a reddish tonality and produced flowers with a pink corolla. Conversely, line '67/3' produced violet flowers. The F_1 hybrid's phenotype was intermediate between the two parents for *adlan*, *toan* and *hyan*. For the remaining traits, the F_1 hybrid was more similar to '67/3' in all environments. Transgressive genotypes among the RILs were limited and only toward '305E40'—more precisely, three RILs for *steanBT*, two RILs for *lveanML* and *adlanBT*) and one RIL for *lveanBT*, and *hyan*. An exception was the speed of plant emergence, which was delayed in 44 RILs with respect to the late female parent '305E40'. The h^2 was overall high, ranging from 0.86 (*lveanBT*) to 0.98 (*sei* and *hyan*) (Table 2).

Significant inter-trait correlations were detected within and across locations (Table 3), and the same traits appeared to be highly correlated in the three locations. No significant correlation was detected for *sei* with other traits as well as between *adlanMT* and traits such as *corcolBT*, *corcolML* or *hyan*. The least correlated traits were *adlan* with *hyan*, *toan* and *corcol* (in all environments), while the most highly correlated were *corcol* and *toan* in both the BT (+0.92), and ML (+0.91) environments.

Table 2. Parameters associated with the framework eggplant genetic map. [a,b] The overall physical start and end positions of all markers of each linkage group. [c] The distance between the linkage group (LG) physical start and LG physical end indicates the overall physical span of all the markers of each linkage group in a particular chromosome.

LG/Chromosome	Size (cM)	Size (Mp)	Markers	LG Physical Start (bp)[a]	LG Physical End (bp)[b]	Physical Span (Mb)[c]	Max Gap (cM)	Gap <5 cM	Markers from Different CH	Markers from CH0	Distorted Markers	Marker Density		Ratio cM/Mb (Estimated Recombination Rate)
												cM	kb	
E01	180.09	136.53	426	1,761,388	136,521,595	134.76	14.89	0.984	0	0	54	0.4	321.3	1.32
E02	326.00	83.34	1454	5796	77,157,371	77.15	6.71	0.999	9	3	1425	0.2	57.4	3.91
E03	148.51	97.01	408	36,214	82,766,928	82.73	18.00	0.985	9	19	7	0.4	238.4	1.53
E04	119.40	105.67	258	1,557,104	105,339,521	103.78	14.46	0.973	0	0	0	0.5	411.2	1.13
E05	157.15	43.85	211	2,017,627	43,756,820	41.74	13.75	0.957	0	13	72	0.7	208.8	3.58
E06	126.94	108.97	838	2187	99,842,165	99.84	12.19	0.995	24	104	8	0.2	130.2	1.16
E07	115.69	142.38	535	16,820	140,599,277	140.58	14.79	0.994	1	16	36	0.2	266.6	0.81
E08	183.94	109.58	730	2123	10,680,425	106.80	8.89	0.999	0	2	1	0.3	150.3	1.68
E09	107.06	36.10	230	11,386	34,244,824	34.23	18.18	0.978	1	15	1	0.5	157.6	2.97
E10	281.23	106.64	386	701,005	106,482,885	105.78	22.29	0.956	0	0	68	0.7	277.0	2.64
E11	141.34	72.29	231	213,621	72,208,518	71.99	12.24	0.974	3	59	39	0.6	314.3	1.96
E12	281.90	100.42	1542	10,639	96,335,130	96.32	22.45	0.995	0	1	33	0.2	65.2	2.81
Total	2169.23	1142.80	7249.00			1095.72			47	232	1744	0.4	216.5	2.12

Table 3. Inter-trait Spearman correlations assessed in the mapping population. In green, significant correlation at $p < 0.05$, in blue at $p < 0.01$.

	adlanML	adlanMT	corcolBT	corcolML	corcolMT	flianBT	flianML	flianMT	sei	hyan	lveanBT	lveanML	lveanMT	steanBT	steanML	steanMT	toanBT	toanML
adlanBT	0.90	0.81	0.21	0.21	0.21	0.43	0.45	0.50	0.02	0.19	0.49	0.42	0.66	0.38	0.38	0.39	0.20	0.16
adlanML		0.83	0.20	0.19	0.18	0.37	0.43	0.45	0.00	0.17	0.40	0.41	0.66	0.32	0.37	0.38	0.19	0.18
adlanMT			0.15	0.15	0.18	0.39	0.42	0.45	0.00	0.11	0.42	0.39	0.65	0.30	0.37	0.36	0.12	0.15
corcolBT				0.85	0.60	0.57	0.53	0.53	0.03	0.69	0.57	0.61	0.42	0.74	0.65	0.69	0.92	0.83
corcolML					0.68	0.51	0.54	0.54	0.05	0.68	0.58	0.58	0.44	0.75	0.68	0.71	0.81	0.91
corcolMT						0.42	0.37	0.41	−0.12	0.48	0.45	0.44	0.34	0.60	0.56	0.55	0.58	0.63
flianBT							0.85	0.81	0.04	0.42	0.80	0.79	0.55	0.79	0.77	0.73	0.50	0.45
flianML								0.85	0.02	0.37	0.83	0.81	0.60	0.75	0.83	0.78	0.42	0.47
flianMT									0.02	0.44	0.82	0.78	0.61	0.77	0.80	0.80	0.45	0.46
sei										0.09	0.07	−0.01	−0.02	0.05	−0.01	0.00	0.07	−0.01
hyan											0.52	0.43	0.27	0.65	0.53	0.54	0.66	0.65
lveanBT												0.83	0.59	0.83	0.82	0.80	0.52	0.50
lveanML													0.52	0.78	0.83	0.83	0.55	0.57
lveanMT														0.54	0.59	0.62	0.35	0.36
steanBT															0.88	0.88	0.70	0.68
steanML																0.91	0.59	0.62
steanMT																	0.62	0.65
toanBT																		0.84

3.3. QTL Analysis

LOD score, percentage of variance explained (PVE), and confidence interval (CI) related to QTLs, are described in Table 4. QTL analyses on all traits and environments yielded a total of 23 major (PVE values >10%) and 11 minor QTLs (Figure 3).

QTL analyses for both *hyan* and *sei*, which were evaluated in a single location, highlighted one major and two minor QTLs. Separate analyses performed on each location for *adlan, lvean, stean, corcol* and *toan* resulted in the identification of a ratio between major and minor QTLs of 8/2, 7/4 and 6/1 in BT, ML and MT, respectively. The majority of QTLs identified can be considered stable, as they had the same genomic position across the three locations, with the exception of the minor QTLs *stean2.1*, confirmed both in MT and ML, but not in BT, and *corcol10.1*, demonstrating a major QTL in BT and a minor QTL in ML, and not detected in MT. Moreover, the major QTL *corcol5.1*, detected both in BT and ML, was mapped in a different position of E05 and was found as playing a minor effect in MT. For the anthocyanin-related QTLs, the positive alleles responsible for an increase in the anthocyanin content and for the presence of a violet vs. reddish pigmentation derived from '67/3'. Concerning the speed of emergence index, for all the QTLs but *sei2.2*, the allele increasing seed vigor derived from '67/3'. The largest single QTL effect was associated with *flian10.1_MT* (69.3% of the PVE). The additive effects of all the QTLs were significant at $p < 0.05$.

All the identified QTLs were distributed over five chromosomes (Table 4), namely E02, E04, E05, E07, E10, and two evident clusters of QTLs were detected (Figure 3). One is located on E05, which also harbors two adjacent sub-clusters of QTLs conserved in the three locations. The former is comprised between 62.45 to 66.5 cM, and contained QTLs for *stean* and *toan*; the other is at 75.48 cM, and included coincident QTLs for *hyan, toan* and *corcol*. The second main cluster is on E10 at 231.77–236.98 cM and included the major QTLs *stean10.1*, *lvean10.1*, *adlan10.1* and *flian10.1*, and the minor QTLs *hyan10.1*, *corcol10.1*, and *toan10.1*.

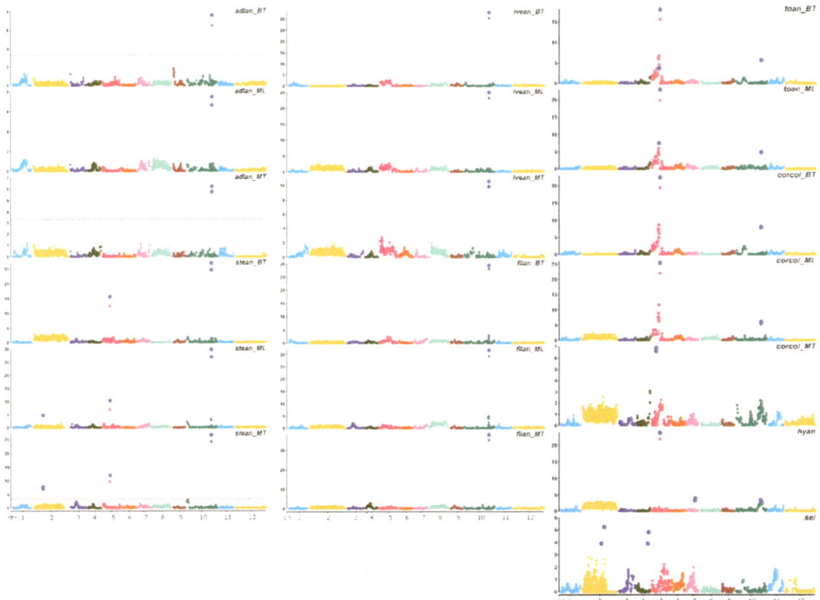

Figure 3. QTLs identified for the traits in study. Blue dots represent markers within the confidence interval of the QTL (LODmax−1 interval), with LOD values plotted against genome locations. Red lines in the Manhattan plots indicate LOD significance threshold.

Table 4. QTL detected in the mapping population. For each trait the chromosomal location (Chr.), the genome-wide thresholds (GW) at $p = 0.05$ (as determined from 1000 permutations) are indicated.

Trait	GW	Battipaglia								GW	Montanaso Lombardo								GW	Monsampolo del Tronto							
		QTL	Chr.	cM	Locus	LOD	CI	PVE	A		QTL	Chr.	cM	Locus	LOD	CI	PVE	A		QTL	Chr.	cM	Locus	LOD	CI	PVE	A
adlan	3.3	adlan 10.1 BT	E10	236.98	CH10_95 003635	7.59	236.98	21.0	−0.6002	3.3	adlan 10.1 ML	E10	236.98	CH10_95 003635	7.47	236.98	20.4	−0.5872	3.3	adlan 10.1 MT	E10	236.98	CH10_95 003635	6.21	236.987	17.7	−0.3465
stean	3.2									3.2	stean 2.1 ML	E02	88.73	CH02_30 555633	4.90	88.73	5.6	−0.2294	3.2	stean2.1 ML	E02	88.73	CH02_30 555713	7.71	88.733	9.5	−0.3255
		stean 5.1 BT	E05	66.39	CH05_36 124744	15.58	62.45–66.51	21.5	−0.4263		stean 5.1 ML	E05	66.39	CH05_36 124744	9.22	66.39	13.0	−0.3156		stean 5.1 ML	E05	64.51	CH05_36 124744	9.70	62.45–64.516	16.8	−0.3882
		stean 10.1 BT	E10	231.46	CH10_94 779014	26.93	231.46	45.1	−0.6000		stean 10.1 ML	E10	231.46	CH10_94 779014	25.92	231.46	48.1	−0.5894		stean 10.1 MT	E10	231.46	CH10_94 779014	20.76	231.465	40.3	−0.5851
lvean	3.2	lvean 10.1 BT	E10	231.46	CH10_94 779014	32.78	231.46	64.2	−0.7883	3.2	lvean 10.1 ML	E10	231.46	CH10_94 779014	24.92	231.46	53.2	−0.6202	3.2	lvean 10.1 MT	E10	231.46	CH10_94 779014	10.52	231.465	28.2	−0.5629
flian	3.1	flian 10.1 BT	E10	231.46	CH10_94 779014	29.26	231.46	59.5	−0.5680	3.1	flian 10.1 ML	E10	231.46	CH10_94 779014	31.64	231.46	61.9	−0.6544	3.1	flian 10.1 MT	E10	231.46	CH10_94 779014	37.70	231.465	69.3	−0.7393
toan	3.4	toan 5.1 BT	E05	75.48	CH05_37 533757	17.89	75.48	35.7	−110.9	3.4	toan 5.1 ML	E05	75.48	CH05_37 533757	22.92	75.48	43.9	−116208									
		toan 5.2 BT	E05	66.39	CH05_36 076199	3.72	61.55–66.39	5.2	−0.6245		toan 5.2 ML	E05	66.39	CH05_36 124744	7.44	61.55–66.39	8.7	−0.7422									
		toan 10.1 BT	E10	232.77	CH10_94 275882	5.67	232.77	9.3	−0.5498		toan 10.1 ML	E10	232.77	CH10_94 275882	4.82	232.77	6.8	−0.4464									
corcol	3.3	corcol 5.1 BT	E05	75.48	CH05_37 533757	22.39	75.484	40.2	−107.477	3.3	corcol 5.1 ML	E05	75.48	CH05_37 533757	25.31	75.48	45.2	−112175	3.3	corcol 5.1 MT	E05	39.39	CH05_17 086140	6.88	39.397	19.3	−0.3440
		corcol 10.1 BT	E10	232.77	CH10_94 275882	7.88	232.77	11.1	−0.5493		corcol 10.1 ML	E10	232.77	CH10_94 281914	5.95	232.77	7.7	−0.4507									
hyan										3.3	hyan5.1	E05	75.48	CH05_37 533757	23.72	75.48	48.1	−0.9234									
											hyan7.1	E07	83.94	CH07_13 2761839	3.93	79.32–83.94	5.8	−0.3223									
											hyan10.1	E10	231.46	CH10_94 684020	3.37	230.12–233.45	4.5	−0.2807									
sei										3.0	sei2.1	E02	204.18	CH02_54 633733	5.23	204.18	10.4	−5.689									
											sei2.2	E02	176.24	CH02_63 996392	3.89	176.24	7.6	5.109									
											sei4.1	E04	108.42	CH04_10 2121728	4.81	102.22–114.85	9.7	−0.8733									

3.3.1. QTL Affecting Plant Pigmentation in Eggplant

Adaxial Leaf Lamina Anthocyanin (*adlan*)

A single major QTL, *adlan10.1*, was mapped on E10 at 236.987 cM next to the marker CH10_95003635 in all the three environments, and explains 21% of PVE in BT, 20.4% in ML and 17.7% in MT.

Stem Anthocyanins (*stean*)

Two major QTLs on E05 and E10 were detected in all the environments. The QTL *stean10.1* on E10 explains from 40% to 48% of the PVE and is located at 231.46 cM (proximal marker: CH10_94779014). The second major QTL *stean5.1*, explains from 13% to 21% of the PVE and maps on E05, within the same confidence interval (62.45–66.51 cM) in all locations (proximal marker: CH05_36124744). A third minor QTL was spotted in ML and MT, but not in BT, on E02 at 88.73 cM (proximal marker: CH02_30555633), and explains 6% and 9% of PVE, respectively.

Leaf Venation Anthocyanins (*lvean*)

One major QTL determining anthocyanin pigmentation in leaf venation (*lvean10.1*), explaining from 29% to 64% of PVE, was mapped on E10 in all locations at 231.46 cM (proximal marker: CH10_94779014).

Flower Anthocyanin Intensity (*flian*)

The unique major QTL *flian10.1* mapped, in all environments, on E10 at 231.46 cM (proximal marker: CH10_94779014), explains from 59% to 69% of the PVE.

Anthocyanin Tonality (*toan*)

Data for *toan* were only available for BT and ML environments. A major QTL (*toan5.1*) explaining 36% and 47% in BT and ML, respectively, was spotted on E05 at 75.48 (proximal marker: CH05_37533757). A minor QTL (*toan5.2*) was identified on E05 within the same CI in both locations at 61.55–66.39 cM. A third QTL with a minor effect (PVE explained from 6.8% to 9.3%) was spotted on E10 at 231.46 cM.

Hypocotyl Anthocyanins (*hyan*)

The QTL analysis for this trait was performed with data from one environment. A major and two minor QTLs were spotted on E05, E07 and E10, respectively. The largest effect locus (*hyan5.1*) explains 48% of PVE and is located on E05 at 75.48 cM (proximal marker: CH05_37533757). A minor QTL (*hyan7.1*), explaining 6% of the PVE, maps on E07 at 83.95 cM (proximal marker: CH07_132671839). The QTL with the lower effect (4.5% PVE explained) was located on E10 at 231.46 cM (proximal marker: CH10_94779014).

Corolla Colour (*corcol*)

A major QTL (*corcol5.1*), explaining from 40% to 44% of the PVE, was spotted on E05 at 75.48 cM (proximal marker: CH05_37533757) in BT and ML, while *corcol5.2* was located at 39.40 cM (proximal marker CH05_17086140) and explains 19% of PVE in MT. A minor QTL (*corcol10.1*), explaining 7.7%–11.1% of the PVE, was spotted on E10 in BT and ML (not in MT) in the same position at 232.77 cM (proximal marker: CH10_94275882).

3.3.2. QTL Affecting Speed of Plant Emergence Index

On E02, we mapped the largest effect locus (*sei2.1*), explaining 10.4% of PVE and located at 204.18 cM (proximal marker: CH02_63996392), together with a minor QTL (*sei 2.2*), at 176 cM (proximal marker: CH02_54633733), which explains 8% of the PVE. A further minor QTL was located on E04, at 108.43 cM (proximal marker: CH04_102121728), which explains 10% of PVE.

3.4. Candidate Genes Identification

To find out candidate genes at the identified QTLs in the confidence interval region, we exploited the annotation of the available eggplant genome sequence by searching for genes, including transcription factors, putatively involved in the genetic control of the traits in study. For each QTL, the position, best candidate genes ID, acronym (abbreviation) and predicted function are reported in Table 5.

Table 5. Candidate genes spotted within the interval of detected QTLs. For each QTL, position, best candidate genes ID, acronym (abbreviation) and putative function are provided.

QTL	Approximative Position	Gene IDs SMEL_	Gene Abbreviation	Predicted Function
sei 2.1	9.2 Mb	002g153950.960	NHL6	2x NDR1/HIN1-like protein 6
	54 Mb	002g158620	PECS–2.1	Pectinesterases 2
		002g158940	SNL2	Paired amphipathic helix protein Sin3-like 2
between sei 2.1 and sei 2.2	~54–60 Mb	002g159100	LAC11	Laccase
		002g159470	ENY	Zinc finger ENHYDROUS
		002g159370.380	PLT6	2x polyol transporter
		002g159480	GBF1	G-box-binding factor 1
sei 2.2	60–63 Mb	002g160070.080	CAR2	2x C2-Domain Abscisic Acid-Related Proteins
		002g160170	GRDP1	Glycine-rich domain-containing protein 1
		002g159720	TCP1	T-complex protein 1 subunit zeta 1
		002g159870	MSR4	Peptide methionine sulfoxide reductase
sei 4.1	102 Mb	004g219910.920		2x Serpins-ZX
		004g220200–220	NPF4.5/NPF4.3	3x NRT1/ PTR protein fam 4.5/4.3
		004g220270	KO	Ent-kaurene oxidase
		004g220280	REM16	B3 domain transcription factor
		004g220780	GAF1	Zinc finger GAI-ASSOCIATED FACTOR 1
	~102 Mb	004g221390		abscisic acid 8' hydrolase 4
		004g221470	BZIP44	bZIP transcription factor 44
toan10.1 corcol 10.1	~94 Mb	010g352310–490	ANS	4x 2-oxoglutarate/Fe(II)-dependent dioxygenase
		010g352500	JRG21	2-oxoglutarate/Fe(II)-dependent dioxygenase
	94 Mb	010g352790		Myb family transcription factor
		010g352650	SAC8	Phosphoinositide phosphatase
lvean 10.1 stean 10.1 flian10.1 hyan 10.1	94.7 Mb	010g352910	BES1/BZR1	BES1/BZR1 transcription factor
		010g352980		Ankyrin repeat-containing protein
		010g352930	DREB2C	Dehydration-Responsive Element-Binding Protein 2C
		010g353040	PPC6–1	protein phosphatase 2C
	~94.7 Mb	010g353090–110	RAPTOR	3x RAPTOR - Regulatory-associated protein of TOR
adlan 10.1	95 Mb	010g353170–200		5x peroxidase
		010g353200		Protein disulfide-isomerase
hyan 7.1	132.7 Mb	007g289310–410	MYBs	6x similar to MYB15/14/58/102
		007g289700.710	NDB	2x NAD(P)H dehydrogenase
		007g289780		F-box/kelch-repeat protein
		007g289410	PPC6–6	protein phosphatase 2C

Table 5. Cont.

QTL	Approximative Position	Gene		Predicted Function
		IDs SMEL_	Abbreviation	
stean 2.1	30.5 Mb	002g155860		Ankyrin repeat-containing protein
		002g155880	LIMYB	L10-interacting MYB domain-containing protein
		002g155890	HSP70	Similar to Heat shock 70 kDa protein
		002g155950.970	LARP1C	2x La-related protein 1C-like
stean 5.1 *toan 5.2*	36.2 Mb	005g235930	HSP	17.3 kDa class II heat shock protein
		005g236210	BHLH93	similar to Transcription factor bHLH93
	~36.2 Mb	005g236240	AAT	Acetyl-CoA-benzylalcohol acetyltransferase
hyan 5.1 *toan 5.1* *corcol 5.1*	~37.5 Mb	005g236840–90		7x Calmodulin-like genes
	37.5 Mb	005g236910.20	BKI1	2x BR1 kinase inhibitors
		005g236720.30	TOGT1	2x Scopoletin glucosyltransferase
	~37.5 Mb	005g236480	BHLH84	Transcription factor bHLH84
		005g236490.00	AZF3,ZAT10	2x Zinc finger protein
		005g236570–620	CYP81Q32,VQ31	cytochrome p450

sei2.1

The major QTL associated to *sei* lies on E02 at 204.18 cM and includes four markers in the confidence interval, whose physical positions are quite distant, i.e., 48–54 Mb and 9.2 Mb. The top-linked marker is located at around 54 Mb, in a region containing 15 genes, among which a Pectinesterase 2 and SNL2, a protein involved in response to hormonal stimulus, appeared to be good candidates. The region at 9.2 Mb contains two colocalizing markers: in the interval around 0.3Mb, five genes are present, including two putative NHL6, putatively involved in the response to abscisic acid (ABA).

sei2.2

The minor QTL *sei2.2* lies on E02 at 176 cM, in a region of six co-localizing markers located at 60–63 Mb. This region contains 70 genes, some of which may be eligible as candidates, such as the ones associated to the response to abscisic acid (GRDP1 and CAR2), two polyol transporters, a TCP1 and peptide methionine sulfoxide reductase. In the region beneath *sei2.1* and *sei2.2*, a laccase, a zinc finger and a G-box-binding factor 1 putatively involved in seed germination are also present.

sei4.1

In the interval around 0.5 Mb from *sei4.1*, several Serpins-ZX and NRT1/PTR genes protein were identified, as well as an Ent-kaurene oxidase and a B3 domain protein, both involved in gibberellin synthesis. By slightly relaxing the confidence interval (until around 1 M), other genes involved in the signaling pathway of ABA degradation (among which an abscisic acid hydroxylase) were found.

stean10.1, lvean10.1, flian10.1, hyan10.1

The major QTLs *stean10.1, lvean10.1, flian10.1,* identified at 231.47 cM on E10 in the three environments, as well as the minor QTL *hyan10.1,* lie at around 94.7 Mb, in the top part of the cluster of QTLs associated to anthocyanin amount/coloration intensity and very close to the QTLs for *corcol* and *toan*. In this region, 17 genetic markers are co-segregating, corresponding to a physical interval comprised between 94.54 and 94.88 Mb. The region includes 19 annotated genes, among which a BES1/BRZ1 transcription factor, a DREB2C (dehydration-responsive element-binding protein 2C), an Ankyrin repeat-containing protein and a PPC6-1 (putative protein phosphatase 2C) are eligible as candidates involved in the anthocyanin synthesis. By slightly relaxing the confidence interval, three proteins annotated as regulatory-associated protein of TOR (RAPTOR) are also present.

corcol10.1-toan10.1

The QTLs *corcol10.1* and L *toan10.1*, both identified in BT and ML in the cluster on E10 at 232.77cM lean in a region located on E10 containing ten co-localizing markers and physically located at 93.8-94.2 Mb. In this region, the most promising candidate genes are a putative MYB family transcription factor (SMEL_010g352790) and a phosphoinositide phosphatase, SAC8 (SMEL_010g352650), a new class of phosphatase playing a role in vacuolar trafficking. By slightly expanding the region of interest, five genes (SMEL_010g352310, SMEL_010g352320, SMEL_010g352330, SMEL_010g352490 and SMEL_010g352500) were identified and annotated as having predicted 2-oxoglutarate/Fe(II)-dependent dioxygenase activity (ANS). In the same region, three sequences with homology with regulatory-associated protein of TOR (RAPTOR) are localized.

adlan10.1

The major QTL *adlan10.1* lies at 236.99 cM on E10, physically located at 94.9–95.08 Mb. A total of seven genes were identified as candidate, five of which were annotated as peroxidases or protein disulfide-isomerases.

stean5.1, toan5.2

The major QTL *stean5.1* identified in the three environments as well as *toan5.2*, specific to BT and ML, underlined a region on E5 in the interval 36.0–36.3 Mb, containing twelve genes. Among them, a class II heat shock protein and a putative BHLH could represent a good candidate. By increasing the confidence interval to the physical position of the marker 3311_PstI_L361, the candidate SMEL_005g236240, annotated as an acetyl-CoA-benzylalcohol acetyltransferase, was found.

toan5.1-corcol5.1-hyan5.1

The major QTL *toan5.1* as well as *corcol5.1* identified in ML and BT and *hyan5.1* in ML lean on E05, in a small region at ~37.5 Mb. Among the seven annotated genes which lie in this interval, two genes annotated as BKI1 (BRI1 kinase inhibitor 1), are eligible as the best candidate. By slightly increasing the region analyzed, two scopoletin glucosyltransferases, as well as a cluster of three genes annotated as encoding cytochrome P450, two zinc fingers and a BHLH-like protein were spotted.

hyan7.1

The minor QTL for *hyan7.1* lies on E07 at 83.94 cM, whose confidence interval physical region spanned between 132.76 and 134.02 Mb. More than 80 annotated genes were found, among which the most interesting are a cluster of putative candidates annotated as similar to MYB14−15−58−102, a PPC6−6 and a F-box/kelch-repeat protein.

stean2.1

The minor QTL *tean2.1* lies on E02 at about 30.5 Mbp. A total of five annotated genes were spotted, among which a L10 interacting MYB domain protein, a Heat shock 70 kDa protein and an AKRP—ankyrin repeat domain-containing protein—are eligible as possible candidates.

4. Discussion

4.1. Genetic Map Construction and Phenotyping

Studies on eggplant genome organization have received an increasing amount of interest in the last few decades, turning it from a "genomic orphan species" to a crop with a high-quality genomic sequence available [36]. As in several other crops, the low level of polymorphism within the cultivated eggplant germplasm required huge efforts in detecting markers exploitable for linkage mapping purposes [61,62].

Several "first generation" inter-specific and intra-specific maps were developed, with the former exploiting a higher genetic polymorphism, but being of minor relevance for marker-assisted breeding [63]. QTLs associated to morphological and plant production traits, as well as parthenocarpy, resistances to fungal and bacterial wilts were identified through bi-parental approaches and genome-wide association (GWA) studies on the basis of the available linkage maps.

An intraspecific F_2 segregating population, obtained from the same cross from which we developed the RIL population in the present study, proved to be a highly efficient tool for the detection of more than 140 QTLs associated to leaf, flower, plant and fruit traits, fruit biochemical composition and resistances to fungal wilts [22,28–31]. Our F_7 RIL population was also previously used for anchoring the "67/3" eggplant genome sequence [36]. Here, we exploited the GBS-derived approach, as applied by Acquadro et al. [64] in eggplant, to develop a new high density linkage map including 7249 SNPs assigned to the 12 chromosomes and spanning 2169.23 cM. Its genetic length is longer than the previously published intra-specific maps [14,22,24,28], as well as the one recently made available by Salgon et al. [34], which spans about 1500 cM and includes 1170 markers.

The newly created map clearly represents a step forward compared to the one we developed for anchoring the genome [36]. This was generated using markers derived from an imputation-based method following low-coverage sequencing of the same mapping population. The pipeline took windows containing 100 SNPs along scaffolds to convert them into genetic markers, which were actually based on the haplotype of 100 SNPs.

This approach was useful in anchoring the scaffolds to pseudomolecules, but some drawbacks are present, especially for QTL analysis and candidate gene identification. Indeed, this contained three chromosomal locations (E02, E08 and E11), which were split into two different portions. The newly developed map actually includes 12 chromosomes, which in turn may increase the efficiency in identifying candidate regions during QTL analysis. Furthermore, the map developed in this study contains a slightly higher number of gaps shorter than 5cM than the previous one, but some chromosomes have a larger gap. On the other hand, the newly created map is shorter (~500 cM) and contains more markers (1285) (Table S2) than the previous one, resulting in a more dense and saturated map.

Finally, in the map used for anchoring the genome sequence, a marker is based on a window of 100 SNPs, whose size is dependent on the polymorphic level of that specific chromosome regions and whose coordinates in the genome are not well defined. On the contrary, in the new map, each marker is based on a single SNP, allowing us to know the precise position of each marker in the genome and make it possible for a breeder to identify the genes located in a QTL region on the basis the available annotation. Furthermore, GBS data provide information on the SNPs that generated the markers, which is of utility for more targeted analysis.

Our map contains about 24% distorted markers, presumably as a result of the genetic distance between the parents as well as possible preferential or gametic/zygotic selection occurring during the development of our RIL population. However, we included these markers in order to increase the genomic coverage of the genetic map, which reached about 96% of the physical sequence. Indeed, if properly handled, these markers do not cause detrimental effects and increase the potential of QTL mapping, as previously reported [65,66]. The breeding line '305E40', used as the female parent, contains scattered introgressed regions from *S. aethiopicum*, with a large portion on chromosome E02, which includes the locus Rfo-sa1 [29,45]. This may justify the reduced recombination observed not only on E02, but also E09 and E12.

All the anthocyanin-related traits showed a high h^2_{BS} (lowest value of 86% for *lvean_BT*) with a high correlation of their phenotypic value among different environments. Similar results were previously reported for some of the traits in the study in the F_2 population developed from the same cross [28,30]. Transgressive genotypes were infrequent and always deviated towards the less pigmented parent '305E40'. The parental line '305E40' produced less vigorous seeds, but about 25% of the RIL population showed a further reduction in the speed of seedling emergence; this is presumably also because we were not able to identify all the QTLs affecting this trait, implying that some other QTLs still remain to be identified.

Conventionally, a 'major' QTL is defined as such when, in addition to justifying a PVE greater than 10% [67], it is conserved in multiple seasons/locations [68–70]. We identified at least one major QTL for all the traits in study including *hyan* and *sei*, although in this case the traits were evaluated only in one environment.

4.2. QTLs and Underlying Candidate Genes

The least and the most convincing LOD scores associated with the major QTLs were 3.89 (*sei2.2*) and 37.70 (*flian10.1_MT*), respectively. The explained PVE varied from 10.4% (*sei2.1*) up to 69.3% (*flian10.1_MT*), and most of the identified QTLs were stable across two/three environments, making them potentially useful for marker-assisted selection. On the other hand, some QTLs were identified in just one (*corcol5.2_MT*) or two (like *corcol5.1* or *stean2.1*) environments. This suggests a strong environmental effect on their expression, but also that the rather limited genetic variation in the mapping population did not allow us to fully dissect the genetic bases of these traits [71].

4.2.1. Seed Emergency Index

Seed germination is the switch from a dormant embryonic state to a highly active phase of growing. It is also defined as the sum of events that begin with seed imbibition and culminate in the emergence of the embryonic axis (usually the radicle) from the seed coat [72,73]. This progression is controlled by several internal factors, such as auxins, abscisic acid (ABA), cytokinins, ethylene, and GA content and balance, as well as environmental factors that include water availability, temperature, and light [74]. Eggplant, as with most of the *Solanum* species, is mainly propagated by seeds, whose vigor influences their germination and seedling emergence performance. Seed dormancy, low uniformity and poor germination rate have been documented in many eggplant accessions as well as in wild and allied species, including those employed as rootstock or those useful for introgression breeding [11,75–79].

Here, we reported, for the first time, three QTLs associated to seed vigor in eggplant, assessed by evaluating the speed of emergence index. Only one major QTL was spotted, suggesting that other key regions controlling this trait are still to be identified.

However, interestingly, both a major and minor QTL (i.e., *sei2.1* and *sei2.2*) were detected on chromosome E02, with *sei2.1* inherited from the female parent '305E40'. As previously pointed out, this breeding line harbors on chromosome E02 an introgressed fragment from *S. aethiopicum*, [28,36], associated to the *Fusarium oxysporum* resistance locus *Rfo-sa1* [45]. Thus, we could speculate that the introgressed portion of *S. aethiopicum* genome might be also involved in the genetic control of this trait, as this allied species usually displays a delayed germination with respect to eggplant.

The introgressed region may also be responsible of an inaccurate positioning of the genomic sequences in this region, and, consequently, in a reduction in the QTL mapping efficiency. Indeed, the candidate genes encompassing *sei2.1* are physically located both at 48–54 Mb and 9.2 Mb on E02. In the first large region, a paired amphipathic helix protein Sin3-like 2 could be a good candidate gene as it belongs to a class of proteins, involved in the response to hormonal stimuli and in the seed dormancy breakdown [80], while the NHL6 genes identified at 9 Mb have been reported to play an important role in the abiotic stress-induced abscisic acid (ABA) signaling and biosynthesis, acting as positive regulator of ABA-mediated seed germination inhibition [81].

Sei2.2 overlies some candidates annotated as similar to previously described genes involved in the ABA response and seed dormancy breakdown: two membrane C2-domain abscisic acid-related proteins (CAR2) [82] and a GRDP1—glycine-rich domain-containing protein 1 [83]. Furthermore, two TCP1 encoding genes and a peptide methionine sulfoxide reductase should also be considered as similar genes are involved in the repair mechanism during seed dormancy release in *Arabidopsis* and the increase in *M. truncatula* seed longevity by reducing the protein oxidation damage [84,85], respectively.

Several Serpins-ZX and NRT1/ PTR coding-genes were identified in the *sei4.1* region. The Serpin gene family has gained attention in wheat and barley for its role in grain development, but these genes could play a possible role in the mobilization of sugars during germination by enhancing β-amylase enzymatic activity and preventing β-amylase aggregation during oxidative stress [86]. The cluster of NPF6/NRT1–1 nitrate or di/tripeptide transporters, also spotted in *sei4.1*, are potentially involved in nitrate sensing and signaling [87,88], and genes belonging to this class are reported to regulate seed development, germination and dormancy cycling in fava bean and *Arabidopsis* [89–91], with a possible involvement in the ABA transport [92]. Nitrate itself is reported as a signal molecule that controls several aspects of plant development including seed dormancy, as higher nitrate accumulation in mother plants leads to lower seed dormancy [93].

Good candidates for future studies might also be other genes in the same region encoding an Ent-kaurene oxidase and two B3 domain proteins, all involved in GA biosynthesis: the latter are regulating factors of the pathway in which the former is a key enzyme, and it is reported that suppressive mutations in the coding region of both genes cause a delay in seed germination and seedling development [94,95].

4.2.2. Anthocyanins

Anthocyanins are among the most represented flavonoid compounds in plants and are responsible for the pigmentation of many flowers and fruits. They have an essential eco-physiological role in attracting pollinators and seed dispersers [96,97] and are also implicated in the response against biotic and abiotic stresses [98,99]

The genetic control of anthocyanin formation, distribution and accumulation has been widely studied in Solanaceae species [38–43,100,101]. This was long thought to be a complex trait in eggplant, involving several loci with assumed epistatic interactions and/or pleiotropic effects [102,103]. More recently, QTL-related studies allowed us to identify the chromosome regions involved in anthocyanin distribution in eggplant tissues and organs, highlighting their synteny with tomato [28,30,32,44] and, thanks to the recent availability of an high quality eggplant genome sequence coupled with metabolomic analyses [37], allowed us to localize putative candidate genes.

As previously observed [28,30,32], our ultra-high density genetic linkage map confirmed that the clusters on E10 and E05 are involved in the pigmentation of eggplant tissues, which may be associated with two different aspects of the anthocyanin synthesis among tissues, but likely control different processes linked to anthocyanin accumulation in diverse tissues. Indeed, the cluster on E10 is mainly prominent for anthocyanin production and accumulation in the vegetative plant organs, except in the corolla of the flower, whose pigmentation is governed by the major QTL on E5 with a smaller contribution by a minor QTL on E10. Conversely, the cluster on E05 contains QTLs more likely associated with the anthocyanin tonality (in flower, with *frucol*, and in general in all the vegetative tissues, represented by *toan*) and with the accumulation of anthocyanins in hypocotyl (*hyan*). Although the phenotypic data available for *hyan* were only collected in one environment, the combined results with *stean* QTLs seem to suggest that both 5 and 10 are involved, but with stronger specific effect on anthocyanin accumulation of *hyan5.1* in the hypocotyl at plantlet stage and of *stean10.1* in the stem of the fully developed plant. Overall, the joint effect of both E05 and E10 QTLs could impact on *hyan*, *stean*, *corcol* and *toan* through an interaction between genes, influencing both tonality and anthocyanin intensity.

4.2.3. Anthocyanin Related Candidate Genes Identifications

The biosynthesis of anthocyanin is one of the most studied pathways in plants, with most of the genes encoding for enzymes and regulatory transcription factors (TFs) identified in several plant species, including Solanaceae [3,104]. The anthocyanin pathway is under the control of many early (EBGs) or late (LBGs) biosynthetic genes, with the former involved in the first steps of biosynthesis of flavonols and other flavonoid compounds and the latter involved in the ensuing steps of the pathway until the final steps of decoration, leading to different anthocyanin compounds. Each enzymatic step of this complex pathway is finely tuned by co-activators independent and functionally redundant R2R3-MYB regulatory proteins which regulate the expression of structural genes, alone or in complexes with other TFs belonging to the basic helix-loop-helix (BHLH) family. The control of the biosynthetic pathway is strongly dependent on tissue, developmental stage and environment, and has only been partially elucidated in eggplant with regard solely to the fruit peel coloration [36,103,105–108].

Cluster on Chromosome E10

The cluster identified on E10 lies in a region of 1.5 Mb, between 93.5 and 95 Mb, containing three clusters of colocalizing QTLs.

We spotted, next to the upper limit of the cluster and close to *toan*, five genes predicted as ANS, which might be involved in the oxidation of leucocyanidin, in the second to last step of anthocyanin biosynthesis [109]. A further characterization highlighted that these genes are located in a genomic region of the parental line '67–3' containing retrotransposon-like sequences, which could alter the expression pattern of nearby genes [110]. Within the QTL for *toan* at 232.77 cM, we identified a putative MYB transcription factor and a phosphoinositide phosphatase, belonging to a class of phosphatases that plays a key role in abiotic stress response, vacuolar trafficking and anthocyanin accumulation [111].

Lying within *stean10.1*, we identified an ankyrin repeat-containing protein coding gene together with a dehydration-responsive element-binding protein 2C and a PPC6–1 (protein phosphatase 2C). Ankyrin repeat proteins were reported to be involved in the anthocyanin synthesis pathway [112], while the other two candidates are putatively involved in the response to abiotic stresses and detoxification [113]. Finally, alongside *adlan10.1*, we identified five peroxidases coding genes which may be involved in the degradation of anthocyanin, influencing the overall coloration of the tissues where they are expressed. Indeed, enzymatic degradation has been considered to be responsible for anthocyanin breakdown in plants, leading to pigment concentration reduction and colour fading [114]. Recent studies have shown that PODs and laccases (LACs) are responsible for anthocyanin catalysis [115, 116], and also, in combination with some environmental factors, such as high temperature and low light density, were reported to enhance the peroxidase activity [104,117].

QTL cluster on Chromosome E05

The region identified on E05 which controls *stean*, *toan*, *hyan* and *corcol* contains two slightly separate clusters. The upper region, including the QTLs for *toan5.2* and *stean5.1*, was already spotted by Barchi et al. [28] as a genomic region involved in the control of several anthocyanin-related traits (such as *stean*) and the corolla color. In the same position, Toppino et al. [30] mapped QTLs associated to peel fruit color as well as to the presence and amount (determined by HPLC) of D3R and nasunin, the two different anthocyanins in the eggplant peel. Analogous QTLs in the distal portion of E05 were also previously identified by GWAS approaches [32,44].

Our candidate gene search highlighted the presence of an acetyl-CoA-benzylalcohol acetyltransferase (AAT), for which we speculate a function in the aromatic group decoration as the last step of the anthocyanin biosynthetic pathway. Furthermore, the distribution and dominance relationships strongly support the hypothesis that AAT is active in '67/3' and inactive in '305E40', and thus responsible of the acetylation of the D3R glucosidic group [118] and the subsequent conversion into nasunin. The comparison of the Illumina sequencing data available for the two parental lines [36] revealed a 1bp indel which could determine a loss-of-function mutation in the *305E40_AAT* CDS sequence, opening the path to a deeper functional study of this gene.

The cluster of QTLs for *corcol*, *hyan* and *toan* on E5 is proximal to *toan5.1* and *stean*; in this region, two scopoletin glucosyltransferase coding genes, involved in the phenylpropanoid pathways [119], were spotted, alongside four genes annotated as cytochrome P450, known as playing pivotal roles in the biosynthesis of plant secondary metabolites, including phenylpropanoids and phytoalexins [120,121].

QTLs for hyan7.1 and stean2.1

In the *hyan7.1* region, a cluster of MYB genes were identified, with homology to sequences known to be involved in the phenylpropanoid pathway and more specifically in the stilbene biosynthesis in *Vitis* [122], lignin in *Arabidopsis* [123] and anthocyanin in forage legumes [124]. In the same region, other interesting candidate genes are a PPC6-6 (probable protein phosphatase 2C) and a F-box/kelch-repeat protein, a class of regulators reported to be associated to phenylpropanoid pathway [125].

Finally, for *stean2.1*, a valid candidate gene was an AKRP—ankyrin repeat domain-containing protein—which was reported to be involved in the anthocyanin synthesis pathway [112].

5. Conclusions

Our results demonstrate that the newly developed map, supported by genome annotation, supplies a key tool to gather valuable information for QTL fine mapping, candidate gene identification, and for the development of molecular markers suitable for identifying favorable alleles, and thus increasing the precision and efficiency of selection in breeding. Our high-density intraspecific map made it possible not only to validate previously reported QTLs, but also to identify new ones associated with the plant anthocyanin pigmentation intensity and tonality, as well as to better define their underlying chromosomal regions. Thanks to the availability of genome annotation, it was also possible to provide a set of relevant candidate genes involved in the anthocyanin biosynthetic process and regulation, some of which are already the subject of ongoing studies.

Finally, the map allowed us to identify the first QTLs affecting seed vigor in eggplant, as measured by the speed of seedling emergence from soil. The identification of the genetic bases of this trait are of key importance, since seed germination and seedling emergence represent two of the most vulnerable phases of a crop cultivation cycle, and less vigorous seeds, other than reducing the crop competitiveness toward weeds, increase the exposure of seedlings to abiotic (drought, heat) and biotic (soil-borne pests) stresses. On the whole, the QTLs we detected provide important knowledge on the genomic region linked physiological and phenotypic properties in eggplant which may be usefully exploited in future breeding programs.

Supplementary Materials: The following are available online at http://www.mdpi.com/2073-4425/11/7/745/s1, Figure S1: Distribution of sequenced reads, after quality cleaning and trimming procedures, across the parental lines and F$_1$ (first three samples) as well as the RIL mapping population (in million reads); Figure S2: Eggplant linkage map. Marker names are shown to the right of each chromosome, with map distances (in cM) shown on the left; Figure S3: Heat map representing pairwise recombination fractions with LOD scores for each marker on all 12 chromosomes; Figure S4: Marey Map plots of SNPs mapped to positions on the 12 *S. melongena* chromosomes vs. their physical positions on the v3. eggplant pseudomolecules from Barchi et al. [36]; Table S1: Markers name, chromosome, genetic position of all markers on the map and RILS graphical haplotypes are reported. The color 'orange' represents homozygous alleles from the "305E40" parent, 'blue' represent homozygous alleles from "67/3" parent, 'green' as heterozygous alleles and white as missing data; Table S2: Parameters associated with the eggplant genetic map compared to the previous developed by Barchi et al. [36].

Author Contributions: Conceptualization, F.M., F.S., S.L. and G.L.R.; Data curation, L.B.; Formal analysis, L.B., M.M. and E.P.; Funding acquisition, L.T., F.C., E.P. and G.L.R.; Investigation, L.T., L.B., F.M., N.A., D.P., S.G., T.S., S.F., A.M., T.C. and G.L.R.; Visualization, L.T., L.B. and E.P.; Writing—original draft, L.T., L.B. and M.M.; Writing—review & editing, F.M., F.C., E.P., F.S., S.L. and G.L.R. All authors have read and agreed to the published version of the manuscript.

Funding: This work was partially funded by the European Union's Horizon 2020 Research and Innovation Program under the Grant Agreement number 677379 (G2P-SOL project: 'linking genetic resources, genomes and phenotypes of solanaceous crops') and by CARIPLO Foundation in the frame of the project Code 2016–0723 (WAKEAPT project: 'Seed Wake-up with Aptamers: a New Technology for Dormancy Release and Improved Seed Priming').

Conflicts of Interest: The authors declare no conflict of interest.

References

1. Fukuoka, H.; Yamaguchi, H.; Nunome, T.; Negoro, S.; Miyatake, K.; Ohyama, A. Accumulation, functional annotation, and comparative analysis of expressed sequence tags in eggplant (*Solanum melongena* L.), the third pole of the genus Solanum species after tomato and potato. *Gene* **2010**, *450*, 76–84. [CrossRef] [PubMed]

2. Hirakawa, H.; Shirasawa, K.; Miyatake, K.; Nunome, T.; Negoro, S.; Ohyama, A.; Yamaguchi, H.; Sato, S.; Isobe, S.; Tabata, S.; et al. Draft genome sequence of eggplant (*Solanum melongena* L.): The representative solanum species indigenous to the old world. *DNA Res. Int. J. Rapid Publ. Rep. Genes Genomes* **2014**, *21*, 649–660. [CrossRef]

3. Albert, V.A.; Chang, T.H. Evolution of a hot genome. *Proc. Natl. Acad. Sci. USA* **2014**, *111*, 5069–5070. [CrossRef] [PubMed]

4. Lester, R.N.; Hasan, S.M.Z. Origin and Domestication of the Brinjal Eggplant, *Solanum Melongena*, from *S. incanum*, in Africa and Asia. In *Solanaceae III: Taxonomy, Chemistry, and Evolution*; Hawkes, J.G., Lester, R.N., Nee, M., Estrada, N., Eds.; Royal Botanic Gardens: Kew, UK, 1991; pp. 369–387.

5. Meyer, R.S.; Karol, K.G.; Little, D.P.; Nee, M.H.; Litt, A. Phylogeographic relationships among Asian eggplants and new perspectives on eggplant domestication. *Mol. Phylogenetics Evol.* **2012**, *63*, 685–701. [CrossRef] [PubMed]

6. Cericola, F.; Portis, E.; Toppino, L.; Barchi, L.; Acciarri, N.; Ciriaci, T.; Sala, T.; Rotino, G.L.G.L.; Lanteri, S. The Population Structure and diversity of eggplant from Asia and the Mediterranean Basin. *PLoS ONE* **2013**, *8*, e73702. [CrossRef] [PubMed]

7. Page, A.; Gibson, J.; Meyer, R.S.; Chapman, M.A. Eggplant domestication: Pervasive gene flow, feralization, and transcriptomic divergence. *Mol. Biol. Evol.* **2019**, *36*, 1359–1372. [CrossRef]

8. FAO. Available online: http://faostat3.fao.org/home/E.org (accessed on 1 June 2020).

9. Zhang, Y.; Liu, H.; Shen, S.; Zhang, X. Improvement of eggplant seed germination and seedling emergence at low temperature by seed priming with incorporation SA into KNO 3 solution. *Front. Agric. China* **2011**, *5*, 534–537. [CrossRef]

10. Gisbert, C.; Prohens, J.; Nuez, F. Treatments for improving seed germination in eggplant and related species. *Acta Hortic.* **2009**, *898*, 45–51. [CrossRef]

11. Forti, C.; Ottobrino, V.; Bassolino, L.; Toppino, L.; Rotino, G.L.; Pagano, A.; Macovei, A.; Balestrazzi, A. Molecular dynamics of pre-germinative metabolism in primed eggplant (*Solanum melongena* L.) seeds. *Hortic. Res.* **2020**, *7*, 87. [CrossRef]

12. Doganlar, S.; Frary, A.; Daunay, M.C.; Lester, R.N.; Tanksley, S.D. A comparative genetic linkage map of eggplant (*Solanum melongena*) and its implications for genome evolution in the Solanaceae. *Genetics* **2002**, *161*, 1697–1711.

13. Doğanlar, S.; Frary, A.; Daunay, M.-C.; Huvenaars, K.; Mank, R.; Frary, A. High resolution map of eggplant (*Solanum melongena*) reveals extensive chromosome rearrangement in domesticated members of the Solanaceae. *Euphytica* **2014**, 1–11. [CrossRef]

14. Wu, F.N.; Eannetta, N.T.; Xu, Y.M.; Tanksley, S.D. A detailed synteny map of the eggplant genome based on conserved ortholog set II (COSII) markers. *Theor. Appl. Genet.* **2009**, *118*, 927–935. [CrossRef]

15. Frary, A.; Doganlar, S.; Daunay, M.C.; Tanksley, S.D. QTL analysis of morphological traits in eggplant and implications for conservation of gene function during evolution of solanaceous species. *Theor. Appl. Genet.* **2003**, *107*, 359–370. [CrossRef]

16. Frary, A.; Frary, A.; Daunay, M.-C.; Huvenaars, K.; Mank, R.; Doğanlar, S. QTL hotspots in eggplant (*Solanum melongena*) detected with a high resolution map and CIM analysis. *Euphytica* **2014**, *197*, 211–228. [CrossRef]

17. Gramazio, P.; Prohens, J.; Plazas, M.; Andújar, I.; Herraiz, F.J.; Castillo, E.; Knapp, S.; Meyer, R.S.; Vilanova, S. Location of chlorogenic acid biosynthesis pathway and polyphenol oxidase genes in a new interspecific anchored linkage map of eggplant. *BMC Plant Biol.* **2014**, *14*, 350. [CrossRef] [PubMed]

18. Sunseri, F.; Sciancalepore, A.; Martelli, G.; Acciarri, N.; Rotino, G.L.; Valentino, D.; Tamietti, G. Development of RAPD-AFLP Map of Eggplant and Improvement of Tolerance to *Verticillium* Wilt. In Proceedings of the Acta Horticulturae, Leuven, Belgium, 30 September 2003; pp. 107–115.

19. Nunome, T.; Ishiguro, K.; Yoshida, T.; Hirai, M. Mapping of fruit shape and color development traits in eggplant (*Solanum melongena* L.) based on RAPD and AFLP markers. *Breed. Sci.* **2001**, *51*, 19–26. [CrossRef]

20. Nunome, T.; Suwabe, K.; Iketani, H.; Hirai, M. Identification and characterization of microsatellites in eggplant. *Plant Breed.* **2003**, *122*, 256–262. [CrossRef]

21. Nunome, T.; Negoro, S.; Kono, I.; Kanamori, H.; Miyatake, K.; Yamaguchi, H.; Ohyama, A.; Fukuoka, H. Development of SSR markers derived from SSR-enriched genomic library of eggplant (*Solanum melongena* L.). *Theor. Appl. Genet.* **2009**, *119*, 1143–1153. [CrossRef]

22. Barchi, L.; Lanteri, S.; Portis, E.; Stàgel, A.; Valè, G.; Toppino, L.; Leonardo Rotino, G. Segregation distortion and linkage analysis in eggplant (*Solanum melongena* L.). *Genome* **2010**, *53*. [CrossRef]

23. Fukuoka, H.; Miyatake, K.; Nunome, T.; Negoro, S.; Shirasawa, K.; Isobe, S.; Asamizu, E.; Yamaguchi, H.; Ohyama, A. Development of gene-based markers and construction of an integrated linkage map in eggplant by using Solanum orthologous (SOL) gene sets. *Theor. Appl. Genet.* **2012**, *125*, 47–56. [CrossRef]

24. Miyatake, K.; Saito, T.; Negoro, S.; Yamaguchi, H.; Nunome, T.; Ohyama, A.; Fukuoka, H. Development of selective markers linked to a major QTL for parthenocarpy in eggplant (*Solanum melongena* L.). *Theor. Appl. Genet.* **2012**, *124*, 1–11. [CrossRef]

25. Miyatake, K.; Saito, T.; Negoro, S.; Yamaguchi, H.; Nunome, T.; Ohyama, A.; Fukuoka, H. Detailed mapping of a resistance locus against *Fusarium* wilt in cultivated eggplant (*Solanum melongena*). *Theor. Appl.* **2016**, *129*, 357–367. [CrossRef] [PubMed]

26. Barchi, L.; Lanteri, S.; Portis, E.; Acquadro, A.; Valè, G.; Toppino, L.; Rotino, G.L.G.L.L.G.L.; Vale, G.; Toppino, L.; Rotino, G.L.G.L.L.G.L.; et al. Identification of SNP and SSR markers in eggplant using RAD tag sequencing. *BMC Genom.* **2011**, *12*. [CrossRef]

27. Baird, N.A.; Etter, P.D.; Atwood, T.S.; Currey, M.C.; Shiver, A.L.; Lewis, Z.A.; Selker, E.U.; Cresko, W.A.; Johnson, E.A. Rapid SNP Discovery and Genetic Mapping Using Sequenced RAD Markers. *PLoS ONE* **2008**, *3*, e3376. [CrossRef] [PubMed]

28. Barchi, L.; Lanteri, S.; Portis, E.; Valè, G.; Volante, A.; Pulcini, L.; Ciriaci, T.; Acciarri, N.; Barbierato, V.; Toppino, L.; et al. A RAD tag derived marker based eggplant linkage map and the location of QTLs determining anthocyanin pigmentation. *PLoS ONE* **2012**, *7*, e43740. [CrossRef] [PubMed]

29. Portis, E.; Barchi, L.; Toppino, L.; Lanteri, S.; Acciarri, N.; Felicioni, N.; Fusari, F.; Barbierato, V.; Cericola, F.; Valè, G.; et al. QTL mapping in eggplant reveals clusters of yield-related loci and orthology with the tomato genome. *PLoS ONE* **2014**, *9*, e89499. [CrossRef] [PubMed]

30. Toppino, L.; Barchi, L.; Lo Scalzo, R.; Palazzolo, E.; Francese, G.; Fibiani, M.; D' Alessandro, A.; Papa, V.; Laudicina, V.A.; Sabatino, L.; et al. Mapping Quantitative Trait Loci Affecting Biochemical and Morphological Fruit Properties in Eggplant (*Solanum melongena* L.). *Front. Plant Sci.* **2016**. [CrossRef]

31. Barchi, L.; Toppino, L.; Valentino, D.; Bassolino, L.; Portis, E.; Lanteri, S.; Rotino, G.L. QTL analysis reveals new eggplant loci involved in resistance to fungal wilts. *Euphytica* **2018**, *214*, 20. [CrossRef]

32. Cericola, F.; Portis, E.; Lanteri, S.; Toppino, L.; Barchi, L.; Acciarri, N.; Pulcini, L.; Sala, T.; Rotino, G.L. Linkage disequilibrium and genome-wide association analysis for anthocyanin pigmentation and fruit color in eggplant. *BMC Genom.* **2014**, *15*, 896. [CrossRef]

33. Portis, E.; Cericola, F.; Barchi, L.; Toppino, L.; Acciarri, N.; Pulcini, L.; Sala, T.; Lanteri, S.; Rotino, G.L. Association mapping for fruit, plant and leaf morphology traits in eggplant. *PLoS ONE* **2015**, *10*, e0135200. [CrossRef]

34. Salgon, S.; Raynal, M.; Lebon, S.; Baptiste, J.M.; Daunay, M.C.; Dintinger, J.; Jourda, C. Genotyping by sequencing highlights a polygenic resistance to ralstonia pseudosolanacearum in eggplant (*Solanum melongena* L.). *Int. J. Mol. Sci.* **2018**, *19*, 357. [CrossRef] [PubMed]

35. Wei, Q.; Wang, W.; Hu, T.; Hu, H.; Wang, J.; Bao, C. Construction of a SNP-based genetic map using SLAF-Seq and QTL analysis of morphological traits in eggplant. *Front. Genet.* **2020**, *11*, 178. [CrossRef]

36. Barchi, L.; Pietrella, M.; Venturini, L.; Minio, A.; Toppino, L.; Acquadro, A.; Andolfo, G.; Aprea, G.; Avanzato, C.; Bassolino, L.; et al. A chromosome-anchored eggplant genome sequence reveals key events in Solanaceae evolution. *Sci. Rep.* **2019**, *9*, 11769. [CrossRef] [PubMed]

37. Sulli, M.; Barchi, L.; Francese, G.; Toppino, L.; Diretto, G.; Mennella, G.; Lanteri, S.; Rotino, G.; Giuliano, G. Metabolic Profiling of a Recombinant Inbred Eggplant Population Reveals Key Metabolic QTLs controlling Fruit Nutritional Quality. In Proceedings of the Plant & Animal Genome XXVII Conference (pe0948), San Diego, CA, USA, 12–16 January 2019.

38. Van Eck, H.J.; Jacobs, J.M.E.; Dijk, J.; Stiekema, W.J.; Jacobsen, E. Identification and mapping of three flower colour loci of potato (*S. tuberosum* L.) by RFLP analysis. *Theor. Appl. Genet.* **1993**, *86*, 295–300. [CrossRef] [PubMed]

39. Chaim, A.B.C.; Borovsky, Y.B.; De Jong, W.D.J.; Paran, I.P. Linkage of the *A* locus for the presence of anthocyanin and *fs10.1*, a major fruit-shape QTL in pepper. *Theor. Appl. Genet.* **2003**, *106*, 889–894. [CrossRef]

40. Van Eck, H.J.; Jacobs, J.M.E.; van den Berg, P.M.M.M.; Stiekema, W.J.; Jacobsen, E. The inheritance of anthocyanin pigmentation in potato (*Solanum tuberosum* L.) and mapping of tuber skin colour loci using RFLPs. *Heredity* **1994**, *73*, 410–421. [CrossRef]

41. Borovsky, Y.; Oren-Shamir, M.; Ovadia, R.; De Jong, W.; Paran, I. The *A* locus that controls anthocyanin accumulation in pepper encodes a *MYB* transcription factor homologous to *Anthocyanin2* of Petunia. *Theor. Appl. Genet.* **2004**, *109*, 23–29. [CrossRef]

42. Bovy, A.; Schijlen, E.; Hall, R. Metabolic engineering of flavonoids in tomato (*Solanum lycopersicum*): The potential for metabolomics. *Metabolomics* **2007**, *3*, 399–412. [CrossRef]

43. Gonzali, S.; Mazzucato, A.; Perata, P. Purple as a tomato: Towards high anthocyanin tomatoes. *Trends Plant Sci.* **2009**, *14*, 237–241. [CrossRef]

44. Ge, H.; Liu, Y.; Jiang, M.; Zhang, J.; Han, H.; Chen, H. Analysis of genetic diversity and structure of eggplant populations (*Solanum melongena* L.) in China using simple sequence repeat markers. *Sci. Hortic.* **2013**, *162*, 71–75. [CrossRef]

45. Toppino, L.; Vale, G.; Rotino, G.L. Inheritance of *Fusarium* wilt resistance introgressed from *Solanum aethiopicum* Gilo and Aculeatum groups into cultivated eggplant (*S.melongena*) and development of associated PCR-based markers. *Mol. Breed.* **2008**, *22*, 237–250. [CrossRef]

46. Doyle, J.J.; Doyle, J.L. Isolation of plant DNA from fresh tissue. *Focus* **1990**, *12*, 13–14.

47. Acquadro, A.; Lanteri, S.; Scaglione, D.; Arens, P.; Vosman, B.; Portis, E. Genetic mapping and annotation of genomic microsatellites isolated from globe artichoke. *Theor. Appl. Genet.* **2009**, *118*, 1573–1587. [CrossRef]

48. Acquadro, A.; Barchi, L.; Gramazio, P.; Portis, E.; Vilanova, S.; Comino, C.; Plazas, M.; Prohens, J.; Lanteri, S. Coding SNPs analysis highlights genetic relationships and evolution pattern in eggplant complexes. *PLoS ONE* **2017**, *12*, e0180774. [CrossRef] [PubMed]

49. Li, H. Aligning Sequence Reads, Clone Sequences and Assembly Contigs with BWA-MEM. *arXiv* **2013**, arXiv:1303.3997.

50. Li, H. A statistical framework for SNP calling, mutation discovery, association mapping and population genetical parameter estimation from sequencing data. *Bioinform.* **2011**, *27*, 2987–2993. [CrossRef]

51. Broman, K.W.; Wu, H.; Sen, S.; Churchill, G.A.; Sen, Ś.; Churchill, G.A.; Sen, S.; Churchill, G.A. R/qtl: QTL mapping in experimental crosses. *Bioinformatics* **2003**, *19*, 889–890. [CrossRef]

52. Van Ooijen, J.W. *JoinMap ®4, Software for the Calculation of Genetic Linkage Maps in Experimental Populations*; Kyazma BV: Wageningen, The Netherlands, 2006.

53. Van Os, H.; Stam, P.; Visser, R.G.F.; van Eck, H.J. SMOOTH: A statistical method for successful removal of genotyping errors from high-density genetic linkage data. *Theor. Appl. Genet.* **2005**, *112*, 187–194. [CrossRef]

54. Voorrips, R.E. MapChart: Software for the graphical presentation of linkage maps and QTLs. *J. Hered.* **2002**, *93*, 77–78. [CrossRef]

55. R Core Team. R: A Language and Environment for Statistical Computing; R Foundation for Statistical Computing. 2020. Available online: https://www.r-project.org/ (accessed on 15 May 2020).

56. Jansen, R.C. Interval mapping of multiple quantitative trait loci. *Genetics* **1993**, *135*, 205–211.

57. Jansen, R.C.; Stam, P. High-resolution of quantitative traits into multiple loci via interval mapping. *Genetics* **1994**, *136*, 1447–1455. [PubMed]

58. Van Ooijen, J.W. *MapQTL 5, Software for the Mapping of Quantitative Trait Loci in Experimental Populations*; Kyazma BV: Wageningen, The Netherlands, 2004; Volume 63.

59. Churchill, G.A.; Do erge, R.W. Empirical Threshold Values for Quantitative Trait Mapping. *Genetics* **1994**, *138*, 963–971. [PubMed]

60. CMplot. Available online: https://github.com/YinLiLin/R-CMplot (accessed on 15 June 2020).

61. Stàgel, A.; Portis, E.; Toppino, L.; Rotino, G.L.; Lanteri, S. Gene-based microsatellite development for mapping and phylogeny studies in eggplant. *BMC Genom.* **2008**, *9*, 357. [CrossRef] [PubMed]

62. Barchi, L.; Acquadro, A.; Alonso, D.; Aprea, G.; Bassolino, L.; Demurtas, O.; Ferrante, P.; Gramazio, P.; Mini, P.; Portis, E.; et al. Single Primer Enrichment Technology (SPET) for high-throughput genotyping in tomato and eggplant germplasm. *Front. Plant Sci.* **2019**, *10*, 1005. [CrossRef] [PubMed]

63. Barchi, L.; Portis, E.; Toppino, L.; Rotino, G.L. Molecular Mapping, QTL Identification, and GWA Analysis. In *The Eggplant Genome*; Springer International Publishing: Cham, Switzerland, 2019; pp. 41–54.

64. Acquadro, A.; Barchi, L.; Portis, E.; Mangino, G.; Valentino, D.; Mauromicale, G.; Lanteri, S. Genome reconstruction in *Cynara cardunculus* taxa gains access to chromosome-scale DNA variation. *Sci. Rep.* **2017**, *7*, 5617. [CrossRef]

65. Zhang, X.F.; Wang, G.Y.; Dong, T.T.; Chen, B.; Du, H.S.; Li, C.B.; Zhang, F.L.; Zhang, H.Y.; Xu, Y.; Wang, Q.; et al. High-density genetic map construction and QTL mapping of first flower node in pepper (*Capsicum annuum* L.). *BMC Plant Biol.* **2019**, *19*, 167. [CrossRef]

66. Hu, X.H.; Zhang, S.Z.; Miao, H.R.; Cui, F.G.; Shen, Y.; Yang, W.Q.; Xu, T.T.; Chen, N.; Chi, X.Y.; Zhang, Z.M.; et al. High-density genetic map construction and identification of QTLs controlling oleic and linoleic acid in peanut using SLAF-seq and SSRs. *Sci. Rep.* **2018**, *8*, 1–10. [CrossRef]

67. Collard, B.; Jahufer, M.; Brouwer, J.; Pang, E. An introduction to markers, quantitative trait loci (QTL) mapping and marker-assisted selection for crop improvement: The basic concepts. *Euphytica* **2005**, *142*, 169–196. [CrossRef]

68. Li, Z.; Jakkula, L.; Hussey, R.S.; Tamulonis, J.P.; Boerma, H.R. SSR mapping and confirmation of the QTL from PI96354 conditioning soybean resistance to southern root-knot nematode. *Theor. Appl. Genet.* **2001**, *103*, 1167–1173. [CrossRef]

69. Lindhout, P. The perspectives of polygenic resistance in breeding for durable disease resistance. *Euphytica* **2002**, *124*, 217–226. [CrossRef]

70. Pilet-Nayel, M.P.-N.; Muehlbauer, F.M.; McGee, R.M.; Kraft, J.K.; Baranger, A.B.; Coyne, C.C. Quantitative trait loci for partial resistance to Aphanomyces root rot in pea. *Theor. Appl. Genet.* **2002**, *106*, 28–39. [CrossRef] [PubMed]

71. Lander, E.S.; Botstein, D. Mapping mendelian factors underlying quantitative traits using RFLP linkage maps. *Genetics* **1989**, *121*, 185–199. [PubMed]

72. Nonogaki, H.; Bassel, G.W.; Bewley, J.D. Germination-still a mystery. *Plant Sci.* **2010**, *179*, 574–581. [CrossRef]

73. Weitbrecht, K.; Müller, K.; Leubner-Metzger, G. First off the mark: Early seed germination. *J. Exp. Bot.* **2011**, *62*, 3289–3309. [CrossRef]

74. Rajjou, L.; Duval, M.; Gallardo, K.; Catusse, J.; Bally, J.; Job, C.; Job, D. Seed Germination and Vigor. *Annu. Rev. Plant Biol.* **2012**, *63*, 507–533. [CrossRef]

75. Adebola, P.O.; Afolayan, A.J. Germination responses of *Solanum aculeastrum*, a medicinal species of the Eastern Cape, South Africa. *Seed Sci. Technol.* **2006**, *34*, 735–740. [CrossRef]

76. Demir, I.; Ermis, S.; Okçu, G.; Matthews, S. Vigour tests for predicting seedling emergence of aubergine (*Solanum melongena* L.) seed lots. *Seed Sci. Technol.* **2005**, *33*, 481–484. [CrossRef]

77. Joshua, A. Seed germination of *Solanum incanum*: An example of germination programs of tropical vegetable crops. *Acta Hortic.* **1977**, *83*, 155–162. [CrossRef]

78. Ibrahim, M.; Munira, M.K.; Kabir, M.S.; Islam, A.K.M.S.; Miah, M.M.U. Seed germination and graft compatibility of wild solanum as rootstock of tomato. *J. Biol. Sci.* **2001**, *1*, 701–703. [CrossRef]

79. Taab, A.; Andersson, L. Seed dormancy dynamics and germination characteristics of *Solanum nigrum*. *Weed Res.* **2009**, *49*, 490–498. [CrossRef]

80. Wang, Z.; Cao, H.; Sun, Y.; Li, X.; Chen, F.; Carles, A.; Li, Y.; Ding, M.; Zhang, C.; Deng, X.; et al. Arabidopsis paired amphipathic helix proteins *SNL1* and *SNL2* redundantly regulate primary seed dormancy via abscisic acid-ethylene antagonism mediated by histone deacetylation. *Plant Cell* **2013**, *25*, 149–166. [CrossRef]

81. Bao, Y.; Song, W.M.; Pan, J.; Jiang, C.M.; Srivastava, R.; Li, B.; Zhu, L.Y.; Su, H.Y.; Gao, X.S.; Liu, H.; et al. Overexpression of the *NDR1/HIN1*-Like gene *NHL6* modifies seed germination in response to abscisic acid and abiotic stresses in *Arabidopsis*. *PLoS ONE* **2016**, *11*. [CrossRef]

82. Lockhart, J. Membrane bound: C2-domain abscisic acid-related proteins help abscisic acid receptors get where they need to go. *Plant Cell* **2014**, *26*, 4566. [CrossRef]

83. Rodríguez-Hernández, A.A.; Ortega-Amaro, M.A.; Delgado-Sánchez, P.; Salinas, J.; Jiménez-Bremont, J.F. *AtGRDP1* Gene Encoding a Glycine-Rich domain protein is involved in germination and responds to ABA Signalling. *Plant Mol. Biol. Report.* **2014**, *32*, 1187–1202. [CrossRef]

84. Arc, E.; Chibani, K.; Grappin, P.; Jullien, M.; Godin, B.; Cueff, G.; Valot, B.; Balliau, T.; Job, D.; Rajjou, L. Cold stratification and exogenous nitrates entail similar functional proteome adjustments during *Arabidopsis* seed dormancy release. *J. Proteome Res.* **2012**, *11*, 5418–5432. [CrossRef]

85. Chatelain, E.; Hundertmark, M.; Leprince, O.; Gall, S.L.; Satour, P.; Deligny-Penninck, S.; Rogniaux, H.; Buitink, J. Temporal profiling of the heat-stable proteome during late maturation of *Medicago truncatula* seeds identifies a restricted subset of late embryogenesis abundant proteins associated with longevity. *Plant Cell Environ.* **2012**, *35*, 1440–1455. [CrossRef] [PubMed]

86. Cohen, M.; Davydov, O.; Fluhr, R. Plant serpin protease inhibitors: Specificity and duality of function. *J. Exp. Bot.* **2019**, *70*, 2077–2085. [CrossRef]

87. Tsay, Y.F.; Chiu, C.C.; Tsai, C.B.; Ho, C.H.; Hsu, P.K. Nitrate transporters and peptide transporters. *FEBS Lett.* **2007**, *581*, 2290–2300. [CrossRef] [PubMed]

88. Ho, C.H.; Lin, S.H.; Hu, H.C.; Tsay, Y.F. *CHL1* functions as a nitrate sensor in plants. *Cell* **2009**, *138*, 1184–1194. [CrossRef]

89. Miranda, M.; Borisjuk, L.; Tewes, A.; Dietrich, D.; Rentsch, D.; Weber, H.; Wobus, U. Peptide and amino acid transporters are differentially regulated during seed development and germination in faba bean. *Plant Physiol.* **2003**, *132*, 1950–1960. [CrossRef]

90. Footitt, S.; Huang, Z.; Clay, H.A.; Mead, A.; Finch-Savage, W.E. Temperature, light and nitrate sensing coordinate Arabidopsis seed dormancy cycling, resulting in winter and summer annual phenotypes. *Plant J.* **2013**, *74*, 1003–1015. [CrossRef] [PubMed]

91. Osuna, D.; Prieto, P.; Aguilar, M. Control of seed germination and plant development by carbon and nitrogen availability. *Front. Plant Sci.* **2015**, *6*, 1023. [CrossRef] [PubMed]

92. Kanno, Y.; Hanada, A.; Chiba, Y.; Ichikawa, T.; Nakazawa, M.; Matsui, M.; Koshiba, T.; Kamiya, Y.; Seo, M. Identification of an abscisic acid transporter by functional screening using the receptor complex as a sensor. *Proc. Natl. Acad. Sci. USA* **2012**, *109*, 9653–9658. [CrossRef]

93. Alboresi, A.; Gestin, C.; Leydecker, M.T.; Bedu, M.; Meyer, C.; Truong, H.N. Nitrate, a signal relieving seed dormancy in Arabidopsis. *Plant Cell Environ.* **2005**, *28*, 500–512. [CrossRef] [PubMed]

94. Guo, X.; Hou, X.; Fang, J.; Wei, P.; Xu, B.; Chen, M.; Feng, Y.; Chu, C. The rice GERMINATION DEFECTIVE 1, encoding a B3 domain transcriptional repressor, regulates seed germination and seedling development by integrating GA and carbohydrate metabolism. *Plant J.* **2013**, *75*, 403–416. [CrossRef] [PubMed]

95. Zhang, H.; Li, M.; He, D.; Wang, K.; Yang, P. Mutations on *ent-kaurene oxidase 1* encoding gene attenuate its enzyme activity of catalyzing the reaction from ent-kaurene to ent-kaurenoic acid and lead to delayed germination in rice. *PLoS Genet.* **2020**, *16*, e1008562. [CrossRef]

96. Harborne, J.B.; Williams, C.A. Advances in flavonoid research since 1992. *Phytochemistry* **2000**, *55*, 481–504. [CrossRef]

97. Schemske, D.W.; Bradshaw, H.D. Pollinator preference and the evolution of floral traits in monkeyflowers (*Mimulus*). *Proc. Natl. Acad. Sci. USA* **1999**, *96*, 11910–11915. [CrossRef]

98. Close, D.C.; Beadle, C.L. The Ecophysiology of foliar anthocyanin. *Bot. Rev.* **2003**, *69*, 149–161. [CrossRef]

99. Gross, J. *Pigments in Vegetables: Chlorophylls and Carotenoids*; Springer Science & Business Media: New York, NY, USA, 2012; ISBN 1461520339.

100. De Jong, W.S.; Eannetta, N.T.; Jong, D.M.; Bodis, M. Candidate gene analysis of anthocyanin pigmentation loci in the Solanaceae. *Theor. Appl. Genet.* **2004**, *108*, 423–432. [CrossRef]

101. Stommel, J.R.; Dumm, J.M. Coordinated regulation of biosynthetic and regulatory genes coincides with anthocyanin accumulation in developing eggplant fruit. *J. Amer. Soc. Hort. Sci.* **2015**, *140*, 129–135. [CrossRef]

102. Tatebe, T. On inheritance of color in *Solanum melongena* L. *Jpn. J. Genet.* **1939**, *15*, 261–271. [CrossRef]

103. Tigchelaar, E.C.; Janick, J.; Erickson, H.T. The genetics of anthocyanin coloration in eggplant (*Solanum melongena* L.). *Genetics* **1968**, *60*, 475–491. [PubMed]

104. Liu, Y.; Tikunov, Y.; Schouten, R.E.; Marcelis, L.F.M.; Visser, R.G.F.; Bovy, A. Anthocyanin biosynthesis and degradation mechanisms in Solanaceous vegetables: A review. *Front. Chem.* **2018**, *6*, 52. [CrossRef]

105. Docimo, T.; Francese, G.; Ruggiero, A.; Batelli, G.; de Palma, M.; Bassolino, L.; Toppino, L.; Rotino, G.L.; Mennella, G.; Tucci, M. phenylpropanoids accumulation in eggplant fruit: Characterization of biosynthetic genes and regulation by a MYB transcription factor. *Front. Plant Sci.* **2015**, *6*, 1233. [CrossRef]

106. Moglia, A.; Florio, F.E.; Iacopino, S.; Guerrieri, A.; Milani, A.M.; Comino, C.; Barchi, L.; Marengo, A.; Cagliero, C.; Rubiolo, P.; et al. Identification of a new R3 MYB type repressor and functional characterization of the members of the MBW transcriptional complex involved in anthocyanin biosynthesis in eggplant (*S. melongena* L.). *PLoS ONE* **2020**, *15*, e0232986. [CrossRef]

107. Zhang, Y.; Hu, Z.; Chu, G.; Huang, C.; Tian, S.; Zhao, Z.; Chen, G. Anthocyanin accumulation and molecular analysis of anthocyanin biosynthesis-associated genes in eggplant (*Solanum melongena* L.). *J. Agric. Food Chem.* **2014**, *62*, 2906–2912. [CrossRef] [PubMed]

108. Xiao, X.O.; Lin, W.Q.; Li, K.; Feng, X.F.; Jin, H.; Zou, H. Transcriptome analyses reveal anthocyanin biosynthesis in eggplants. *PeerJ Prepr.* **2018**, *6*, e27289v1.

109. Saito, K.; Kobayashi, M.; Gong, Z.; Tanaka, Y.; Yamazaki, M. Direct evidence for anthocyanidin synthase as a 2-oxoglutarate-dependent oxygenase: Molecular cloning and functional expression of cDNA from a red forma of Perilla frutescens. *Plant J.* **1999**, *17*, 181–189. [CrossRef]

110. Hirsch, C.D.; Springer, N.M. Transposable element influences on gene expression in plants. *Biochim. Et Biophys. Acta—Gene Regul. Mech.* **2017**, *1860*, 157–165. [CrossRef]

111. Williams, M.E.; Torabinejad, J.; Cohick, E.; Parker, K.; Drake, E.J.; Thompson, J.E.; Hortter, M.; DeWald, D.B. Mutations in the Arabidopsis phosphoinositide phosphatase gene *SAC9* lead to overaccumulation of PtdIns(4,5)P2 and constitutive expression of the stress-response pathway. *Plant Physiol.* **2005**, *138*, 686–700. [CrossRef] [PubMed]

112. Yoo, J.; Ho Shin, D.; Cho, M.-H.; Kim, T.-L.; Hee Bhoo, S.; Hahn, T.-R. An ankyrin repeat protein is involved in anthocyanin biosynthesis in Arabidopsis. *Physiol. Plant.* **2011**, *142*, 314–325. [CrossRef] [PubMed]

113. Shen, X.F.; Zhao, Y.; Jiang, J.P.; Guan, W.X.; Du, J.F. Phosphatase *Wip1* in immunity: An overview and update. *Front. Immunol.* **2017**, *8*, 8. [CrossRef] [PubMed]

114. Luo, H.; Li, W.; Zhang, X.; Deng, S.; Xu, Q.; Hou, T.; Pang, X.; Zhang, Z.; Zhang, X. In planta high levels of hydrolysable tannins inhibit peroxidase mediated anthocyanin degradation and maintain abaxially red leaves of Excoecaria Cochinchinensis. *BMC Plant Biol.* **2019**, *19*, 1–20. [CrossRef]

115. Fang, F.; Zhang, X.L.; Luo, H.H.; Zhou, J.J.; Gong, Y.H.; Li, W.J.; Shi, Z.W.; He, Q.; Wu, Q.; Li, L.; et al. An intracellular Laccase is responsible for epicatechin-mediated Anthocyanin degradation in litchi fruit Pericarp. *Plant Physiol.* **2015**, *169*, 2391–2408. [CrossRef] [PubMed]

116. Zipor, G.; Duarte, P.; Carqueijeiro, I.; Shahar, L.; Ovadia, R.; Teper-Bamnolker, P.; Eshel, D.; Levin, Y.; Doron-Faigenboim, A.; Sottomayor, M.; et al. *In planta* anthocyanin degradation by a vacuolar class III peroxidase in *Brunfelsia calycina* flowers. *New Phytol.* **2015**, *205*, 653–665. [CrossRef]

117. Sytar, O.; Zivcak, M.; Bruckova, K.; Brestic, M.; Hemmerich, I.; Rauh, C.; Simko, I. Shift in accumulation of flavonoids and phenolic acids in lettuce attributable to changes in ultraviolet radiation and temperature. *Sci. Hortic.* **2018**, *239*, 193–204. [CrossRef]

118. Ino, I.; Yamaguchi, M.A. Acetyl-coenzyme A: Anthocyanidin 3-glucoside acetyltransferase from flowers of *Zinnia elegans*. *Phytochemistry* **1993**. [CrossRef]

119. Chong, J.; Baltz, R.; Schmitt, C.; Beffa, R.; Fritig, B.; Saindrenan, P. Downregulation of a pathogen-responsive tobacco UDP-Glc:phenylpropanoid glucosyltransferase reduces scopoletin glucoside accumulation, enhances oxidative stress, and weakens virus resistance. *Plant Cell* **2002**. [CrossRef]

120. Tanaka, Y.; Brugliera, F. Flower colour and cytochromes P450. *Philos. Trans. R. Soc. B Biol. Sci.* **2013**, *368*, 20120432. [CrossRef]

121. Su, V.; Hsu, B.D. Transient expression of the cytochrome p450 *CYP78A2* enhances anthocyanin production in flowers. *Plant Mol. Biol. Report.* **2010**. [CrossRef]

122. Höll, J.; Vannozzi, A.; Czemmel, S.; Donofrio, C.; Walker, A.R.; Rausch, T.; Lucchin, M.; Boss, P.K.; Dry, I.B.; Bogsa, J. The *R2R3-MYB* transcription factors *MYB14* and *MYB15* regulate stilbene biosynthesis in *Vitis vinifera*. *Plant Cell* **2013**. [CrossRef]

123. Ali, M.B.; McNear, D.H. Induced transcriptional profiling of phenylpropanoid pathway genes increased flavonoid and lignin content in Arabidopsis leaves in response to microbial products. *BMC Plant Biol.* **2014**. [CrossRef] [PubMed]

124. Albert, N.W. Subspecialization of R2R3-MYB repressors for anthocyanin and proanthocyanidin regulation in forage legumes. *Front. Plant Sci.* **2015**. [CrossRef] [PubMed]
125. Zhang, X.; Gou, M.; Liu, C.J. Arabidopsis kelch repeat F-Box proteins regulate phenylpropanoid biosynthesis via controlling the turnover of phenylalanine ammonia-lyase. *Plant Cell* **2013**. [CrossRef]

Article

Natural Variation Uncovers Candidate Genes for Barley Spikelet Number and Grain Yield under Drought Stress

Samar G. Thabet [1], Yasser S. Moursi [1], Mohamed A. Karam [1], Andreas Börner [2] and Ahmad M. Alqudah [2,*]

[1] Department of Botany, Faculty of Science, University of Fayoum, Fayoum 63514, Egypt; sgs03@fayoum.edu.eg (S.G.T.); ysm01@fayoum.edu.eg (Y.S.M.); mak04@fayoum.edu.eg (M.A.K.)
[2] Research Group Resources Genetics and Reproduction, Department Genebank, Leibniz Institute of Plant Genetics and Crop Plant Research, 06466 Seeland OT Gatersleben, Germany; boerner@ipk-gatersleben.de
* Correspondence: alqudah@ipk-gatersleben.de or ahqudah@gmail.com

Received: 22 March 2020; Accepted: 5 May 2020; Published: 11 May 2020

Abstract: Drought stress can occur at any growth stage and can affect crop productivity, which can result in large yield losses all over the world. In this respect, understanding the genetic architecture of agronomic traits under drought stress is essential for increasing crop yield potential and harvest. Barley is considered the most abiotic stress-tolerant cereal, particularly with respect to drought. In the present study, worldwide spring barley accessions were exposed to drought stress beginning from the early reproductive stage with 35% field capacity under field conditions. Drought stress had significantly reduced the agronomic and yield-related traits such as spike length, awn length, spikelet per spike, grains per spike and thousand kernel weight. To unravel the genetic factors underlying drought tolerance at the early reproductive stage, genome-wide association scan (GWAS) was performed using 121 spring barley accessions and a 9K single nucleotide polymorphisms (SNPs) chip. A total number of 101 significant SNPs, distributed over all seven barley chromosomes, were found to be highly associated with the studied traits, of which five genomic regions were associated with candidate genes at chromosomes 2 and 3. On chromosome 2H, the region between 6469300693-647258342 bp includes two candidate drought-specific genes (*HORVU2Hr1G091030* and *HORVU2Hr1G091170*), which are highly associated with spikelet and final grain number per spike under drought stress conditions. Interestingly, the gene expression profile shows that the candidate genes were highly expressed in spikelet, grain, spike and leaf organs, demonstrating their pivotal role in drought tolerance. To the best of our knowledge, we reported the first detailed study that used GWAS with bioinformatic analyses to define the causative alleles and putative candidate genes underlying grain yield-related traits under field drought conditions in diverse barley germplasm. The identified alleles and candidate genes represent valuable resources for future functional characterization towards the enhancement of barley cultivars for drought tolerance.

Keywords: GWAS; drought; barley; spikelet development; candidate gene

---

1. Introduction

Barley (*Hordeum vulgare* L.) ranks as one of the most important cereal crops worldwide. Globally, barley is the fourth most important cereal crop in terms of production after maize (*Zea mays* L.), rice (*Oryza sativa* L.), and wheat (*Triticum spp.*) (Faostat 2017, http://www.fao.org/faostat/en/#home). One limitation in achieving the production target is abiotic stress which limits the quality and nutritional value of the grain in cereal crops worldwide [1]. Among all abiotic stresses, drought is the most important environmental stress which limits crop production and yield [2–4], and is becoming more common particularly in the arid and semiarid regions [2].

Crops can be exposed to drought during their entire life cycle from vegetative to reproductive stages [5]. Drought stress affects crop growth and yield during all developmental stages [6]. Water shortage at early growth stages can cause severe problems for seedlings, restricting the emergence, growth and development of seedlings, and thus affecting grain yield [7]. Furthermore, the developing plants will have poor tillering capacity, leading to fewer tillers per unit area and thus lower yield potential. Moreover, drought in the period of stem elongation causes a decrease in the number of grains per unit area because it has a negative impact on floret formation and fertility [8].

At post-anthesis, water insufficiency reduces the grain filling rate and duration leading to shriveled grains [2]. Moreover, the effect of drought on yield is highly complex and involves processes as diverse as reproductive organs, gametogenesis, fertilization, embryogenesis and seed development [6,9]. Reproductive and seed development phases are especially sensitive to drought stress [2,10]. In barley, a reduction in the number of grains per spike, grain filling duration and dry matter accumulation have already been reported to decrease grain yield [2,10]. Several studies have reported that early growth stage parameters (e.g., tiller number, biomass formation, etc.) are highly correlated with yield potential and grain quality at harvest under both normal and drought stress conditions in various cereal crops, including barley [11,12]. Accordingly, understanding the genetic basis for drought tolerance in crop plants by identifying the genetic loci and the candidate genes associated with these traits is useful for developing new varieties with more drought-tolerant characters.

The Genome-Wide Association Scan (GWAS) approach is widely employed to reveal associations between genomic loci and advantageous traits in a given population based on linkage disequilibrium [13]. These loci then become targets for improving new genotypes by the breeder. The GWAS is very effective in identifying major candidate genes regulating mono- or oligogenic agricultural traits. Recently, GWAS has been successfully used to identify genes for yield-related traits [14,15]. Barley has a high-level population structure such as two-rowed and six-rowed cultivars, spring and winter barley [16]. This may lead to spurious marker–trait associations in GWAS [17]. Therefore, it is important to use strong statistical methods and strategies to control the population structure [17]. A mixed-linear model (MLM) approach has been developed to control spurious associations through account multiple levels of relatedness leading to better performance [18].

GWAS has successfully yielded genomic locations for quantitative trait loci (QTL) in crop cereals [13]. Identification of genomic locations for QTL using linked segregating markers is considered to be highly useful for marker-assisted breeding. Nevertheless, the ultimate goal of GWAS in crop species is to detect new QTL through genetic dissection of complex traits.

The present study aimed at identifying the genetic basis of drought tolerance at different developmental and growth phases in 121 spring barley accessions under field conditions using GWAS. In total, 101 SNPs showed association with different traits that were distributed across the seven chromosomes of barley. The identified QTL colocalized with several genes that are exclusively distributed on chromosomes 2H and 3H. The annotation and expression of these genes demonstrated their roles in drought tolerance.

2. Materials and Methods

2.1. Experimental Setup and Phenotyping

In total, 121 diverse spring barley accessions from different geographical origins were grown under the field conditions in the 2017/2018 growing season at the Experimental Station of University of Fayoum. The collection included 83 cultivars, 29 landraces, and 9 breeding lines. They originated from Europe (EU, 62), West Asia and North Africa (WANA, 24), East Asia (EA, 22), and the Americas (AM, 13). The row types were two-rowed (72) and six-rowed (49). The population structure and genetic diversity using the genotypic information of the accession in the collection are shown in Figure S1 that demonstrated there is no clear structure on our collection.

Five seeds from each accession were sown in plastic pots (40 cm × 26 cm × 26 cm) filled with field-soil on the 1st of December 2017 under field conditions. The soil texture was classified as clay-loam with pH = 8.0, total N% = 9.4 and available P = 58.0 ppm. Manual irrigation was performed as required, and 5 g (17:11:10/N:P:K) fertilizer was added to each pot. In field-grown plants, each accession was replicated in three pots of each treatment (control and drought). At the beginning of early reproductive (spikelet development phase) phase (~25 days after sowing), the plants were thinned into three plants per pot, standing with a border to eliminate positional and environmental effects on growth and development. Weeds were controlled manually.

Plants were irrigated until the onset of early reproductive, then the plants were then exposed to two watering treatments: (1) well-watered treatment (soil maintained at ~75% of field capacity (FC)); and (2) severe drought stress (at 35% FC). To maintain the targeted (~75% FC) and drought (~35% FC), eight randomly selected accessions were used as a reference. Before irrigation, eight reference pots were weighed and watered to adjust the corresponding field capacity, and the rest of the experiment was watered accordingly. Irrigation and drought treatment continued until maturity after that, irrigation withholds until harvest. Nine morphological, developmental and grain yield-related traits were measured from at least three biological replicates for each accession under each treatment. More information about the phenotypic trait measurements are explained in Table 1. Respective drought tolerance indices were calculated from the recorded data as described in (Table 1).

Table 1. The name and abbreviation of measured traits and respective description of measurements.

Trait	Abbreviation		Description
	Control	Drought	
Awn Tipping	AT_C	AT_D	The number of days from planting up to awn tipping.
Spike Heading	SH_C	SH_D	The number of days from planting up to spike heading
Anther Extrusion	AE_C	AE_D	The number of days from planting up to anther extrusion.
Plant Height	PH_C	PH_D	The distance between the ground level to the tip of the highest spikelet (excluding awns) in cm.
Spike Length	SL_C	SL_D	Distance from the base of the spike to the tip of the highest spikelet (excluding awns) in cm.
Awn Length	AL_C	AL_D	Distance from the tip of the spike to the end of the awn in cm.
No of Spikelets per Spike	NSS_C	NSS_D	The actual count of the number of spikelets.
No of Grains per Spike	NGS_C	NGS_D	The actual count of the number of the grains.
Thousand Grain Weight	TGW_C	TGW_D	The weight of 1000 grains randomly taken from each plot in gram (g).
Drought tolerance index (Awn Tipping)	DTI_AT		$DTI\,(AT) = \dfrac{AT \text{ under drought}}{AT \text{ under control}} \times 100$
Drought tolerance index (Spike Heading)	DTI_SH		$DTI\,(SH) = \dfrac{SH \text{ under drought}}{SH \text{ under control}} \times 100$
Drought tolerance index (Anther Extrusion)	DTI_AE		$DTI\,(AE) = \dfrac{AE \text{ under drought}}{AE \text{ under control}} \times 100$
Drought tolerance index (Plant Height)	DTI_PH		$DTI\,(PH) = \dfrac{PH \text{ under drought}}{PH \text{ under control}} \times 100$
Drought tolerance index (Spike Length)	DTI_SL		$DTI\,(SL) = \dfrac{SL \text{ under drought}}{SL \text{ under control}} \times 100$
Drought tolerance index (Awn Length)	DTI_AL		$DTI\,(AL) = \dfrac{AL \text{ under drought}}{AL \text{ under control}} \times 100$
Drought tolerance index (No of Spikelet per Spike)	DTI_NSS		$DTI\,(NSS) = \dfrac{NSS \text{ under drought}}{NSS \text{ under control}} \times 100$
Drought tolerance index (No of Grain per Spike)	DTI_NGS		$DTI\,(NGS) = \dfrac{NGS \text{ under drought}}{NGS \text{ under control}} \times 100$
Drought tolerance index (Thousand Grain Weight)	DTI_TGW		$DTI\,(TGW) = \dfrac{TGW \text{ under drought}}{TGW \text{ under control}} \times 100$

The average of temperature and humidity during the growing season of this experiment (2017–2018) was 16.62 °C and 55.16%, respectively. The maximum temperature was recorded in November and April (24.33 and 27.66 °C, respectively) with the minimum temperature from January to February

(8.29 and 9.12 °C, respectively). The minimum humidity was 30.37% and 30.41% in November and February, respectively, whereas the maximum humidity was recorded in April (78.33%) (Figure S2).

2.2. Data Analysis

Analysis of variance [19] was conducted to compare the controlled and drought stress conditions at $p < 0.05$ for all measured traits using GENSTAT 18 [20]. Data were analyzed as a randomized complete block design (RCBD) with three replications. The treatments were considered as main plots and the accessions were considered as sub-plots. Broad-sense heritability (H^2) for the measured traits under each condition separately was calculated using GENSTAT 18. The phenotypic data were subjected to Residual Maximum Likelihood (REML) to analyze it in mixed linear model (MLM). Mean estimation of each measured trait in each accession under each treatment was calculated as Best Linear Unbiased Estimates (BLUEs) using GENSTAT 18. Correlation matrix analysis among the traits in each treatment was separately calculated by GENSTAT 18. Comparison between treatments at $p < 0.05$ for each trait including boxplots were calculated using R-studio [21].

2.3. Genome-Wide Marker–Trait Associations

The accessions were genotyped by a 9K Illumina^{TM} SNPs chip. In the analysis, we only used the markers which passed the quality control as minor allele frequency (MAF) ≥ 0.05 with their physical positions. A mixed linear model (MLM) was performed to determine marker–trait associations between the estimated phenotypic traits (BLUEs) and genotypic data. Different statistical models, e.g., general-linear model (GLM), mixed linear model (MLM), and compressed MLM (CMLM)) were tested in GWAS using GAPIT R package [22]. Finally, we used MLM as a powerful model considering the population structure, including kinship and PCA, to control the population structure influence. False discovery rate (FDR) at 0.001 was calculated for each trait under each treatment separately and association signals passed the threshold of FDR at 0.001 ($-\log_{10} p$-values $\geq$ FDR) were used in further analyses. To be sure that our associations were true, we followed the GWAS and post-GWAS protocol published recently [13].

2.4. SNP-Gene-Based Association and Haplotype Analysis

At each chromosome, linkage disequilibrium (LD, r^2) among the significant SNPs within highly associated genomic region was calculated and presented as heatmap plot. This allowed us to define the most important physical position that had been used for candidate gene identification. The physical positions of SNPs exceeding FDR within the linkage disequilibrium interval were used in annotation for high-confidence (HC) candidate gene with other respective information using the barley genome explorer web-based with recent barley genome dataset (BARLEX; http://apex.ipk-gatersleben.de).

SNPs within the candidate gene physical position were used for further validation of SNP-Gene-based haplotype analyses and expression analyses. *T*-test at $p < 0.05$ was used to calculate the significant differences between alleles on the associated trait(s) [13]. RNA-seq datasets were derived from 16 different tissues of barley cv. 'Morex' cultivar, each with three biological replicates. In total 48 samples were used for generation of RNA-seq data.

From seven vegetative, six inflorescence, two developing grain and one germinating grain tissues, more details about RNA-seq experiments was published by Mascher et al. [23]. We used BARLEX; an expression database for barley that presented as FPPM (fragments per kilobase per million mapped reads).

3. Results

3.1. Phenotypic Characteristics and Natural Variation

In total, nine traits were recorded under control and drought treatments. Additionally, respective drought tolerance indices were calculated and used as a derived trait for GWAS. A wide range of phenotypic variation with normal distribution was detected for all traits (Figure 1 and Figure S3). The means of most of the studied traits showed a significant reduction under drought treatment compared to control conditions (Table 2 and Figure S3). There were no significant differences between the treatments in phase transition i.e., AT, SH and AE developmental stages (Table 2 and Figure S3). Drought treatment influenced significantly other developmental and yield traits such as plant height and spikelet number per spike (Table 2 and Figure S3). Notably, the genotypes showed wide range of variation in all drought tolerance indices (Table 2, Figure 2 and Figure S4).

Furthermore, heritability values were relatively high under drought, ranging from 0.64 for TGW to 0.84 for AT, whereas they ranged from 0.72 to 0.80 for AT and SH, respectively under control conditions. Additionally, the heritability values varied for tolerance indices from 0.68 for AE_DTI and NGS_DTI to 0.78 for AT_DTI (Table 2).

Table 2. Analysis of variance and heritability for the measured traits under control and drought treatments.

Trait	Control			H^2	Drought			H^2
	T	G	T × G		T	G	T × G	
Awn Tipping	ns	***	**	0.72	ns	**	***	0.84
Spike Heading	ns	***	**	0.80	ns	**	***	0.78
Anther Extrusion	ns	***	*	0.79	ns	*	***	0.81
Plant Height	***	***	***	0.75	***	***	***	0.71
Spike Length	***	***	***	0.74	***	***	***	0.69
Awn Length	***	***	***	0.79	***	***	***	0.76
No of Spikelet per Spike	***	***	***	0.76	***	***	***	0.65
No of Grain per Spike	***	***	***	0.77	***	***	***	0.68
Thousand Grain Weight	***	***	***	0.72	***	***	***	0.64
Awn Tipping_DTI	–	***	–	0.78				
Spike Heading_DTI	–	***	–	0.71				
Anther Extrusion_DTI	–	***	–	0.68				
Plant Height_DTI	–	***	–	0.73				
Spike Length_DTI	–	***	–	0.74				
Awn Length_DTI	–	***	–	0.71				
No of Spikelet per Spike _DTI	–	***	–	0.74				
No of Grain per Spike _DTI	–	***	–	0.68				
Thousand Grain Weight_DTI	–	***	–	0.77				

H^2—Heritability; T—Treatment; G—Genotype; T × G—Treatment by Genotype interaction; DTI—Drought Tolerant Index; The degree of significance is indicated as * p, 0.05; ** p, 0.01; *** p, 0.001; ns: not significant.

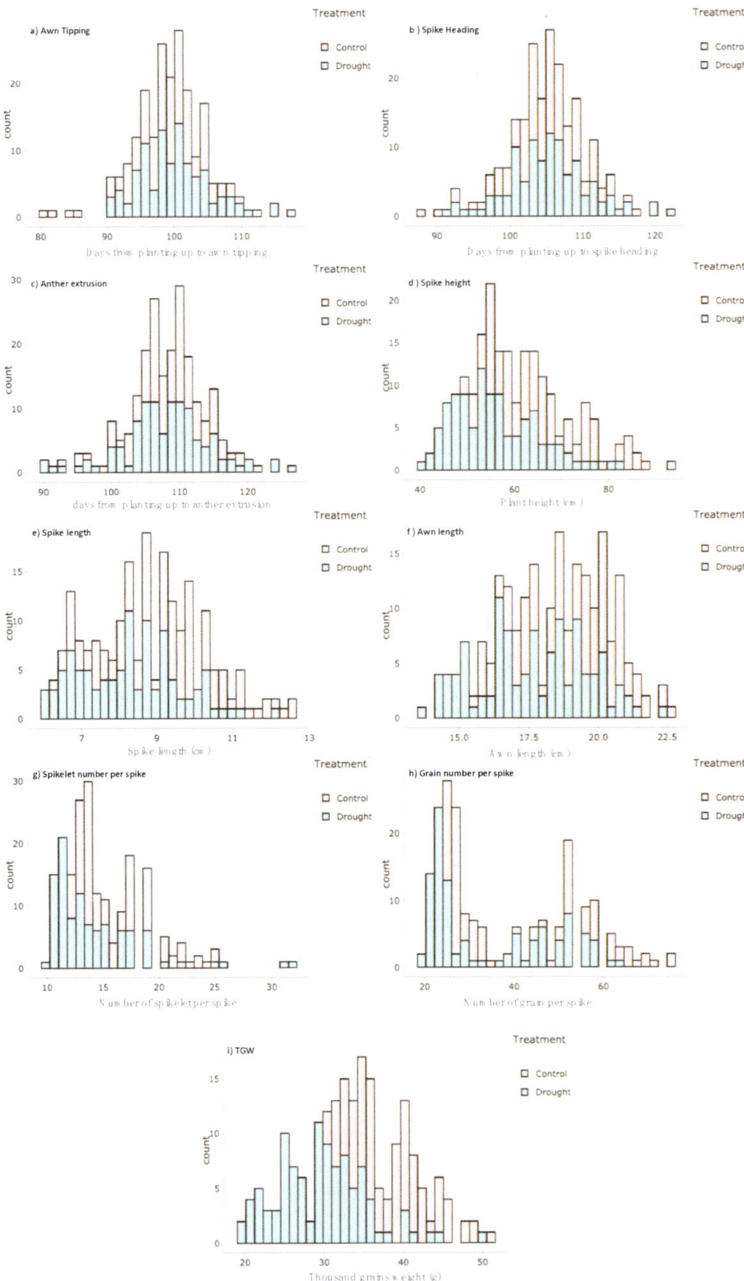

Figure 1. Histogram of phenotypic values distribution analysis of the studied traits; (**a**) Awn Tipping, (**b**) Spike Heading, (**c**) Anther Extrusion, (**d**) Plant Height, (**e**) Spike length, (**f**) Awn Length, (**g**) Spikelet per Spike, (**h**) Grain per Spike and (**i**) Thousand Grain Weight (TGW) in 121 spring barley accessions under control and drought stress.

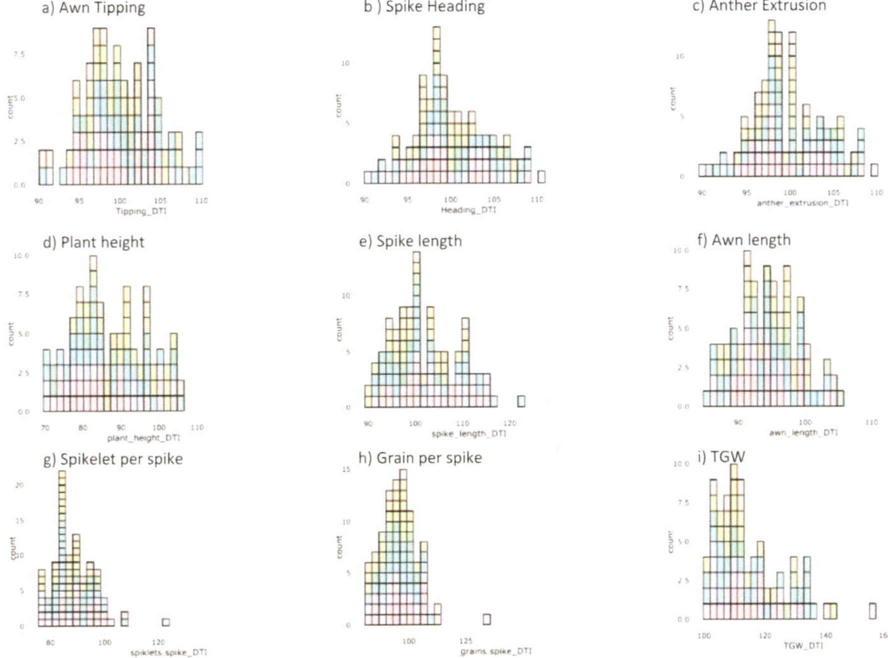

Figure 2. Histogram of phenotypic values distribution analysis of drought tolerance index variation of the studied traits in 121 spring barley accessions under control and drought stress. (**a**) Awn Tipping, (**b**) Spike Heading, (**c**) Anther Extrusion, (**d**) Plant Height, (**e**) Spike length, (**f**) Awn Length, (**g**) Spikelet per Spike, (**h**) Grain per Spike and (**i**) Thousand Grain Weight (TGW).

3.2. Correlations Analysis

Significant positive correlations were observed among various traits within both treatments. For example, AE showed a significantly high positive correlation with AT and SH ($r − 0.98$ *** and $r = 1.00$ ***, respectively) under control conditions and ($r = 0.93$ *** and $r = 1.00$ ***, respectively) under drought treatment (Figure 3A,B and Figure S5). On the contrary, some traits showed high significant negative correlations under both conditions. Interestingly, TGW showed negative correlation with NGS under control and drought conditions ($r = −0.21$ and $−0.44$ ***, respectively; Figure 3A,B).

For DTI, AT_DTI showed high significant positive correlation with SH_DTI and AE_DTI ($r = 0.92$ *** and $r = 0.91$ ***, respectively). Additionally, SH_DTI exhibited high positive correlation with AE_DTI ($r = 0.99$ ***). The SL_DTI with AL_DTI showed a significant negative correlation ($r = −0.27$ ***; Figure S5).

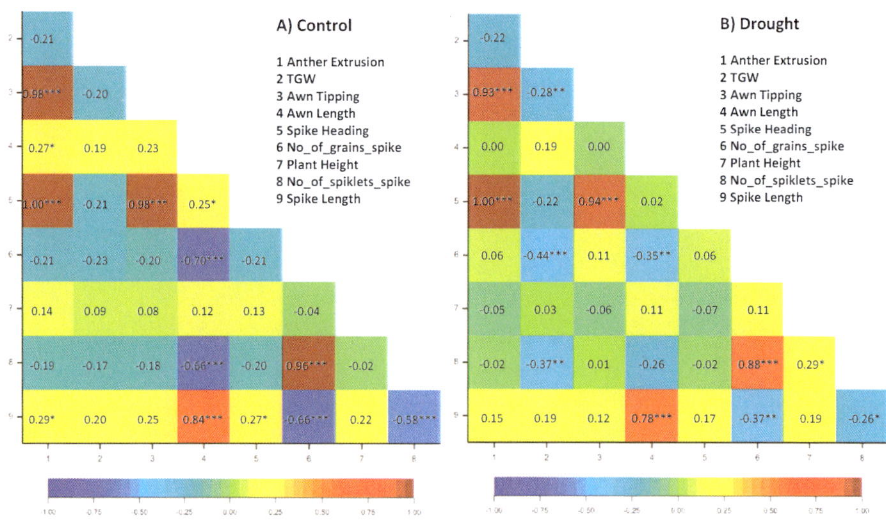

Figure 3. Correlations matrix among the studied traits in barley genotypes (**A**) under control, and (**B**) under drought stress. The degree of significance is indicated as * p, 0.05; ** p, 0.01; *** p, 0.001; ns: not significant.

3.3. Natural Genetic Variation and Candidate Genes Potentially Underlying Drought Tolerance

GWAS analysis of 121 selected accessions was performed to find out the natural genetic variation of the studied traits. We detected a total number of 101 significant marker–trait associations (with $-\log_{10}$ p-value ≥ 3) distributed over the seven barley chromosomes (Figure 4 and Table S1). There was plenty of natural genetic variation of all studied traits in this collection. Through GWAS analysis, we found twelve interesting genomic regions for genetic variation of all studied traits distributed on only two chromosomes (2H and 3H).

Highly significant LD was found among significant SNPs within these genomic regions (Figure 4b), indicating that these significant SNPs are potentially harboring important candidate genes in addition to their useful for marker-assisted selection. On chromosome 2H, two adaptive genes were identified, whereas on chromosome 3H, ten genes that represent a combination of both adaptive and constitutive genes were identified. Adaptive genes might be control-specific, i.e., genes that regulate trait variation under control only, or drought-specific (genes that regulate trait variation under drought only). Constitutive genes regulate trait variation under both control and drought conditions (Table 3).

In total, eight SNPs were associated ($-\log_{10}$ p-value ≥ 3) with AT parameters. Out of these, five SNPs were adaptive: three were control-specific and two were drought-specific; the remaining three SNPs were constitutive. The constitutive SNPs were identified on chromosomes 1H, 3H and 4H. The most significant one (with $-\log_{10}$ p-value $= 8.7$) was observed on chromosome 3H at 126.69 cM. Only one constitutive gene was identified *HORVU3Hr1G098200* (Chr. 3; 126.69 cM) (Table 3).

Sixteen SNPs were associated ($-\log_{10}$ p-value ≥ 3) with SH parameters. Out of these, seven SNPs were adaptive: four SNPs were control-specific and three drought-specific; the remaining nine SNPs were constitutive. The constitutive SNPs were identified on chromosomes 1H, 2H, 3H, 4H, 5H and 6H where the most significant one (with $-\log_{10}$ p-value $= 4.7$) was observed on chromosome 6H at 53.54 cM. Only seven SNPs showed association with candidate genes. Four genes were control-specific; *HORVU3Hr1G088300*, *HORVU3Hr1G089160*, *HORVU3Hr1G089080* and *HORVU3Hr1G098200*. Three constitutive genes were identified: *HORVU3Hr1G098200*, *HORVU3Hr1G116790* and *HORVU3Hr1G115810* (Table 3).

In total, 18 SNPs showed association (with $-\log_{10}$ p-value $\geq$ 3) with AE parameters. Ten SNPs were adaptive: six were control-specific and four were drought-specific; the remaining SNPs were constitutive. These constitutive SNPs were mapped on chromosomes 2H, 3H, 4H, 5H and 6H. The most significant one (with $-\log_{10}$ p-value = 4.6) was detected on chromosomes 6H at 53.54 cM. Out of these, twelve SNPs showed association with candidate genes. Eight genes were control-specific: *HORVU3Hr1G018650*, *HORVU3Hr1G020430*, *HORVU3Hr1G020660*, *HORVU3Hr1G019590*, *HORVU3Hr1G088300*, *HORVU3Hr1G089160*, *HORVU3Hr1G089080* and *HORVU3Hr1G098200* and four genes are constitutive: *HORVU3Hr1G116790*, *HORVU3Hr1G098200* and *HORVU3Hr1G115810* (Table 3).

For the trait PH, six SNPs were associated. Out of these, five SNPs were adaptive: four SNPs were control-specific and one SNP was drought-specific. Only one constitutive SNP was found on chromosome 7H at 131.59 cM. There were no SNPs associated with candidate genes for PH (Table 3).

GWAS analysis showed that six SNPs were associated with SL parameters. Only one SNP was associated with a candidate gene that was control-specific, *HORVU3Hr1G098200* (Table 3).

In total, eight SNPs showed association (p-value $\leq$ 0.001) with AL. Out of these, six SNPs were adaptive: two were control-specific and four were drought-specific; the remaining two SNPs were constitutive and mapped on chromosomes 2H and 3H. Six SNPs were associated with candidate genes. Only one gene was control-specific, *HORVU3Hr1G098200*; and one was constitutive, *HORVU3Hr1G098200* (Table 3).

For NSS, 15 SNPs showed significant association with $-\log_{10}$ p-value $\geq$ 3. Six of them were associated with adaptive candidate genes. Three SNPs were control-specific: *HORVU3Hr1G018650*, *HORVU3Hr1G020430* and *HORVU3Hr1G020660* and another three were drought-specific: *HORVU2Hr1G091030*, *HORVU2Hr1G091170* and *HORVU3Hr1G019590*. No constitutive genes were identified (Table 3).

In total, 15 SNPs were associated with NGS parameters. Seven SNPs showed association with candidate genes. Five genes were control-specific: *HORVU2Hr1G091030*, *HORVU3Hr1G018650*, *HORVU3Hr1G020430*, *HORVU3Hr1G020660* and *HORVU3Hr1G019590* and two drought-specific genes: *HORVU2Hr1G091030* and *HORVU2Hr1G091170* (Table 3).

For TGW parameters, seven SNPs showed significant association with $-\log_{10}$ p-value $\geq$ 3. Only two SNPs showed association with candidate genes. One was control-specific, *HORVU3Hr1G098200*, and the other drought-specific one was *HORVU3Hr1G098200* (Table 3).

Overall, all the identified genes revealed a pleiotropic effect, i.e., each gene controlled more than one trait. The gene *HORVU3Hr1G098200*, for instance, regulates the variation of ten traits (Table 3). On the contrary, they differ in their mode of action, and some of them are adaptive genes such as *HORVU2Hr1G091170* and *HORVU3Hr1G018650*. The first gene modulates the variation of NGS and NSS under drought; whereas the second controls them under control. Other genes are constitutive, such as *HORVU3Hr1G116790* and *HORVU3Hr1G115810* (Table 3). Surprisingly, the gene *HORVU3Hr1G098200* showed a constitutive/adaptive mode of action. It controls the variation of (TGW) constitutively, while regulating the variation of (AE, AL, SH and SH) in an adaptive manner, i.e., under control only.

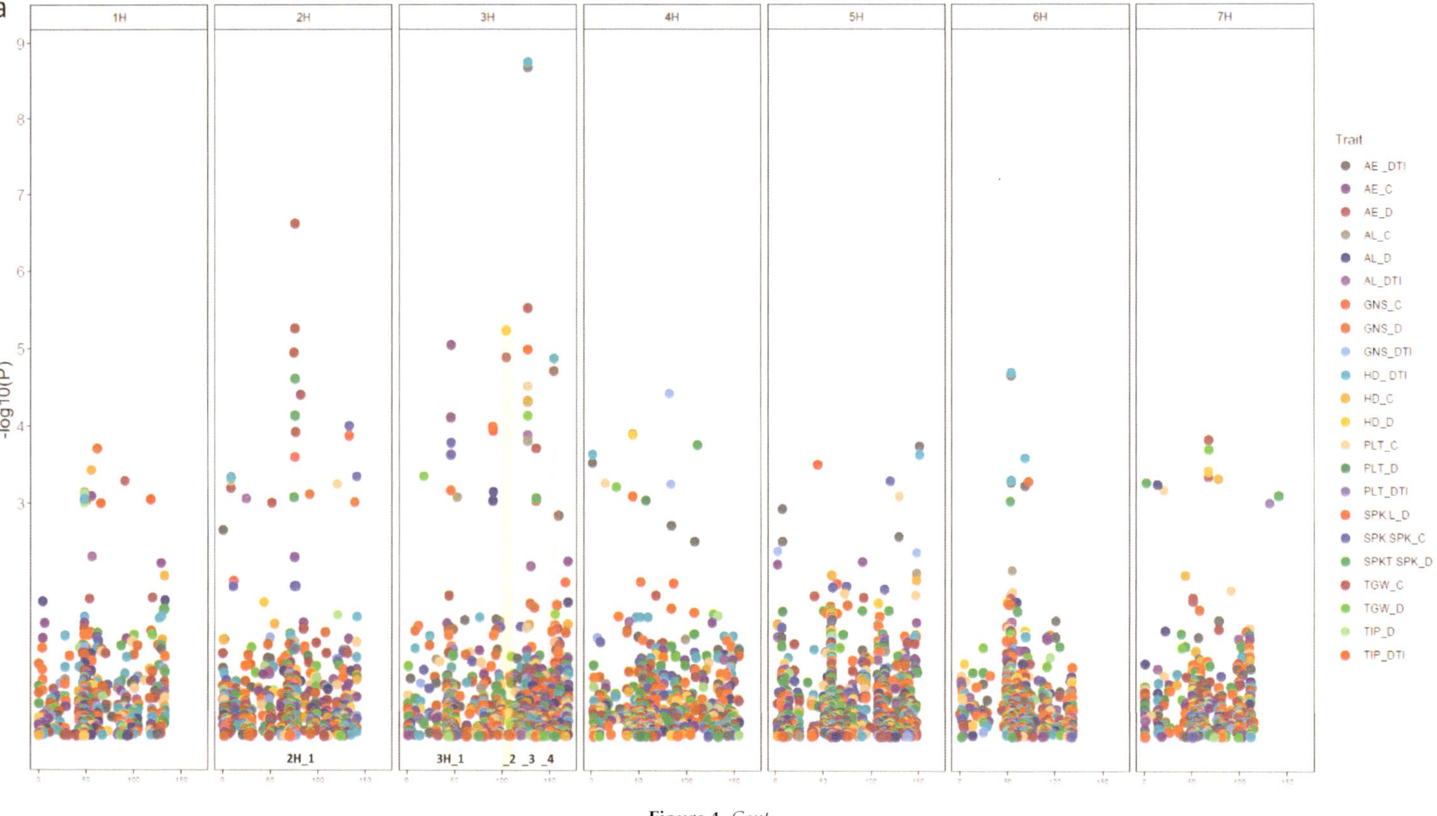

Figure 4. *Cont.*

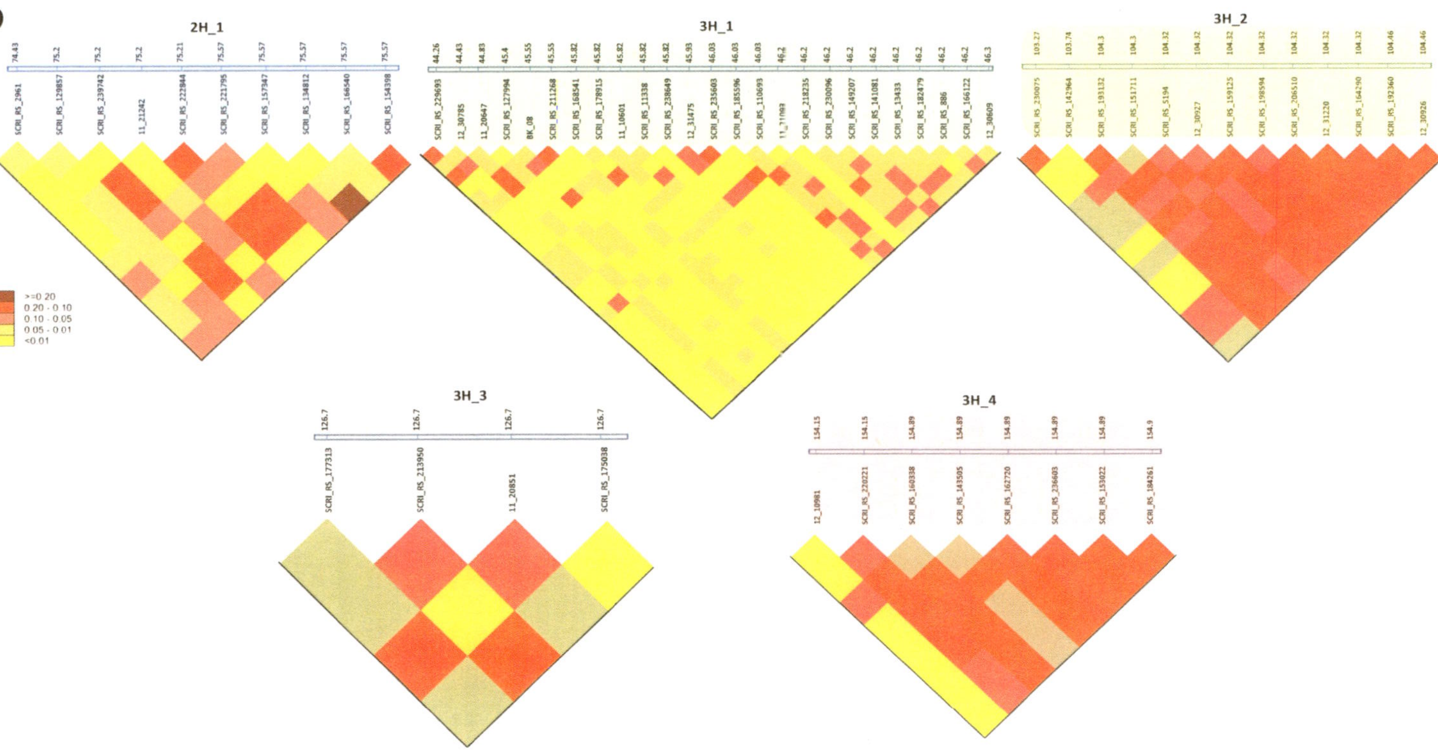

Figure 4. (**a**) The significant SNPs (101 SNPs, $-\log_{10} \geq 3$) associated with all traits under control and drought stress conditions. The x-axis shows the chromosomes and the SNP positions. The y-axis shows the $-\log_{10}$ (*P*-value) for each SNP marker. (**b**) Heatmap linkage disequilibrium to detect the region of candidate genes, which show a consistent effect on traits and associated with SNPs passing $\geq$ FDR using their physical position.

Table 3. The functional annotation of the putative candidate genes associated with the studied traits under drought and control growth conditions.

Trait	Marker	Chr	Pos.	SNP Pos.	Gene	Gene Pos.	GO ID	Annotation in Barley	Orthologs
NGS_C NGS_D NSS_D	SCRI_RS_166540	2H	75.56	646934425	HORVU2Hr1G091030	646930069-646939693	None	RNA polymerase II C-terminal domain phosphatase-like 1	AT4G21670
NGS_D NSS_D	SCRI_RS_157347	2H	75.56	647255135	HORVU2Hr1G091170	647252087-647258342	None	expansin B3	AT4G28250
AE_C NGS_C NSS_C	SCRI_RS_229693	3H	44.26	48643412	HORVU3Hr1G018650	48643412-48647662	GO:0000287 GO:0003824 GO:0030976	pyruvate decarboxylase-2	AT5G54960
AE_C NGS_C NSS_C	BOPA2_12_30737	3H	45.55	63626447	HORVU3Hr1G020430	63623393-63623651	None	Protein HASTY 1	AT3G05040
AE_C NGS_C NSS_C	BOPA1_7045-950	3H	45.55	64618663	HORVU3Hr1G020660	64617802-64621872	GO:0005515	Chromosome 3B, genomic scaffold, cultivar Chinese Spring	AT2G36270
AE_C NGS_C NSS_D	BOPA1_2391-566	3H	46.25	55590274	HORVU3Hr1G019590	55915087-55917096	GO:0003677	myb domain protein 37	AT5G23000
AE_C SH_C	SCRI_RS_230075	3H	103.27	624309908	HORVU3Hr1G088300	624309585-624311169	None	Chromosome 3B, genomic scaffold, cultivar Chinese Spring	AT2G36270
AE_C SH_C	BOPA2_12_30223	3H	104.32	627767267	HORVU3Hr1G089160	627749990-627754917	GO:0003677 GO:0003700 GO:0006355	AP2-like ethylene-responsive transcription factor	AT2G28550
AE_C SH_C	SCRI_RS_192360	3H	104.26	627260224	HORVU3Hr1G089080	627258327-627265762	None	undescribed protein	
AE_DTI AL_DTI AE_C AL_C SH_C SL_C TGW_C TGW_D SH_DTI AT_DTI	SCRI_RS_177313	3H	126.69	657713242	HORVU3Hr1G098200	657712024-657713975	GO:0005515	Chromosome 3B, genomic scaffold, cultivar Chinese Spring	AT2G36270
AE_DTI SH_DTI	BOPA2_12_10981	3H	154.15	696452271	HORVU3Hr1G116790	696450874-696453390	GO:0005215 GO:0006810 GO:0016020	Aquaporin-like superfamily protein	AT2G36830
AE_DTI SH_DTI	SCRI_RS_237738	3H	154.6	694105354	HORVU3Hr1G115810	694103839-694106776	None	Kinetochore protein spc25	

3.4. SNP-Gene-Based Analysis

Twelve SNPs at chromosomes 2H and 3H were physically co-located inside the candidate genes (Table 3). Two SNPs at 2H (SCRI_RS_166540 and SCRI_RS_157347) were detected within the genes *HORVU2Hr1G091030* and *HORVU2Hr1G091170*, respectively. Meanwhile, ten SNPs at 3H were co-located within the physical position of candidate genes. For example, SNP numbers BOPA1_2391-566 and SCRI_RS_177313 were located within *HORVU3Hr1G019590* and *HORVU3Hr1G098200* genes, respectively. Interestingly, these SNPs were mostly associated with NGS, NSS, and TKW under drought stress. The rest of the genes at 3H were associated with traits under drought stress or with DTI for SH and AE.

The allelic analysis of the SNPs that are associated with traits under drought showed that alleles A, G and A from markers SCRI_RS_166540, SCRI_RS_157347 and BOPA1_2391-566, respectively, have a highly significant impact on NSS (Figures 5a and 6). The genes *HORVU2Hr1G091030* and *HORVU2Hr1G091170* at 2H controlled NGS under drought via markers SCRI_RS_166540 and SCRI_RS_157347, where alleles G and A, respectively, increased NGS significantly (Figures 5b and 6). Only one marker (SCRI_RS_177313 from *HORVU3Hr1G098200* gene) showed a significant effect on TKW by allele A that increased the value under drought (Figure 5b).

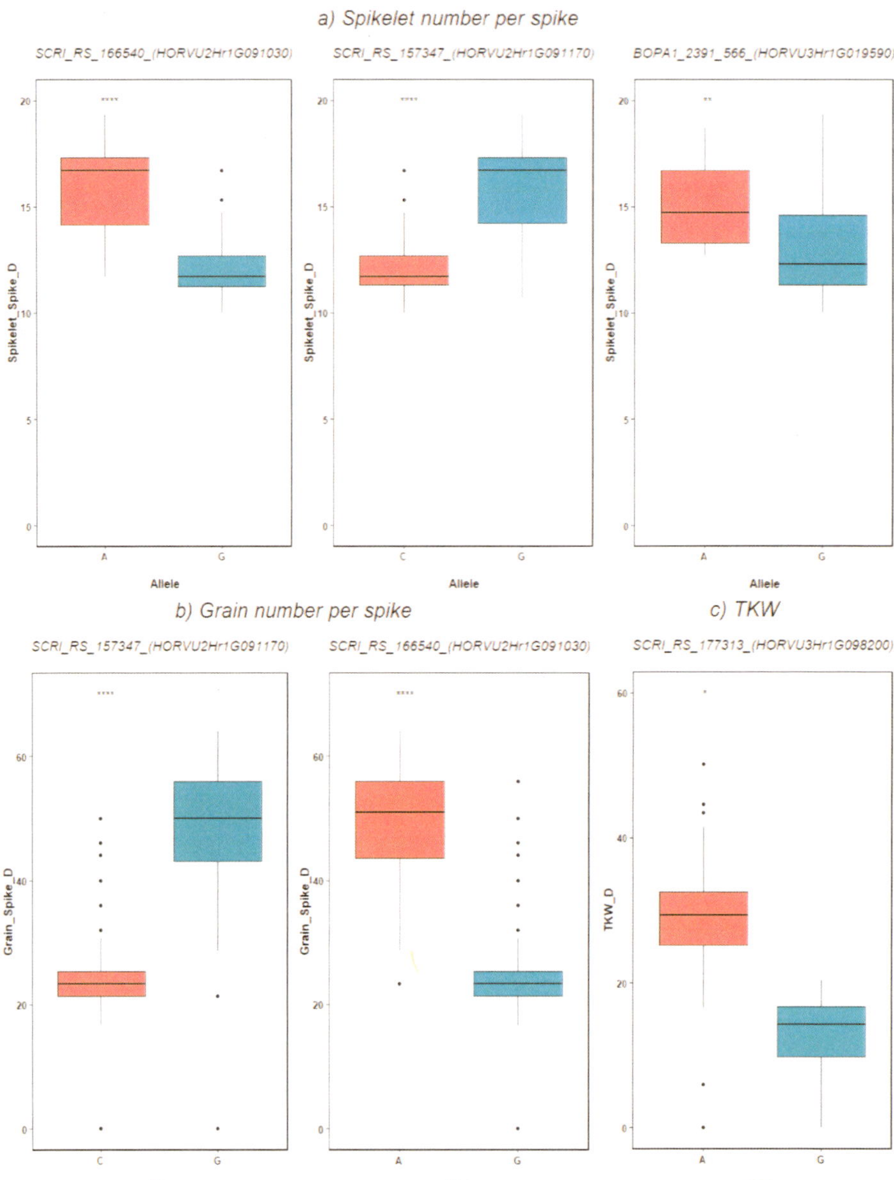

Figure 5. SNP-gene-based analysis for (**a**) Spikelet number per spike, (**b**) Grain number per spike in barley genotype, and (**c**) TKW (gram). The degree of significance indicated as * p, 0.05; ** p, 0.01; *** p, 0.001; **** p, 0.0001.

47

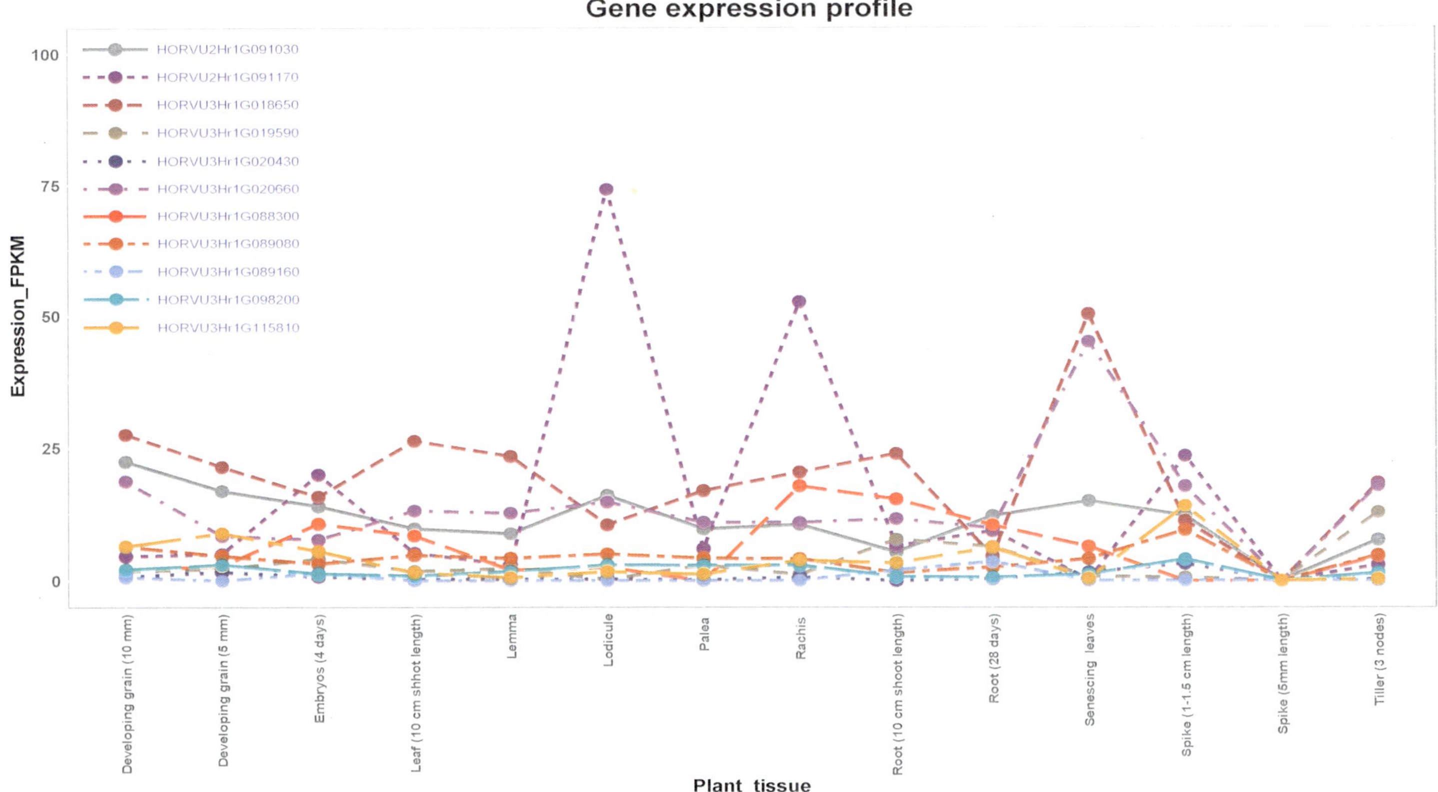

Figure 6. Expression analysis of the candidate genes in different plant organs at different developmental stages.

3.5. Expression Analysis of Candidate Genes

The expression analysis of candidate genes in different organs showed a wide range of expression for the genes (Figure 6). Notably, the associated genes with spikelet and grain number per spike under drought stress (Figure 5) showed a high expression for most of the organs (Figure 6). Gene *HORVU2Hr1G091170* at 2H in particular that had high impact on spikelet and grain numbers under drought showed highest expression in the respective grain organs, e.g., lodicule and rachis in addition to spike at 1–1.5 cm length (Figure 6). The second highest expressed gene from the highly associated ones was *HORVU2Hr1G091030*. This gene was highly expressed in developing grain and lodicule (Figure 6). The expression of these genes demonstrated their biological roles in the spikelet and grain development under drought conditions. Other genes, e.g., *HORVU3Hr1G020660* and *HORVU3Hr1G018650*, showed high expression particularly in senescing leaves, suggesting their roles in leaf development (Figure 6).

4. Discussion

Studying drought stress tolerance under field conditions in cereals such as barley is very limited, as it requires complex and laborious experiments for population characterization, in addition to being influenced by environmental factors [3,4]. Nevertheless, the present study focused on investigating the natural variation in diverse spring barley collections and on identifying candidate genes associated with the traits of interest under field conditions.

In the present study, there was a considerable reduction in most traits under drought stress compared to control conditions. These results indicated that drought stress reduced grain yield by decreasing NSS, NGS, and TGW. Our results are supported by the findings of [2,24], who examined the response of spring barley to pre- and post-anthesis drought and reported a yield reduction due to pre-anthesis water deficit on several fertile spikes and grains per plant. Our results are in agreement with those of Samarah, Alqudah, Amayreh and McAndrews [2], who found that drought stress reduced grain yield by reducing the number of tillers, spikes and grains per plant and individual grain weight in barley (*Hordeum vulgare L.*) as a result of early maturity and shortened grain filling duration at 25% field capacity compared to control. In conclusion, drought stress negatively influenced barley yield through impairing grain development, size and grain filling duration.

Agronomic traits such as grain yield and its components (NSS, NGS, and TGW) are the major selection criteria for drought tolerance in barley breeding [25]. Therefore, understanding the interplay among these parameters is of high importance. The correlation between TGW and NGS was always negative and significant under control and drought, respectively. The positive and significant correlation between TGW and SL indicates that yield mainly depends on spike length. Zhou et al. [26] suggested that the grain yield traits interacting with each other, increase in one of them (e.g., TGW) can be correlated with a reduction in another (e.g., NGS), which is in agreement with our results. In support of our findings, in a European spring barley collection [27], it was found that NGS showed negative correlations with all other yield parameters except grain length, concluding that the yield was mainly dependent on grain size and SL rather than NGS. Several reports showed that drought-tolerant genotypes implement high productivity under both stressed and unstressed conditions [2,28]. Therefore, the comparative analysis of the yield components under drought and well-watered conditions can be used for predicting stress tolerance of genotypes, and then in selection of more tolerant barley lines for future breeding purposes [24,29].

The Role of Putative Candidate Genes in Drought Tolerance

Promising genomic regions are located at position 75.56 cM on chromosome 2H, harboring two important candidate genes. The first one is *HORVU2Hr1G091030*, which encodes RNA polymerase II C-terminal domain phosphatase-like 1 (*CPL1*), for NGS_C, NGS_D, and NSS_D. *CPL1* is a negative regulator of stress-responsive gene transcription, ABA, and stress responses [30]. In Arabidopsis, *CPL1* regulates gene expression under various osmotic stresses through ABA signaling [30]. Loss of *CPL1* function in the mutants enhances tolerance to oxidative stress including drought and salt stresses [31]. In rice, *OsCPL1* is expressed in the young spikelets [32]. Most likely, this gene is expressed during the development of barley spikelets under drought. Additionally, it controls the NGS under drought and control conditions in a constitutive manner. The second candidate, *HORVU2Hr1G091170*, encodes expansin B3 for NSS_D and NGS_D. Expansins have been implicated in the responses of various plant species to water stress. In maize, increased expansin activity was found to be involved in maintaining the growth of primary roots at a low water potential [33]. The expression of a root β-expansin gene, *GmEXPB2*, is remarkably participated in root system architecture responses to several abiotic stresses, such as Fe, P, and water deficiency [34]. *RhEXPA4* is a rose expansin gene that is up-regulated in rose petals after dehydration [35], and it confers salt and drought tolerance to transgenic Arabidopsis [36]. In potato, most of expansin-like B genes have a potential role in multistress tolerance and upregulated under stresses including drought [37]. These two genes—*HORVU2Hr1G091030* and *HORVU2Hr1G091170*—showed different expression profiles (Figure 6). The noisy expression profile of *HORVU2Hr1G091170*, suggesting a stress-responsive gene, exclusively regulates the variation of NGS_D and NSS_D under drought (Table 3). While the slightly constant expression profile of *HORVU2Hr1G091030* indicates a constitutive gene that regulates the variation of NGS under control and drought. These findings are in accordance with the results of several authors who found that the stress-responsive genes exhibiting noisy expression profiles, while the constitutive ones showing constant expression patterns reviewed in Lopez-Maury et al. [38].

Notably, the allelic diversity analysis shows that allele A and G from *HORVU2Hr1G091030* and *HORVU2Hr1G091170* genes, respectively, have a significant positive impact on NSS and then NGS under drought stress. The expression of these genes during spike, spikelet and grain developmental stages demonstrates their influence on agronomic traits under drought stress conditions. In terms of molecular breeding, the above-mentioned alleles were the highest alleles explained the natural variation under drought stress suggesting their usefulness in breeding programs. Taken together, this provides evidence that these genes are drought-specific and involved in the drought stress-tolerance pathway. These findings indicate that both genes are of high importance for enhancing barley grain yield under drought stress.

Interestingly, ten candidate genes were detected on chromosome 3H. Out of these, three genes at positions 44.26 and 45.55 cM were identified as candidates for various traits such as AE, NGS, and NSS under control conditions. The first one is *HORVU3Hr1G018650* encoding pyruvate decarboxylase-2 (*PDC2*), which belongs to pyruvate decarboxylase (*PDC*) gene family. Pyruvate has been involved in the ethanolic fermentation pathway that is associated with flooding tolerance when plant cells switch from respiration to anaerobic fermentation [39]. Additionally, fermentation has important functions in the presence of oxygen, mainly in germinating pollen and during abiotic stress. This indicates the interdependency between pyruvate decarboxylase (PDC) and AE, NGS, and NSS under control conditions. Furthermore, PDC, which catalyzes the first step in this pathway, is thought to be a main regulatory enzyme [40]. In Arabidopsis, the expression of PDC genes during abiotic stresses has been reported [41]. In maize and Arabidopsis, strong induction of fermentation genes takes place in anaerobic conditions [42]. Thus, it is conceivable that ethanolic fermentation is part of a general response to environmental stress, e.g., drought stress.

For traits such as AE_C, NGS_C, and NSS_D at position 46.25 cM, *HORVU3Hr1G019590* encodes myb domain protein 37. MYB (myeloblastosis) has a regulatory role in ABA signaling by activating some stress-inducible genes [43]. In Arabidopsis, MYB, namely *AtMYB60*, *AtMYB44*, and *AtMYB15*, have been implicated in the regulation of stomatal closure and ABA-mediated response to drought and salt stresses [44]. Agarwal et al. [45] detected that *AtMYB15* was expressed in both vegetative and reproductive organs and up-regulated by cold and salt stresses. The differential expression of numerous MYB TFs in the *Triticeae* was shown to be involved in the response of abiotic stress conditions such as drought and salt stresses [46,47]. These findings suggest that the genes in question are constitutive genes and may have a role in drought tolerance in barley during heading and maturation.

For SH_C and AE_C, *HORVU3Hr1G089160* encodes AP2-like ethylene-responsive transcription factor at position 104.32 cM. The AP2/ERF superfamily regulates diverse developmental responses such as flower pedicel abscission [48], leaf senescence [49], cell proliferation and shoot branching [50]. Houston et al. [51] observed that mutants of *HvAP2* internode lead to the reduction of elongation in both the culm and the spike in barley, suggesting that the *HvAP2* alleles increase grain yield by controlling spike density. In the current study, the allelic variation was only observed under control condition indicating that this gene might not be involved directly in drought tolerance.

Additionally, *HORVU3Hr1G116790* encodes Aquaporin-like superfamily protein which is a candidate for SH_DTI and AE_DTI. Aquaporins (AQPs) are a class of water channel proteins that belong to the major intrinsic protein (MIP) superfamily of membrane proteins [52]. These proteins regulate the movement of water and other small molecules across plant vacuolar and plasma membranes [53]. Aquaporins have also been suggested to have an essential role in plant tolerance of biotic and abiotic stresses [54], and extension growth [52]. Furthermore, it was reported in [55] that various aquaporin homologs are involved in plant stress responses against a variety of environmental stresses that disturb plant cell osmotic balance and nutrient homeostasis.

The most effective gene *HORVU3Hr1G098200* was mapped at 126.69 cM. This gene orchestrates the variation of ten traits both in constitutive and adaptive manner (Table 3). This explains the significant correlations between these traits, either under control or drought (Table 3 and Figure 3A,B). Shi et al. [56] reported that Chromosome 3B harbors genes that may be significant in controlling agronomical important traits, such as yield and resistance to biotic and abiotic stress in wheat. The QTL for these traits maps quite close to semidwarf1 (sdw1) on chromosome 3H at 126.69 cM. The *semi dwarf 1 (sdw1)* gene has previously been found to control the most desired agronomic traits barley reviewed by Hedden [57]. The pleiotropic effect of *sdw1* gene was evidenced in wheat [58,59].

Additionally, the gene *HORVU3Hr1G098200* showed an interplay between the constitutive and adaptive control pattern. For example, it constitutively controls the variation of thousand grain weight (TGW) under both control and drought stress. At the same time, it controls the spike heading only under control in an adaptive manner (Table 3). This pattern indicates that *HORVU3Hr1G098200* is partially constitutive and partially stress-responsive gene. Therefore, we considered it an adaptive gene when it controlled a trait(s) under control or drought. On the other hand, we considered it constitutive when it controlled trait(s) under both conditions. This gene, nevertheless, exhibited constant expression profile in different plant tissues, spanning the reproductive period from anther extrusion until the seed set, suggesting a key role in controlling different traits and more likely to be constitutive (Figure 6). This finding is consistent with the results of [60], who found that cells express some genes constantly to maintain the concentration of some proteins tuned with the cell physiological needs. The constant expression pattern characterizes the constitutive genes rather than the stress responsive (adaptive) ones (reviewed in Lopez-Maury, Marguerat and Bahler [38]. According to Blum [61], drought stress when expressed as a final yield is affected by constitutive and adaptive plant traits (genes). These constitutive QTL/genes represent an instrumental tool for selection as they show stability across different environments compared to the adaptive ones. Moreover, the selection for drought tolerance based on these QTL/genes does not require drought stress.

The last candidate gene is *HORVU3Hr1G115810*, which encodes Kinetochore protein spc25 and affects AE_DTI and SH_DTI. Kinetochore proteins may have a pivotal role for centromere and kinetochore functioning [62–64], and chromosome segregation mediating [65]. Specifically, kinetochore protein MIS12 is required for the co-orientation of sister kinetochores during meiosis I in maize [66]. NDC80 kinetochore protein serves as a contact point for chromosome–spindle interaction [67]. Interestingly, QTL at 3H 126 and 154 cM have previously been reported to be associated with grain number and yield in barley under drought stress conditions [4,68,69].

On chromosome 3H, three genes (*HORVU3Hr1G020660*, *HORVU3Hr1G088300*, and *HORVU3Hr1G098200*) are counterparts of the ortholog in Arabidopsis, *AT2G36270*. The corresponding genes are expressed during the reproductive stage in the different flower and seed organs (Figure 6), indicating their significance in flowering and seed set. Our results are in accordance with the findings of Klepikova et al. [70], who compared the spatiotemporal expression and stability of a lot of genes in 79 organs and developmental stages. Additionally, they found that *AT2G36270* had been significantly expressed in senescent organs (leaves and silique); this is similar to expression profile of the gene *HORVU3Hr1G020660* that highly expressed in senescent leave (Figure 6). Taking these findings together, the similarity of spatiotemporal expression patterns, as well as functions of these genes in both barley and Arabidopsis, suggests that they might be are homologous for AT2G36270.

The significant role of *AT2G36270* (*ABI5*, *AtABI5* and *BZIP39*) during different growth stages, as well as in drought tolerance, has been evidenced in several studies. Finkelstein et al. [71] reported that it encodes a member of the basic leucine zipper transcription factor family, involved in ABA-regulated gene expression during germination, seed development and subsequent reproductive stage. In particular, *ABI5* regulates a set of the late embryogenesis-abundant genes during both seed and ABA-inducible vegetative gene expression in wild-type and abi5-1 plants [71]. Mittal et al. [72] reported that overexpression of *AtABI5* in transgenic cotton (*Gossypium hirsutum*) showed resistance to the imposed drought stress through ROS scavenging and osmotic adjustment, enhancing photosynthesis, as well as traits of drought avoidance (bigger root and leaf systems) and tolerance (longer internode length and higher stem weight) leading to better establishment under water shortage. In rice (*Oryza sativa*), overexpression of *OsbZIP46CA1* significantly increased tolerance to drought and osmotic stresses at flowering stage, and suggested that *OsbZIP46* is a positive regulator of ABA signaling and drought stress tolerance by modulating many stress-related genes [73].

In summary, the present study showed the value of using field experiments to investigate natural phenotypic and genetic variation in a worldwide panel of barley accessions underlying agronomic traits such as grain yield and its components which can be exploited for crop improvement. Drought stress negatively influences most of the studied traits. In addition, we observed significant positive correlations among various traits within both control and drought treatments. Candidate genes associated with drought response were detected on two chromosomes, notably 2H and 3H. Interestingly, most the candidate genes are described to be involved in responses to abiotic stresses such as drought and salt. Interestingly, the genomic regions at 75 cM on 2H and 126.69 cM on 3H harbor three candidate gene *HORVU2Hr1G091030 HORVU2Hr1G091170* and *HORVU3Hr1G098200*, which are highly associated with spikelet and final grain number per spike, suggesting the crucial role in controlling grain yield under drought conditions. The discovered SNPs and candidate genes for drought response will be helpful for breeding drought tolerant barley cultivars.

5. Conclusions

Conclusively, the present study showed that drought negatively affects the yield-related traits. Despite of the reduction in most traits under drought, the heritability estimates for all respective traits were high, indicating the potential of this collection to conduct a GWAS analysis looking for drought-controlling alleles/genes. Additionally, the present study revealed that combining GWAS and bioinformatics is a very instrumental approach to identify candidate genes even for polygenic traits such as the yield-related components. Our results confirmed that the yield-related components are under polygenetic control; under contrasting growth conditions (control and drought). The candidate genes exhibited different patterns of traits control; some genes were adaptive (*HORVU2Hr1G091170*), while other genes were constitutive (*HORVU2Hr1G091030* and *HORVU3Hr1G098200*). The constitutive pleiotropic genes are of high importance to improve drought tolerance because they can be employed to improve several traits at a time with no need to test under drought. The causative genes showed different expression patterns; the constitutive genes showed constant expression profiles, while the adaptive genes showed a noisy expression profile. Most of the causative genes were expressed in spikelet organs (palea, lema and lodicules), as well as in grain, spike and leaf, indicating their potential role in drought tolerance, in particular, during the reproductive stage. To get more comprehensive and clear answers, a new detailed experiment is underway to study the gene expression of the candidate genes identified in this study especially, the gene *HORVU3Hr1G098200* because it regulates the variation of ten traits, and because of its constitutive/adaptive mode of action.

Supplementary Materials: The following are available online at http://www.mdpi.com/2073-4425/11/5/533/s1, Figure S1: Population structure based on their row type (two-rowed vs. six-rowed), geographical of origin; Figure S2: Weather data at Fayoum station that indicates Max-Min Temperature and Humidity; Figure S3: Boxplot analysis of variation of the traits; Figure S4: Boxplot analysis of drought tolerance index variation of the traits in barley genotypes under control and drought stress; Figure S5: Correlations matrix among drought tolerant index; Table S1: Summary of the most significant SNP associated with studied traits.

Author Contributions: Conceptualization, S.G.T., Y.S.M., M.A.K. and A.M.A.; methodology, S.G.T., Y.S.M., M.A.K. and A.M.A.; software, A.M.A.; validation, S.G.T., Y.S.M., M.A.K. and A.M.A.; formal analysis, A.M.A.; investigation, S.G.T., Y.S.M. and A.M.A.; resources, A.B.; data curation, S.G.T., Y.S.M. and A.M.A.; writing—original draft preparation, S.G.T.; writing—review and editing, S.G.T., Y.S.M. and A.M.A.; visualization, S.G.T., Y.S.M. and A.M.A.; supervision, Y.S.M., M.A.K. and A.M.A.; software, A.M.A. All authors have read and agreed to the published version of the manuscript.

Funding: This research received no external funding.

Acknowledgments: The authors would like to thank all who contributed to this work.

Conflicts of Interest: The authors declare no conflict of interest.

References

1. Lobell, D.B.; Schlenker, W.; Costa-Roberts, J. Climate trends and global crop production since 1980. *Science* **2011**, *333*, 616–620. [CrossRef]
2. Samarah, N.H.; Alqudah, A.M.; Amayreh, J.A.; McAndrews, G.M. The effect of late-terminal drought stress on yield components of four barley cultivars. *J. Agron. Crop Sci.* **2009**, *195*, 427–441. [CrossRef]
3. Thabet, S.G.; Alqudah, A.M. Crops and drought. In *eLS*; John Wiley & Sons, Ltd., Ed.; John Wiley & Sons, Ltd.: Hoboken, NJ, USA, 2019; pp. 1–8. [CrossRef]
4. Sallam, A.; Alqudah, A.M.; Dawood, M.F.A.; Baenziger, P.S.; Borner, A. Drought stress tolerance in wheat and barley: Advances in physiology, breeding and genetics research. *Int. J. Mol. Sci.* **2019**, *20*, 3137. [CrossRef] [PubMed]
5. Shavrukov, Y.; Kurishbayev, A.; Jatayev, S.; Shvidchenko, V.; Zotova, L.; Koekemoer, F.; de Groot, S.; Soole, K.; Langridge, P. Early flowering as a drought escape mechanism in plants: How can it aid wheat production? *Front. Plant Sci.* **2017**, *8*, 1950. [CrossRef] [PubMed]
6. Alqudah, A.M.; Samarah, N.H.; Mullen, R.E. Drought stress effect on crop pollination, seed set, yield and quality. In *Alternative Farming Systems, Biotechnology, Drought Stress and Ecological Fertilization*; Springer: Dordrecht, The Netherlands, 2011; pp. 193–213. [CrossRef]
7. Thabet, S.G.; Moursi, Y.S.; Karam, M.A.; Graner, A.; Alqudah, A.M. Genetic basis of drought tolerance during seed germination in barley. *PLoS ONE* **2018**, *13*, e0206682. [CrossRef]
8. Fischer, R.; Turner, N.C. Plant productivity in the arid and semiarid zones. *Annu. Rev. Plant Physiol.* **1978**, *29*, 277–317. [CrossRef]
9. Barnabas, B.; Jager, K.; Feher, A. The effect of drought and heat stress on reproductive processes in cereals. *Plant Cell Environ.* **2008**, *31*, 11–38. [CrossRef]
10. Samarah, N.; Alqudah, A. Effects of late-terminal drought stress on seed germination and vigor of barley (Hordeum vulgareL.). *Arch. Agron. Soil Sci.* **2011**, *57*, 27–32. [CrossRef]
11. Botwright, T.; Condon, A.; Rebetzke, G.; Richards, R. Field evaluation of early vigour for genetic improvement of grain yield in wheat. *Aust. J. Agric. Res.* **2002**, *53*, 1137–1145. [CrossRef]
12. Cowley, R.; Luckett, D.; Moroni, J.; Diffey, S. Use of remote sensing to determine the relationship of early vigour to grain yield in canola (*Brassica napus* L.) germplasm. *Crop Pasture Sci.* **2014**, *65*, 1288–1299. [CrossRef]
13. Alqudah, A.M.; Sallam, A.; Stephen Baenziger, P.; Borner, A. GWAS: Fast-forwarding gene identification and characterization in temperate Cereals: Lessons from Barley—A review. *J. Adv. Res.* **2020**, *22*, 119–135. [CrossRef] [PubMed]
14. Reig-Valiente, J.L.; Marques, L.; Talon, M.; Domingo, C. Genome-wide association study of agronomic traits in rice cultivated in temperate regions. *BMC Genom.* **2018**, *19*, 706. [CrossRef] [PubMed]

15. Mazaheri, M.; Heckwolf, M.; Vaillancourt, B.; Gage, J.L.; Burdo, B.; Heckwolf, S.; Barry, K.; Lipzen, A.; Ribeiro, C.B.; Kono, T.J.Y.; et al. Genome-wide association analysis of stalk biomass and anatomical traits in maize. *BMC Plant Biol.* **2019**, *19*, 45. [CrossRef] [PubMed]

16. Teulat, B.; Merah, O.; Sirault, X.; Borries, C.; Waugh, R.; This, D. QTLs for grain carbon isotope discrimination in field-grown barley. *Theor. Appl. Genet.* **2002**, *106*, 118–126. [CrossRef] [PubMed]

17. Hou, S.; Zhu, G.; Li, Y.; Li, W.; Fu, J.; Niu, E.; Li, L.; Zhang, D.; Guo, W. Genome-wide association studies reveal genetic variation and candidate genes of drought stress related traits in cotton (*Gossypium hirsutum* L.). *Front. Plant Sci.* **2018**, *9*, 1276. [CrossRef] [PubMed]

18. De Ronde, J.A.; Cress, W.A.; Kruger, G.H.; Strasser, R.J.; Van Staden, J. Photosynthetic response of transgenic soybean plants, containing an Arabidopsis P5CR gene, during heat and drought stress. *J. Plant Physiol.* **2004**, *161*, 1211–1224. [CrossRef]

19. Haberle, J.; Svoboda, P.; Raimanová, I. The effect of post-anthesis water supply on grain nitrogen concentration and grain nitrogen šeld of winter wheat. *Plant Soil Environ.* **2018**, *54*, 304–312. [CrossRef]

20. VSN-International. *VSN GenStat for Windows*, 18th ed.; VSN International: Hemel Hempstead, UK, 2016; Available online: https://genstat.kb.vsni.co.uk/knowledge-base/hcitegen (accessed on 15 August 2019).

21. RStudio-Team. *RStudio: Integrated Development for R*; RStudio, Inc.: Boston, MA, USA, 2015; Available online: http://www.r-studio.com/ (accessed on 15 August 2019).

22. Lipka, A.E.; Tian, F.; Wang, Q.; Peiffer, J.; Li, M.; Bradbury, P.J.; Gore, M.A.; Buckler, E.S.; Zhang, Z. GAPIT: Genome association and prediction integrated tool. *Bioinformatics* **2012**, *28*, 2397–2399. [CrossRef]

23. Mascher, M.; Gundlach, H.; Himmelbach, A.; Beier, S.; Twardziok, S.O.; Wicker, T.; Radchuk, V.; Dockter, C.; Hedley, P.E.; Russell, J.; et al. A chromosome conformation capture ordered sequence of the barley genome. *Nature* **2017**, *544*, 427–433. [CrossRef]

24. Al-Ajlouni, Z.; Al-Abdallat, A.; Al-Ghzawi, A.; Ayad, J.; Abu Elenein, J.; Al-Quraan, N.; Baenziger, P. Impact of pre-anthesis water deficit on yield and yield components in barley (*Hordeum vulgare* L.) plants grown under controlled conditions. *Agronomy* **2016**, *6*, 33. [CrossRef]

25. Hossain, A.; Teixeira da Silva, J.A.; Lozovskaya, M.V.; Zvolinsky, V.P. High temperature combined with drought affect rainfed spring wheat and barley in South-Eastern Russia: I. Phenology and growth. *Saudi J. Biol. Sci.* **2012**, *19*, 473–487. [CrossRef] [PubMed]

26. Zhou, H.; Liu, S.; Liu, Y.; Liu, Y.; You, J.; Deng, M.; Ma, J.; Chen, G.; Wei, Y.; Liu, C.; et al. Mapping and validation of major quantitative trait loci for kernel length in wild barley (Hordeum vulgare ssp. spontaneum). *BMC Genet.* **2016**, *17*, 130. [CrossRef] [PubMed]

27. Xu, X.; Sharma, R.; Tondelli, A.; Russell, J.; Comadran, J.; Schnaithmann, F.; Pillen, K.; Kilian, B.; Cattivelli, L.; Thomas, W.T.B.; et al. Genome-wide association analysis of grain yield-associated traits in a pan-european barley cultivar collection. *Plant Genome* **2018**, *11*, 0073. [CrossRef]

28. Sharafi, S.; Ghassemi-Golezani, K.; Mohammadi, S.; Lak, S.; Sorkhy, B. Evaluation of drought tolerance and yield potential in winter barley (*Hordeum vulgare*) genotypes. *J. Food Agric. Environ.* **2011**, *9*, 419–422.

29. Perlikowski, D.; Kosmala, A.; Rapacz, M.; Koscielniak, J.; Pawlowicz, I.; Zwierzykowski, Z. Influence of short-term drought conditions and subsequent re-watering on the physiology and proteome of Lolium multiflorum/Festuca arundinacea introgression forms, with contrasting levels of tolerance to long-term drought. *Plant Biol. Stuttg* **2014**, *16*, 385–394. [CrossRef] [PubMed]

30. Koiwa, H.; Barb, A.W.; Xiong, L.; Li, F.; McCully, M.G.; Lee, B.H.; Sokolchik, I.; Zhu, J.; Gong, Z.; Reddy, M.; et al. C-terminal domain phosphatase-like family members (AtCPLs) differentially regulate Arabidopsis thaliana abiotic stress signaling, growth, and development. *Proc. Natl. Acad. Sci. USA* **2002**, *99*, 10893–10898. [CrossRef]

31. Thatcher, L.F.; Foley, R.; Casarotto, H.J.; Gao, L.L.; Kamphuis, L.G.; Melser, S.; Singh, K.B. The Arabidopsis RNA Polymerase II Carboxyl Terminal Domain (CTD) Phosphatase-Like1 (CPL1) is a biotic stress susceptibility gene. *Sci. Rep.* **2018**, *8*, 13454. [CrossRef]

32. Ji, H.; Kim, S.R.; Kim, Y.H.; Kim, H.; Eun, M.Y.; Jin, I.D.; Cha, Y.S.; Yun, D.W.; Ahn, B.O.; Lee, M.C.; et al. Inactivation of the CTD phosphatase-like gene OsCPL1 enhances the development of the abscission layer and seed shattering in rice. *Plant J.* **2010**, *61*, 96–106. [CrossRef]

33. Wu, Y.; Thorne, E.T.; Sharp, R.E.; Cosgrove, D.J. Modification of expansin transcript levels in the maize primary root at low water potentials. *Plant Physiol.* **2001**, *126*, 1471–1479. [CrossRef]

34. Guo, W.; Zhao, J.; Li, X.; Qin, L.; Yan, X.; Liao, H. A soybean beta-expansin gene GmEXPB2 intrinsically involved in root system architecture responses to abiotic stresses. *Plant J.* **2011**, *66*, 541–552. [CrossRef]

35. Dai, F.; Nevo, E.; Wu, D.; Comadran, J.; Zhou, M.; Qiu, L.; Chen, Z.; Beiles, A.; Chen, G.; Zhang, G. Tibet is one of the centers of domestication of cultivated barley. *Proc. Natl. Acad. Sci. USA* **2012**, *109*, 16969–16973. [CrossRef] [PubMed]

36. Lu, P.; Kang, M.; Jiang, X.; Dai, F.; Gao, J.; Zhang, C. RhEXPA4, a rose expansin gene, modulates leaf growth and confers drought and salt tolerance to Arabidopsis. *Planta* **2013**, *237*, 1547–1559. [CrossRef] [PubMed]

37. Chen, Y.; Li, C.; Yi, J.; Yang, Y.; Lei, C.; Gong, M. Transcriptome response to drought, rehydration and re-dehydration in potato. *Int. J. Mol. Sci.* **2019**, *21*, 159. [CrossRef] [PubMed]

38. Lopez-Maury, L.; Marguerat, S.; Bahler, J. Tuning gene expression to changing environments: From rapid responses to evolutionary adaptation. *Nat. Rev. Genet.* **2008**, *9*, 583–593. [CrossRef]

39. Kürsteiner, O.; Dupuis, I.; Kuhlemeier, C. The pyruvate decarboxylase1 gene of Arabidopsis is required during anoxia but not other environmental stresses. *Plant Physiol.* **2003**, *132*, 968–978. [CrossRef]

40. Uga, Y.; Kitomi, Y.; Ishikawa, S.; Yano, M. Genetic improvement for root growth angle to enhance crop production. *Breed Sci.* **2015**, *65*, 111–119. [CrossRef]

41. Dolferus, R.; Ellis, M.; De Bruxelles, G.; Trevaskis, B.; Hoeren, F.; Dennis, E.; Peacock, W. Strategies of gene action in Arabidopsis during hypoxia. *Ann. Bot.* **1997**, *79*, 21–31. [CrossRef]

42. Wahid, A.; Gelani, S.; Ashraf, M.; Foolad, M.R. Heat tolerance in plants: An overview. *Environ. Exp. Bot.* **2007**, *61*, 199–223. [CrossRef]

43. Abe, H.; Urao, T.; Ito, T.; Seki, M.; Shinozaki, K.; Yamaguchi-Shinozaki, K. Arabidopsis AtMYC2 (bHLH) and AtMYB2 (MYB) function as transcriptional activators in abscisic acid signaling. *Plant Cell* **2003**, *15*, 63–78. [CrossRef]

44. Jaradat, M.R.; Feurtado, J.A.; Huang, D.; Lu, Y.; Cutler, A.J. Multiple roles of the transcription factor AtMYBR1/AtMYB44 in ABA signaling, stress responses, and leaf senescence. *BMC Plant Biol.* **2013**, *13*, 192. [CrossRef]

45. Agarwal, M.; Hao, Y.; Kapoor, A.; Dong, C.-H.; Fujii, H.; Zheng, X.; Zhu, J.-K. A R2R3 type MYB transcription factor is involved in the cold regulation of CBF genes and in acquired freezing tolerance. *J. Biol. Chem.* **2006**, *281*, 37636–37645. [CrossRef] [PubMed]

46. Rahaie, M.; Xue, G.P.; Naghavi, M.R.; Alizadeh, H.; Schenk, P.M. A MYB gene from wheat (*Triticum aestivum* L.) is up-regulated during salt and drought stresses and differentially regulated between salt-tolerant and sensitive genotypes. *Plant Cell Rep.* **2010**, *29*, 835–844. [CrossRef] [PubMed]

47. Zhang, L.; Zhao, G.; Jia, J.; Liu, X.; Kong, X. Molecular characterization of 60 isolated wheat MYB genes and analysis of their expression during abiotic stress. *J. Exp. Bot.* **2012**, *63*, 203–214. [CrossRef] [PubMed]

48. Nakano, T.; Fujisawa, M.; Shima, Y.; Ito, Y. The AP2/ERF transcription factor SlERF52 functions in flower pedicel abscission in tomato. *J. Exp. Bot.* **2014**, *65*, 3111–3119. [CrossRef]

49. Koyama, T.; Nii, H.; Mitsuda, N.; Ohta, M.; Kitajima, S.; Ohme-Takagi, M.; Sato, F. A regulatory cascade involving class II ETHYLENE RESPONSE FACTOR transcriptional repressors operates in the progression of leaf senescence. *Plant Physiol.* **2013**, *162*, 991–1005. [CrossRef]

50. Mehrnia, M.; Balazadeh, S.; Zanor, M.I.; Mueller-Roeber, B. EBE, an AP2/ERF transcription factor highly expressed in proliferating cells, affects shoot architecture in Arabidopsis. *Plant Physiol.* **2013**, *162*, 842–857. [CrossRef]

51. Houston, K.; McKim, S.M.; Comadran, J.; Bonar, N.; Druka, I.; Uzrek, N.; Cirillo, E.; Guzy-Wrobelska, J.; Collins, N.C.; Halpin, C.; et al. Variation in the interaction between alleles of Hv APETALA2 and microRNA172 determines the density of grains on the barley inflorescence. *Proc. Natl. Acad. Sci. USA* **2013**, *110*, 16675–16680. [CrossRef]

52. Johansson, I.; Karlsson, M.; Johanson, U.; Larsson, C.; Kjellbom, P. The role of aquaporins in cellular and whole plant water balance. *Biochim. Biophys. Acta* **2000**, *1465*, 324–342. [CrossRef]

53. Maurel, C.; Verdoucq, L.; Luu, D.T.; Santoni, V. Plant aquaporins: Membrane channels with multiple integrated functions. *Annu. Rev. Plant Biol.* **2008**, *59*, 595–624. [CrossRef]

54. Li, J.; Ban, L.; Wen, H.; Wang, Z.; Dzyubenko, N.; Chapurin, V.; Gao, H.; Wang, X. An aquaporin protein is associated with drought stress tolerance. *Biochem. Biophys. Res. Commun.* **2015**, *459*, 208–213. [CrossRef]

55. Afzal, Z.; Howton, T.C.; Sun, Y.; Mukhtar, M.S. The roles of aquaporins in plant stress responses. *J. Dev. Biol.* **2016**, *4*, 9. [CrossRef] [PubMed]

56. Shi, X.; Ling, H.-Q. Current advances in genome sequencing of common wheat and its ancestral species. *Crop J.* **2018**, *6*, 15–21. [CrossRef]

57. Hedden, P. The genes of the green revolution. *Trends Genet.* **2003**, *19*, 5–9. [CrossRef]

58. Borner, A.; Worland, A.; Plaschke, J.; Schumann, E.; Law, C. Pleiotropic effects of genes for reduced height (Rht) and day-length insensitivity (Ppd) on yield and its components for wheat grown in middle Europe. *Plant Breed.* **1993**, *111*, 204–216. [CrossRef]

59. Youssefian, S.; Kirby, E.; Gale, M. Pleiotropic effects of the GA-insensitive Rht dwarfing genes in wheat. 2. Effects on leaf, stem, ear and floret growth. *Field Crops Res.* **1992**, *28*, 191–210. [CrossRef]

60. Geisel, N. Constitutive versus responsive gene expression strategies for growth in changing environments. *PLoS ONE* **2011**, *6*, e27033. [CrossRef]

61. Blum, A. Drought resistance and its improvement. In *Plant Breeding for Water-Limited Environments*; Springer: Berlin/Heidelberg, Germany, 2011; pp. 53–152.

62. Dawe, R.K.; Reed, L.M.; Yu, H.-G.; Muszynski, M.G.; Hiatt, E.N. A maize homolog of mammalian CENPC is a constitutive component of the inner kinetochore. *Plant Cell* **1999**, *11*, 1227–1238. [CrossRef]

63. Zhong, C.X.; Marshall, J.B.; Topp, C.; Mroczek, R.; Kato, A.; Nagaki, K.; Birchler, J.A.; Jiang, J.; Dawe, R.K. Centromeric retroelements and satellites interact with maize kinetochore protein CENH3. *Plant Cell* **2002**, *14*, 2825–2836. [CrossRef]

64. Cheeseman, I.M.; Desai, A. Molecular architecture of the kinetochore-microtubule interface. *Nat. Rev. Mol. Cell. Biol.* **2008**, *9*, 33–46. [CrossRef]

65. Zheng, H.; Wu, H.; Pan, X.; Jin, W.; Li, X. Aberrant meiotic modulation partially contributes to the lower germination rate of pollen grains in maize (*Zea mays* L.) under low nitrogen supply. *Plant Cell Physiol.* **2017**, *58*, 342–353. [CrossRef]

66. Li, X.; Dawe, R.K. Fused sister kinetochores initiate the reductional division in meiosis I. *Nat. Cell Biol.* **2009**, *11*, 1103–1108. [CrossRef] [PubMed]

67. Powers, A.F.; Franck, A.D.; Gestaut, D.R.; Cooper, J.; Gracyzk, B.; Wei, R.R.; Wordeman, L.; Davis, T.N.; Asbury, C.L. The Ndc80 kinetochore complex forms load-bearing attachments to dynamic microtubule tips via biased diffusion. *Cell* **2009**, *136*, 865–875. [CrossRef] [PubMed]

68. Jabbari, M.; Fakheri, B.A.; Aghnoum, R.; Mahdi Nezhad, N.; Ataei, R. GWAS analysis in spring barley (*Hordeum vulgare* L.) for morphological traits exposed to drought. *PLoS ONE* **2018**, *13*, e0204952. [CrossRef]

69. Varshney, R.K.; Paulo, M.J.; Grando, S.; van Eeuwijk, F.A.; Keizer, L.C.P.; Guo, P.; Ceccarelli, S.; Kilian, A.; Baum, M.; Graner, A. Genome wide association analyses for drought tolerance related traits in barley (*Hordeum vulgare* L.). *Field Crops Res.* **2012**, *126*, 171–180. [CrossRef]

70. Klepikova, A.V.; Kasianov, A.S.; Gerasimov, E.S.; Logacheva, M.D.; Penin, A.A. A high resolution map of the Arabidopsis thaliana developmental transcriptome based on RNA-seq profiling. *Plant J.* **2016**, *88*, 1058–1070. [CrossRef] [PubMed]

71. Finkelstein, R.; Gampala, S.S.L.; Lynch, T.J.; Thomas, T.L.; Rock, C.D. Redundant and distinct functions of the ABA response loci ABA-INSENSITIVE(ABI)5 and ABRE-BINDING FACTOR (ABF)3. *Plant Mol. Biol.* **2005**, *59*, 253–267. [CrossRef]

72. Mittal, A.; Gampala, S.S.L.; Ritchie, G.L.; Payton, P.; Burke, J.J.; Rock, C.D. Related to ABA-Insensitive3(ABI3)/Viviparous1 and AtABI5 transcription factor coexpression in cotton enhances drought stress adaptation. *Plant Biotechnol. J.* **2014**, *12*, 578–589. [CrossRef]

73. Tang, N.; Zhang, H.; Li, X.; Xiao, J.; Xiong, L. Constitutive activation of transcription factor OsbZIP46 improves drought tolerance in rice. *Plant Physiol.* **2012**, *158*, 1755–1768. [CrossRef]

 *genes*

Article

Characteristics of Microsatellites Mined from Transcriptome Data and the Development of Novel Markers in *Paeonia lactiflora*

Yingling Wan [1], Min Zhang [1], Aiying Hong [2], Yixuan Zhang [1] and Yan Liu [1,3,4,5,*]

[1] College of Landscape Architecture, Beijing Forestry University, Beijing 100083, China; wan_yingling@bjfu.edu.cn (Y.W.); 18310257817@163.com (M.Z.); zyx_cyclone@163.com (Y.Z.)
[2] Management Office of Caozhou Peony Garden, Heze 274000, Shandong, China; m18811581510@163.com
[3] Beijing Key Laboratory of Ornamental Plants Germplasm Innovation & Molecular Breeding, Beijing Forestry University, Beijing 100083, China
[4] National Engineering Research Center for Floriculture, Beijing Forestry University, Beijing 100083, China
[5] Beijing Laboratory of Urban and Rural Ecological Environment, Beijing Forestry University, Beijing 100083, China
* Correspondence: yanwopaper@yahoo.com

Received: 28 January 2020; Accepted: 17 February 2020; Published: 19 February 2020

Abstract: The insufficient number of available simple sequence repeats (SSRs) inhibits genetic research on and molecular breeding of *Paeonia lactiflora*, a flowering crop with great economic value. The objective of this study was to develop SSRs for *P. lactiflora* with Illumina RNA sequencing and assess the role of SSRs in gene regulation. The results showed that dinucleotides with AG/CT repeats were the most abundant type of repeat motif in *P. lactiflora* and were preferentially distributed in untranslated regions. Significant differences in SSR size were observed among motif types and locations. A large number of unigenes containing SSRs participated in catalytic activity, metabolic processes and cellular processes, and 28.16% of all transcription factors and 21.74% of hub genes for inflorescence stem straightness were found to contain SSRs. Successful amplification was achieved with 89.05% of 960 pairs of SSR primers, 55.83% of which were polymorphic, and most of the 46 tested primers had a high level of transferability to the genus *Paeonia*. Principal component and cluster dendrogram analyses produced results consistent with known genealogical relationships. This study provides a set of SSRs with abundant information for future accession identification, marker-trait association and molecular assisted breeding in *P. lactiflora*.

Keywords: herbaceous peony; molecular marker; next-generation sequencing; pedigree

1. Introduction

Herbaceous peony, which has many varieties with distinct flower types and colors, provides great commercial benefits in the form of cut flowers and potted plants. It has a long juvenile period before flowering, which slows the development of new cultivars with specific and stable characteristics by traditional hybridization breeding [1]. Based on appropriate DNA markers, molecular-assisted breeding can be employed to select a target genotype and detect whether hybrids have the expected trait at an early stage; thus, it improves breeding efficiency and accuracy and saves time, labor and material resources [2]. The molecular breeding of herbaceous peony is not currently well developed due to a lack of foundational research; hence a large number of highly polymorphic and stable molecular markers of herbaceous peony should be developed to further identify associations with target traits.

Microsatellites, also known as simple sequence repeats (SSRs), are widely used for plant fingerprinting, genetic diversity assessment and association analysis between target traits and quantitative trait loci (QTLs) [3–5] due to their abundance in the genome, high polymorphism, codominant inheritance and good reproducibility [6,7]. SSRs can be developed from DNA or complementary DNA (cDNA) reverse transcribed from RNA [8]. In four previous studies on herbaceous peony, there were fewer than 16 SSR primers in each SSR-enriched genomic library or magnetic bead enrichment dataset [9–11]. Previous researchers synthesized 384 pairs of SSR primers from barcoded Illumina sequencing libraries of several species from the genus *Paeonia*; the researchers utilized 12 pairs of these SSRs and nine other SSR pairs to successfully identify 93 genotypes of *Paeonia* [12]. The numbers and repeat motif types of expressed sequence tag SSRs (EST-SSRs) from two sets of transcriptome data were previously reported by bioinformatics analysis, but additional PCR experiments or further validation were not performed [13,14]. Moreover, based on the transferability of SSRs among congeners of dicotyledonous plants [15], several primers from *Paeonia* were selected and used to successfully identify cultivars of *Paeonia lactiflora* [16]. Therefore, neither the number nor the application range of SSRs in herbaceous peony was not sufficient in these previous studies.

SSRs within genes or ESTs are more likely than genic SSRs obtained from SSR-enriched libraries or random DNA sequences to be effectively linked to target traits [17]. In *Populus tomentosa*, the genic SSRs selected from candidate genes related to wood formation were successfully used in family-based linkage mapping [18]. Twenty-four SSR primers of *Pisum sativum* were successfully mapped to several existing linkage groups [19]. Similar methods were also used in raspberry, in which SSRs were associated with several developmental traits [20]. For *Paeonia rockii*, SSR markers were used to perform association mapping, and 2.68–23.97% of flowering trait variance was explained [21]. Moreover, a genetic linkage map of tree peony covering five linkage groups was constructed by 124 EST-SSR primers [22]. Thus, development SSRs from herbaceous peony transcriptome may be effective for future use. Further studies showed the genomic distribution of SSRs is nonrandom. SSRs in genes may influence gene transcription or translation and gene activity [6,23], and recent studies showed a higher abundance of SSRs in response to environmental stress [24]. The polymorphism levels and potential functions of SSRs differ among the 5′ untranslated region (5′ UTR), the 3′ UTR and coding sequences (CDs) are different; SSRs in 5′ UTR may affect transcription or translation, SSRs in CDS may inactivate or activate genes, or truncate proteins, and SSRs in 3′ UTR may cause silencing or slippage [25]. However, no such information has been reported in herbaceous peony, which is not conducive to developing desirable SSR markers.

In this study, we mined SSRs from herbaceous peony transcriptome data and analyzed the distribution and location of the SSRs and the function of unigenes containing them. Initially, a total of 960 pairs of SSR primers were developed and amplified in eight cultivars from a core collection to initially validate the polymorphism level. Then, 46 pairs of primers were used to analyze transferability among nine species in *Paeonia*. Finally, we constructed a phylogenetic tree containing seven species and 24 varieties. This study provides a number of efficient and informative SSR primers for future molecular-assisted breeding of herbaceous peony.

2. Materials and Methods

2.1. SSR Identification, Annotation from Transcriptome Data and SSR Primer Design

All the SSR sequences used in this study were obtained from transcriptome data from inflorescence stems of *P. lactiflora* 'Da Fugui' and 'Chui Touhong' at five developmental stages (i.e., stages representing every seven days from stem elongation to flowering). The transcriptome data have been deposited in the Sequence Read Archive (SRA) database as described previously (accession number: PRJNA528693) [26,27] and were assembled by Trinity 3.0. MISA-web (http://pgrc.ipk-gatersleben.de/misa/) was used to search for SSRs in the unigenes [28]. The parameters were set as follows: dinucleotide (Di-) repeats had to be repeated at least 6 times, trinucleotide (Tri-) repeats had to be repeated at least five times, tetranucleotide (Tetra-) repeats had to be repeated at least four times, pentanucleotide (Penta-) repeats had to be repeated at least four times, and hexanucleotide

(Hexa-) repeats had to be repeated at least four times. Interruptions were set to 100 to merge two SSR sequences into one SSR when the distance was shorter than 100 bp. Notably, mononucleotide repeats were not analyzed in this study. To identify possible SSR functions for future use, unigenes that contained SSRs were mapped to terms in the Gene Ontology (GO) database, gene numbers were calculated for each term, and the Kyoto Encyclopedia of Genes and Genomes (KEGG) database was used for pathway analysis. TransDecoder v3.0.1 (http://transdecoder.github.io) was used to identify candidate coding regions, dividing transcript sequences into 5′ UTR, CDS and 3′ UTR sections. Primer 3 (http://bioinfo.ut.ee/primer3) [29] was used to design primers on both sides of the microsatellite sequences, following previous product size, primer length, GC content and annealing temperature principles [8].

2.2. Plant Materials

To evaluate the specificity and polymorphism of primers, fresh young leaves of *P. lactiflora* 'Qihua Lushuang', 'Jinxing Shanshuo', 'Lian Tai', 'Fu Shi', 'Da Fugui', 'Dongfang Shaonu', 'Yangfei Chuyu' and 'Hong Fushi' were obtained from Caozhou Peony Garden, Heze, Shandong Province, China, in April 2019. To evaluate transferability, fresh leaves of seven species of *Paeonia* were collected from different habitats in China: i.e., *P. lactiflora* was collected from Xilin Gol, Inner Mongolia (115°13′–117°06′ E, 43°02′–44°52′ N), *Paeonia emodi* and *Paeonia sterniana* were collected from Tibet (84°35′–86°20′ E, 28°3′–29°3′ N), *Paeonia obovata* was collected from Pingquan, Hebei Province (118°21′–119°15′ E, 40°24′–40°40′ N), *Paeonia anomala* was collected from Altay city (85°31′–91°04′ E, 45°00′–49°10′ N), *Paeonia intermedia* was collected from Yumin, Xinjiang Province (82°12′–83°30′ E, 45°24′–46°3′ N), and *Paeonia veitchii* was collected from Lanzhou, Gansu Province (103°40′ E, 36°03′ N). Twenty-four cultivars of *P. lactiflora* used for phylogenetic analysis were also collected from Caozhou Peony Garden. These leaves were bagged with silica gel and transported to Beijing, ground to powder with liquid nitrogen and stored at −80 °C in the laboratory of the National Engineering Research Center for Floriculture, Beijing, China. A DNAsecure plant kit (TIANGEN Biotech, Co., Ltd., Beijing, China) was used for DNA extraction. The quality and quantity of total DNA were estimated by a NanoDrop 2000 spectrophotometer (Thermo Scientific, Waltham, MA, USA). DNA was diluted to 30 ng/μL in preparation for polymerase chain reaction (PCR).

2.3. SSR Primer Evaluation in Eight Cultivars

A total of 960 pairs of primers were selected for synthesis (Ruibiotech, Co., Ltd., Beijing, China). To improve the efficiency of primer fluorescence labeling, the thermocycler amplification protocol was conducted in two rounds. First, the primers synthesized for the DNA of the eight cultivars were used for amplification. The 10 μL PCR mixture consisted of 0.1 μL of 10 μmol/μL forward primer containing the M13(-21) tail at its 5′ end and reverse primer, 1 μL of 30 ng/μL DNA, 5 μL of 0.1 U/μL 2×Taq PCR MasterMix (containing 0.05 units/μL Taq DNA polymerase (recombinant), 4 mM $MgCl_2$ and 0.4 mM dNTPs, Aidlab Biotechnologies Co., Ltd., Beijing, China) and 3.8 μL of ddH_2O. After an initial denaturation step of 95 °C for 5 min, 20 cycles of 95 °C for 30 s, 60 °C for 30 s and 72 °C for 30 s, as well as extension at 72 °C for 10 min, were performed. Second, to efficiently and economically analyze the length of PCR products, fluorescently labeled (i.e., FAM, HEX, TAMRA or ROX) M13(-21) universal primers were added to the PCR mix [30]. The 10 μL PCR mixture contained 0.15 μL of 10 μmol/μL M13(-21) universal primer and reverse primer, 2 μL of the PCR product from the first round, 5 μL of 0.1 U/μL 2×Taq PCR MasterMix and 2.7 μL of ddH_2O. In the thermocycler, amplification was performed at 95 °C for 5 min and followed by 35 cycles of 95 °C for 30 s, 52 °C for 30 s and 72 °C for 30 s, as well as extension at 72 °C for 10 min. After 3% agarose gel electrophoresis, the amplified loci of the final PCR product were detected by a 3730xl DNA Analyzer with 96 capillaries (Applied Biosystems, Foster City, CA, USA) and sized with GS500 LIZ. The amplified loci were analyzed by GeneMarker V2.2.0.

2.4. Phylogenetic Analysis of Seven Species of Paeonia and 24 Cultivars of P. lactiflora

As shown in Table 1, the 46 forward primers showing the most abundant polymorphic loci were resynthesized by adding a fluorescent label to the 5′ tail. The 10 µL PCR mixture consisted of 0.2 µL of 10 µmol/µL forward primer and reverse primer, 1 µL of 30 ng/µL DNA template, 5 µL of 0.1 U/µL 2×Taq PCR MasterMix and 3.6 µL of ddH$_2$O. PCR was performed at 95 °C for 5 min, followed by 35 cycles of 94 °C for 30 s, annealing at an appropriate temperature (as shown in Table S3) for 30 s, and 72 °C for 30 s and a final extension at 72 °C for 7 min.

Table 1. Size range of amplification products, sample size (N), the frequency of null allele at locus (Null allele), number of different alleles (Na), number of effective alleles (Ne), Shannon's information index (I), observed heterozygosity (Ho), expected heterozygosity (He), fixation index (F) and polymorphism information content (PIC)./means we did not calculated the frequency of null allele at the locus because there was no significant difference in the observed and expected value in this locus and it followed Hardy-Weinberg equilibrium (Chi-Square tests, $p < 0.05$).

Locus	Repeat Motif	Size Range (bp)	N	Na	Ne	I	Ho	He	Null allele	F	PIC
T125	(TC)9	128–170	31	15	7.84	2.29	0.77	0.87	0.073	0.11	0.86
T163	(TA)8	146–160	31	8	3.96	1.63	0.65	0.75	0.085	0.14	0.71
T237	(CT)10	95–109	31	8	4.59	1.76	0.68	0.78	0.095	0.13	0.75
T241	(GA)15	127–157	31	12	6.77	2.11	0.81	0.85	/	0.05	0.84
C160	(AT)9ctcctt(CTC)5	211–227	29	8	4.58	1.79	0.66	0.78	0.117	0.16	0.76
TA564	(TA)8	172–201	26	9	5.43	1.90	0.77	0.82	0.058	0.06	0.79
T317	(CT)10	157–183	30	9	5.94	1.92	0.83	0.83	/	0.00	0.81
S024	(AAT)16	161–201	31	10	5.60	1.94	0.77	0.82	0.042	0.06	0.80
T040	(GA)7	228–246	31	7	5.02	1.73	0.68	0.80	0.106	0.15	0.77
T179	(AT)7	235–269	31	10	4.93	1.83	0.71	0.80	0.070	0.11	0.77
T192	(CT)8	212–230	31	8	4.50	1.75	0.68	0.78	0.093	0.13	0.75
T300	(AT)10	218–238	31	9	4.45	1.80	0.48	0.78	0.171	0.38	0.75
TA673	(CA)18	190–288	31	15	6.74	2.20	0.81	0.85	0.058	0.05	0.84
T210	(GA)10	253–285	31	12	4.44	1.86	0.81	0.77	0.044	−0.04	0.75
TA038	(CT)9	259–299	31	13	6.74	2.18	0.13	0.85	0.391	0.85	0.84
S033	(TAT)7	137–181	31	10	2.47	1.39	0.42	0.59	0.123	0.29	0.57
T304	(GA)10	147–163	30	9	4.75	1.77	0.63	0.79	0.144	0.20	0.76
T863	(AG)10	140–160	31	10	6.26	2.03	0.74	0.84	0.070	0.12	0.82
TA028	(AC)6	101–123	31	8	3.88	1.69	0.48	0.74	0.177	0.35	0.72
T239	(CT)9	185–221	31	11	2.98	1.59	0.58	0.66	0.069	0.13	0.64
TA144	(CT)6	186–286	31	9	4.00	1.64	0.68	0.75	/	0.10	0.71
T852	(AT)8	158–367	31	13	5.95	2.06	0.35	0.83	0.268	0.57	0.81
F106	(TATG)5	200–290	31	13	5.02	2.03	0.55	0.80	0.145	0.32	0.78
TA082	(AG)6	242–280	30	13	6.00	2.03	0.73	0.83	0.084	0.12	0.81
TA079	(AT)8	232–264	29	8	2.93	1.45	0.52	0.66	0.161	0.21	0.62
T356	(AG)9	247–269	30	9	4.76	1.81	0.33	0.79	0.296	0.58	0.76
TA133	(AT)8	258–286	28	9	4.28	1.72	0.75	0.77	0.095	0.02	0.73
SA010	(AAT)5	264–294	31	9	4.98	1.85	0.71	0.80	0.079	0.11	0.77
W75	(CTCAC)5	257–293	31	8	4.65	1.70	0.65	0.79	0.092	0.18	0.75
T205	(TC)9	275–301	31	13	6.94	2.18	0.71	0.86	/	0.17	0.84
S853	(TTG)5	158–186	29	8	4.67	1.71	0.69	0.79	/	0.12	0.75
T859	(AG)19	166–200	29	11	7.61	2.15	0.90	0.87	0.000	-0.03	0.85
TA566	(GA)6	167–185	25	7	4.83	1.71	0.80	0.79	/	−0.01	0.76
TA695	(AC)15	164–180	30	8	3.00	1.49	0.57	0.67	0.119	0.15	0.64
S237	(CTG)8	193–219	31	13	9.96	2.41	0.74	0.90	0.081	0.18	0.89
TA464	(TC)8	200–214	30	8	5.56	1.85	0.73	0.82	0.062	0.11	0.80
T160	(CT)10	205–227	30	10	5.47	1.92	0.70	0.82	/	0.14	0.80
TA134	(GA)16	142–224	29	13	5.43	2.09	0.93	0.82	0.019	−0.14	0.80
SA061	(GAA)7	257–277	29	6	3.43	1.46	0.38	0.71	0.281	0.46	0.67
TA074	(CT)8	260–306	30	13	5.79	2.03	0.97	0.83	0.000	−0.17	0.81
TA704	(TC)17	237–277	30	7	5.19	1.73	0.70	0.81	0.111	0.13	0.78
TA610	(TC)8	257–291	28	11	6.25	2.05	0.68	0.84	0.117	0.19	0.82
S025	(ATT)7	268–350	29	13	5.14	2.02	0.55	0.81	0.200	0.32	0.79
TA022	(TC)6	276–318	30	16	6.87	2.23	0.90	0.85	0.000	−0.05	0.84
TA086	(AG)10	393–419	31	10	5.62	1.94	0.74	0.82	0.077	0.10	0.80
TA700	(AG)9	278–312	29	13	5.76	2.05	0.62	0.83	0.155	0.25	0.81
Mean			30.07	10.26	5.26	1.88	0.67	0.80	0.11	0.16	0.77
SE			0.20	0.36	0.21	0.03	0.02	0.01	0.01	0.03	0.01

Polymorphism information content (PIC) was calculated by the Microsatellite Toolkit according to the methods of a previous study [31]. GenAlEx 6.51b2 was used for genetic analysis and principal coordinates analysis (PCoA), and parameters including the number of alleles (*Na*), tests for Hardy-Weinberg equilibrium, the number of effective alleles (*Ne*), Shannon's information index (*I*), heterozygosity (*Ho*), heterozygosity (*He*) and PIC were calculated [32,33]. The frequency of null alleles at loci were estimated by maximum likelihood method [34]. A dendrogram of the accessions of *Paeonia* was generated according to Bruvo's distance with 1000 bootstrap replicates by the R package poppr [35].

3. Results

3.1. Numbers and Distribution of SSRs in Transcriptome Data

A total of 122,670 unigenes with a total length of 1.06E+08 bp were searched by MISA-web, and 10,468 SSRs (including 825 compound formations) were found. These SSRs were distributed among 8837 unigenes (7.20%), 1321 of which contained more than one SSR. As shown in Figure 1A, Di- repeats were the most abundant (63.52%) type of repeat motif, followed by Tri- repeats (22.70%), Tetra- repeats (7.73%) and all the other types of repeat motifs (6.05%). Di- repeats were of four types, namely, AG/CT (36.55%), AT/AT (14.78%), AC/GT (11.97%) and CG/CG (0.23%). The number of each type of Di- repeat (except CG/CG) exceeded the number of SSRs. TransDecoder analysis showed that these SSRs involved 4482 CDSs and 3958 unigenes. As shown in Figure 1B, the positions of many motifs were not putative, and in known positions of unigenes, different motifs exhibited distinct preferences. Di- repeats were mostly located in the 5′ UTR, followed by the 3′ UTR. In CDSs, Tri- repeats were the most abundant motif, and Tetra- repeats were mostly located in the 3′ UTR and 5′ UTR.

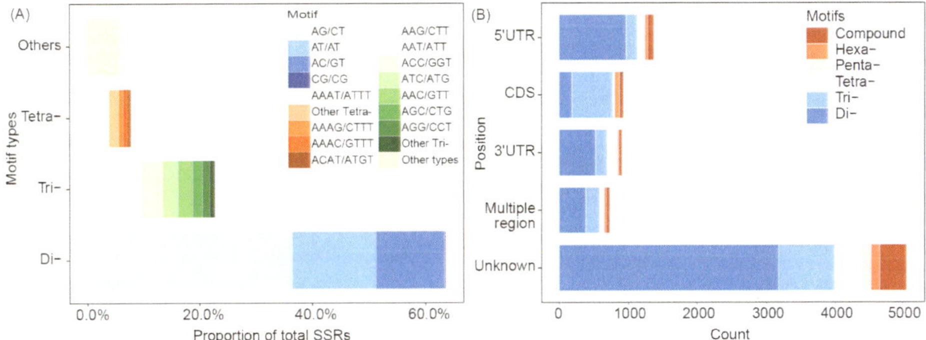

Figure 1. Distribution of different repeat motifs and positions of simple sequence repeats (SSRs) in unigenes. (**A**) Proportion and distribution of each type of motif in dinucleotide (Di-), trinucleotide (Tri-), tetranucleotide (Tetra-) and other (i.e., Penta-, Hexa- and compound) repeats. In the legend, 'other Tri-' consists of ACG/CGT (0.26%), ACT/AGT (0.26%) and CCG/CGG (0.42%), and 'Other Tetra-' consists of 26 types of Tetra- repeats, the most abundant of which are AATC/ATTG (0.22%) and AGGG/CCCT (0.22%). (**B**) Abundances of six motifs in different unigene positions. Two types of SSRs were located in multiple regions. One of these SSRs was located across two 5′ UTRs, a coding sequences (CDS) and a 3′ UTR; another was located in multiple CDSs in one unigene. The SSR locations differed from each other. Unknown refers to the SSRs without matching locations.

SSR size was analyzed as shown in Figures S1 and S2 and Table S1. For each type of repeat motif, the smallest SSRs were the most abundant, with sizes of 12 for Di- repeats, 15 for Tri- repeats, 16 for Tetra- repeats, 20 for Penta- repeats and 24 for Hexa- repeats, all of which had distinct discrete values with the increase in repeat motifs. Different types of repeat motifs exhibited significantly distinct sizes according to pairwise comparisons (Wilcoxon rank sum test, with Bonferroni adjustment). For all the SSRs, the most common size was 12 for Di- repeats (frequency of 1894), 15 for Tri- repeats (frequency of 1309) and 14 for Di- repeats (frequency of 1200). For each location (including the unknown positions) of

SSRs, a large number of discrete values were observed to have high repetitions. Significant differences in the sizes of SSRs among the 5′ UTR, 3′ UTR and CDSs appeared according to pairwise comparisons (Wilcoxon rank sum test, with Bonferroni adjustment). Only the size of SSRs in the 5′ UTR differed from that in multiple regions. The size of SSRs in unknown regions was not obviously different from that in the 3′ UTR and multiple regions.

3.2. Annotation of the Unigenes with SSRs

To understand the potential functions of the unigenes containing SSRs, we classified these unigenes using GO annotation. A total of 8837 unigenes were blasted against a protein database and annotated with GO terms. The results showed that 8522 unigenes (96.44%) were involved in three functional groups and 42 putative processes or functions, as shown in Figure 2. These unigenes participated in 17 types of biological processes, 11 types of molecular functions and 14 types of cellular components. Catalytic activity (972, 11.40% of the total blasted genes), metabolic process (959, 11.25%) and cellular process (930, 10.91%) were the three most abundant terms for putative gene functions. Fewer than ten unigenes were involved in each of the two types of biological processes, six types of molecular functions and four types of cellular components, and rhythmic process (1), transcription factor (TF) activity and protein binding (1) and extracellular region part (1) were the terms assigned the fewest unigenes containing SSRs.

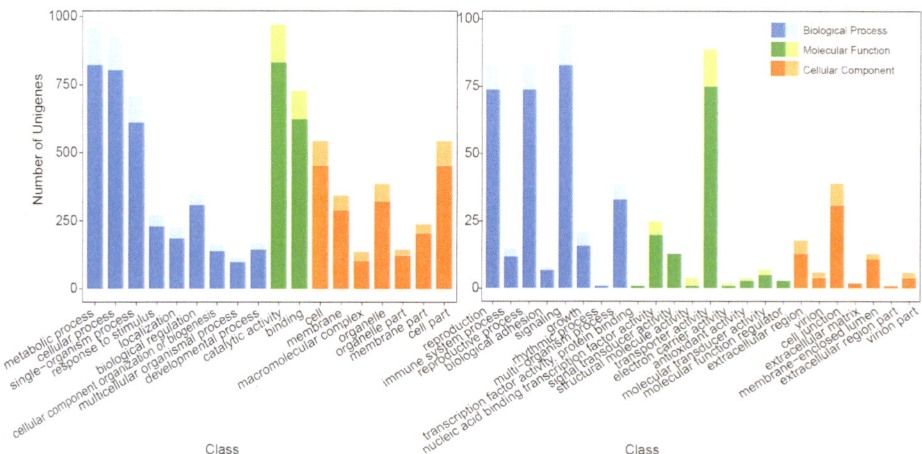

Figure 2. Gene Ontology (GO) analysis of unigenes containing SSRs. The lighter color of each bar represents the number of unigenes without matching coding sequences.

To further categorize the unigenes, KEGG annotation was used. As shown in Figure 3, unigenes with SSRs were involved in 126 pathways, which were divided into five classes and 18 subclasses. In total, unigenes participating in metabolism were most abundant. At the subclass level, the global and overview pathway (38.22%) had the largest number of unigenes, followed by the transcription pathway (11.44%), and the carbohydrate metabolism pathway (9.79%). The top five unigenes belonged to global and overview subgroups, and they were involved in metabolic pathways (15.38%), biosynthesis of secondary metabolites (8.63%), biosynthesis of antibiotics (4.11%), microbial metabolism in diverse environments (3.98%) and carbon metabolism (2.73%).

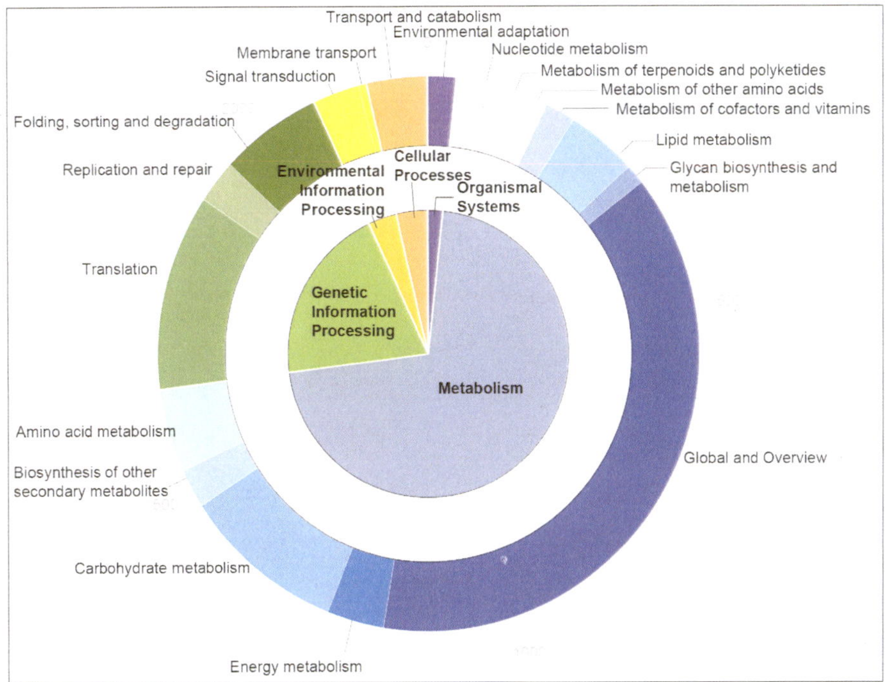

Figure 3. Kyoto Encyclopedia of Genes and Genomes (KEGG) pathway annotation of unigenes with embedded SSRs. The numbers outside the circle represent the cumulative number of unigenes beginning at zero and moving in a clockwise manner.

As shown in Table S2, the TFs containing SSRs accounted for 28.16% of the total TFs (1218) in the transcriptome of herbaceous peony, including 51 kinds of TF families (e.g., ERF, MYB, MYB related and ARF). Regarding the hub genes (46) for inflorescence stem straightness, as described in a previous study [27], 21.74% of the genes contained SSRs and were involved in lignin monolignol biosynthesis (*4CL1*, *CCoAOMT2*, *HST* and *CAD2*), xylan synthesis and metabolic process (*IRX-15 L*), auxin signaling transduction (*IAA26*, *IAA31* and *SAUR20*) and lateral organ boundary domain TF (*LBD15* and *LBD36*) pathways.

3.3. Initial Amplification of SSR Primers

Primer3 was used to design primers for the 9643 SSRs. Appropriate primers could not be designed for 2369 of these SSRs due to short or missing flanking sequences. A total of 7274 pairs of primers were designed, 3721 of which were able to identify CDSs. We further selected 960 pairs of primers considering SSR types and locations for synthesis and used them for amplification in eight distinct cultivars of *P. lactiflora*. As shown in Figure S3, 89.05% of the primers resulted in successful amplification in these cultivars, and 55.83% (i.e., 62.72% of the total with successful amplification) of the SSR marker primers had polymorphic amplification products. The total number of polymorphic loci decreased with an increasing number of alleles per locus. Appropriately 30% of the primers amplified only two or three types of products, and almost 16% of the primers amplified more than five alleles in the eight DNA templates.

3.4. Polymorphism in P. lactiflora

The 46 primers (listed in Table S3) with the most abundant amplified loci were used to reveal the information and transferability of the primers, as shown in Table 1.

Forty-four pairs of primers were amplified in the accessions; however, TA564 (172–201 bp) and TA566 (167–185 bp) were successful in only 26 and 25 accessions, respectively. The product size range varied among accessions. The products of T852 (209 bp) and TA144 (100 bp) had the maximum size difference among accessions, and T163 (14 bp), T237 (14 bp) and TA464 (14 bp) presented the smallest size differences among accessions. A total of 472 different alleles (Na) were amplified. The Na ranged from 6 to 16, with an average value of 10.26 ± 0.36; the Ne ranged from 2.47 to 9.96, with an average value of 5.26 ± 0.21. I varied from 1.39 to 2.41, with an average value of 1.88. The ranges of Ho and He were 0.13–0.97 (the average was 0.67) and 0.59–0.90 (the average was 0.80), respectively. SSR typing data of seven locus followed Hardy-Weinberg equilibrium, and the frequency of null alleles in other locus were 0.000–0.391 (the average was 0.11). The PIC ranged from 0.57–0.89 with an average value of 0.77.

3.5. Transferability of SSR Markers among Paeonia Species

PCoA of 31 accessions was conducted according to the amplified alleles, as shown in Figure 4A. Eigen values by axis and sample eigen vectors are shown in Table S4. A total of 30 dimensions were extracted, and dimension 1 (10.7%) and dimension 2 (7.1%) represented 17.8% of the total information. The points representing the 24 cultivars of *P. lactiflora* were close to each other and separated from points of the other seven species. Of these species, *P. lactiflora* was nearest to the 24 cultivars in the PCoA plot, and *P. intermedia* was significantly separated from all the accessions. A cluster dendrogram was drawn according to Bruvo's distances calculated by the amplified alleles, as shown in Figure 4B. All the accessions were divided into two groups, and the species *P. lactiflora* and its cultivars were tightly clustered and separated from the other *Paeonia* species, which was consistent with the results of PCoA. Furthermore, at a height of 0.85, *P. obovata* and *P. emodi* were further clustered and separate from the other four species. Among the cultivars of *P. lactiflora*, 'Yinxian Xiuhongpao' had the maximum distance from the other cultivars and was separated at a height of 0.65.

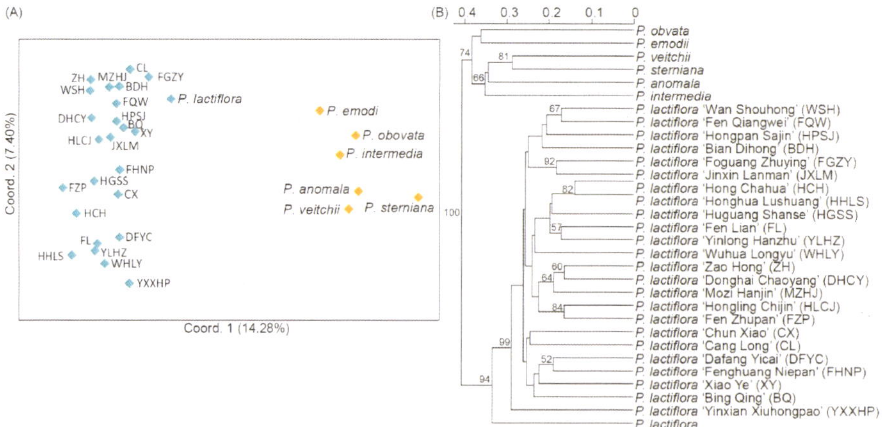

Figure 4. Principal coordinates analysis (PCoA) of amplified loci (**A**) and dendrogram generated by Bruvo's distances (**B**) of 31 accessions, including seven species and 24 cultivars of *Paeonia lactiflora*. The cultivar names are abbreviated with capitalized letters in (**A**), and their full names are shown in (**B**). The UPGMA tree was produced with 1000 bootstrap replicates, and the node values greater than 50 are shown in the tree.

4. Discussion

Next-generation sequencing makes it possible to develop microsatellites efficiently and inexpensively [8]. In this study, we identified a total of 10,468 SSRs, covering 7.20% of the transcripts assembled using our transcriptome data. This method was significantly more convenient and effective than the use of SSR-enriched genomic libraries or magnetic bead enrichment, which were the primary methods used in previous herbaceous peony studies [9–11,36,37]. The coverage of SSRs in ESTs reported in the present study was higher than that reported in five cereals (average of 3.2%) [38] and similar (6.6%) to that generated by Trinity for *P. lactiflora* 'Hang Baishao' [13]. The distribution of motif types varies among plants. The most frequent motif type for *Parrotia subaequalis* was Di- [39], while the most frequent motif type for *Lychnis kiusiana* and *Dendrocalamus hamiltonii* was Tri- [40,41]. In this study, Di- repeats were the most abundant motif (mononucleotides were not considered) in *P. lactiflora*, and AG/CT accounted for 36.55% of all SSRs, followed by AT/AT (14.78%) and AC/GT (11.97%). The observation that AG/CT was the most frequent repeat was consistent with the finding of a previous study in the genus *Paeonia* regardless of assemble methods [13], while the proportions of TC/GA and AC/GT repeats significantly differed [14,42]. Differences in the transcriptomic SSR motifs can explain the relatively low transferability of SSR primers (approximately 26%) from the genus *Paeonia* to *P. lactiflora* [16].

SSR size varied significantly among unigene locations and motif types in this study. Furthermore, our results suggested distinct preferences in the distribution of motifs among different gene parts; in annotated positions, a large number of Di- repeat motifs were distributed in the 5' UTR and 3' UTR, and most of the Tri- repeat motifs were distributed in CDSs. Polymorphism level is affected by location; notably, a large proportion of SSRs in the 3' UTR were polymorphic in *Hordeum vulgare* [43]. Further studies in grape showed that the most polymorphic SSR position differed at three levels, that is, among cultivars, among cultivars and species, and among species and genera [25]. As shown in Figure S3, in the initial screening of 960 SSR primers in eight cultivars, we found that at the cultivar level, SSRs from the 5' UTR (64.05%) were the most polymorphic, followed by those from the 3' UTR (60.61%). This result suggested that the polymorphism level of SSR locations was related to species.

SSRs from the transcribed sequence may be directly related to phenotypic variation and thus related to functional trait. SSR alleles associated with biotic or abiotic stress, such as heat, cold, salt and resistant to multiple diseases have been reported [44–47]. SSRs from specific organ are likely to associate with corresponding morphological traits; SSRs obtained from flower bud transcriptome in *P. rockii* have been demonstrated significantly associating with flower colors and shapes [21]. In this study, SSRs were investigated from transcriptome that was obtained from two cultivars with distinct straightness of inflorescence stem, and a large quantity of unigenes with SSRs were annotated with the catalytic activity, metabolic process and cellular process terms. Furthermore, 28.16% of all TFs and 21.74% of the hub genes for inflorescence stem straightness contained SSRs. We speculate that these SSRs likely associated with straightness characteristics of the herbaceous peony inflorescence stem, while association analysis and QTL mapping are needed in further study.

In our experiment, 89.05% of 960 pairs of primers were validated by PCR, and 55.83% of the primers were polymorphic, which was approximately the percentage (59.90%) previously reported in peony [10], higher than the percentage (36.67%) in *Amentotaxus argotaenia* [48], and lower than the percentage for SSR markers (77.2%) generated from the soybean genome [49]. These amplification differences may be due to the number of individuals used for amplification or locus mutations (e.g., insertions, deletions and translocations) among species or cultivars [50]. To identify the reason, more individuals should be subjected to PCR amplification, and cloning experiments and sequencing should be carried out.

The mean Na in this study was 10.26, and Ho and He were 0.67 and 0.80, respectively, which were higher than the Ho and He reported in previous studies on tree peony and herbaceous peony [9,10,51]. The mean PIC value was 0.77, showing a high level of high informativeness [52]; compared to mean PIC value (0.4149–0.678) revealed by previous SSR development [9,10,36,37], our results were significantly higher, suggesting our contribution for future effective genetic analysis or QTL mapping of *Paeonia* with fewer SSRs. Previous studies suggested that the presence of null alleles is common, and it has influence on evaluation of genetic diversity of population, even causes misunderstanding in parentage analysis [53,54]. Literature showed that the frequencies of null alleles were almost fewer than 0.40 and most of them were fewer than 0.20 [55]. Our results showed the frequencies of null alleles of 22 SSRs were less than 0.08 (or no presence), and only that of four SSRs were between 0.20 and 0.40. The future use of these SSRs should carefully consider the influence of null alleles according to research objective and choose the appropriate SSRs.

EST-SSRs developed for one species can be transferred to related species, with transferability varying depending on the plant and SSR source. In *Magnolia wufengensis* and *Elymus sibiricus*, the transferability of EST-SSRs to related species was 50–68.1% and 49.1%, respectively. In SSRs from candidate genes of *Oryza sativa*, transferability ranged from 70.37% to 77.78% according to different complexes [56]. In this study, 52.17% of the 46 pairs of SSR primers selected from the initial screening could be completely transferred to seven species of the genus *Paeonia*, 39.13% of the pairs had high transferability (six or five of seven accessions were successfully amplified), and 4.35% of the pairs had moderate transferability and could be amplified in four of seven species.

The diversity of dimensions extracted from PCoA and the low explanatory power of one dimension suggested that the genetic background of the *Paeonia* accessions involved in this study varied greatly. Combining the PCoA plot and the dendrogram of 31 accessions, the genetic relationships between these accessions were almost consistent with their recognized morphological classification [57,58]. These SSR markers can be used in genetic variance analysis and to initially evaluate the value of breeding parents selected according to genetic distance in the genus *Paeonia*.

5. Conclusions

In this study, a large quantity of informative SSRs were conveniently identified from transcriptome data of *P. lactiflora*, and the distribution and location of motifs were defined. SSR containing genes associated with TFs and inflorescence stem straightness were identified, providing a foundation for future marker-trait association research. To the best of our knowledge, this is the first study to comprehensively reveal the characteristics and functional annotations of EST-SSRs in *P. lactiflora*. In future studies, more herbaceous peony accessions should be tested to further evaluate the polymorphism of markers, and more functional markers potentially associated with traits should be developed to advance the molecular breeding of *P. lactiflora*.

Supplementary Materials: The following are available online at http://www.mdpi.com/2073-4425/11/2/214/s1, Figure S1: Size distribution of SSRs with different motifs and different unigene positions. (A) SSR size of different motifs. Considering the data coverage and to improve the readability of the graph, the figure shows 99% of the data. The distribution of all data is shown in Figure S2. Size differences between motifs were analyzed with pairwise comparisons using the Wilcoxon rank sum test (after Bonferroni correction), and all the outputs were less than 2.2×10^{-16}. (B) Sizes of SSRs in different unigene positions. The figure shows 98% of the data. The distribution of all data is shown in Table S1. Size differences between positions were compared using the same method as in (A), and the figure shows combinations with p-values less than 0.05, Figure S2: Size distribution of each type of repeat motif in all data, Figure S3: Polymorphism rate of primers designed for different positions of SSRs in unigenes, Table S1: Numbers of SSRs with distinct sizes in different regions, Table S2: Count of SSRs in TFs, Table S3: Sequence, position, repeat motif, annotation, annealing temperature and amplified product size of 46 SSR primers, Table S4: Eigen values by axis and sample eigen vectors.

Author Contributions: Conceptualization, Y.W. and Y.L.; methodology, Y.W. and M.Z.; software, Y.W.; validation, Y.W., M.Z. and Y.Z.; formal analysis, Y.W.; investigation, A.H. and Y.Z.; resources, A.H. and Y.Z.; data curation, M.Z.; writing—original draft preparation, Y.W.; writing—review and editing, M.Z., A.H., Y.Z., and Y.L.; visualization, Y.W.; supervision, Y.L.; project administration, Y.L.; funding acquisition, Y.L. All authors have read and agreed to the published version of the manuscript.

Funding: This research was funded by the Beijing Municipal Science & Technology Commission, grant number D161100001916004.

Conflicts of Interest: The authors declare no conflict of interest.

References

1. Kamenetsky, R.; Dole, J. Herbaceous peony (*Paeonia*): Genetics, physiology and cut flower production. *Floric. Ornam. Biotechnol.* **2012**, *6*, 62–77.
2. Collard, B.C.; Mackill, D.J. Marker-assisted selection: An approach for precision plant breeding in the twenty-first century. *Philos. Trans. R. Soc. B Biol. Sci.* **2008**, *363*, 557–572. [CrossRef]
3. Agarwal, M.; Shrivastava, N.; Padh, H. Advances in molecular marker techniques and their applications in plant sciences. *Plant Cell Rep.* **2008**, *27*, 617–631. [CrossRef]
4. Bhattarai, G.; Mehlenbacher, S.A. In silico development and characterization of tri-nucleotide simple sequence repeat markers in hazelnut (*Corylus avellana* L.). *PLoS ONE* **2017**, *12*, e0178061. [CrossRef] [PubMed]
5. Nie, G.; Tang, L.; Zhang, Y.; Huang, L.; Ma, X.; Cao, X.; Pan, L.; Zhang, X.; Zhang, X. Development of SSR markers based on transcriptome sequencing and association analysis with drought tolerance in perennial grass *Miscanthus* from China. *Front. Plant Sci.* **2017**, *8*, 801. [CrossRef] [PubMed]
6. Varshney, R.K.; Graner, A.; Sorrells, M.E. Genic microsatellite markers in plants: Features and applications. *Trends Biotechnol.* **2005**, *23*, 48–55. [CrossRef] [PubMed]
7. Liu, S.; An, Y.; Li, F.; Li, S.; Liu, L.; Zhou, Q.; Zhao, S.; Wei, C. Genome-wide identification of simple sequence repeats and development of polymorphic SSR markers for genetic studies in tea plant (*Camellia sinensis*). *Mol. Breed.* **2018**, *38*, 59. [CrossRef]
8. Taheri, S.; Abdullah, T.L.; Yusop, M.R.; Hanafi, M.M.; Sahebi, M.; Azizi, P.; Shamshiri, R.R. Mining and development of novel SSR markers using next generation sequencing (NGS) data in plants. *Molecules* **2018**, *23*, 399. [CrossRef]
9. Cheng, Y.; Kim, C.-H.; Shin, D.-I.; Kim, S.-M.; Koo, H.-M.; Park, Y.-J. Development of simple sequence repeat (SSR) markers to study diversity in the herbaceous peony (*Paeonia lactiflora*). *J. Med. Plants Res.* **2011**, *5*, 6744–6751. [CrossRef]
10. Li, L.; Cheng, F.Y.; Zhang, Q.X. Microsatellite markers for the Chinese herbaceous peony *Paeonia lactiflora* (*Paeoniaceae*). *Am. J. Bot.* **2011**, *98*, e16–e18. [CrossRef]
11. Sun, J.; Yuan, J.; Wang, B.; Pan, J.; Zhang, D. Development and characterization of 10 microsatellite loci in *Paeonia lactiflora* (*Paeoniaceae*). *Am. J. Bot.* **2011**, *98*, e242–e243. [CrossRef] [PubMed]
12. Gilmore, B.; Bassil, N.; Nyberg, A.; Knaus, B.; Smith, D.; Barney, D.L.; Hummer, K. Microsatellite marker development in peony using next generation sequencing. *J. Am. Soc. Hortic. Sci.* **2013**, *138*, 64–74. [CrossRef]
13. Zhang, J.; Wu, Y.; Li, D.; Wang, G.; Li, X.; Xia, Y. Transcriptomic analysis of the underground renewal buds during dormancy transition and release in 'Hangbaishao' Peony (*Paeonia lactiflora*). *PLoS ONE* **2015**, *10*, e0119118.
14. Ma, Y.; Cui, J.; Lu, X.; Zhang, L.; Chen, Z.; Fei, R.; Sun, X. Transcriptome analysis of two different developmental stages of *Paeonia lactiflora* seeds. *Int. J. Genom.* **2017**. [CrossRef]
15. Barbara, T.; PALMA-SILVA, C.; Paggi, G.M.; Bered, F.; Fay, M.F.; Lexer, C. Cross-species transfer of nuclear microsatellite markers: Potential and limitations. *Mol. Ecol.* **2007**, *16*, 3759–3767. [CrossRef]
16. Zhang, J.; Liu, A.; Zhang, S.; Xie, Y.; Yan, L. Using the SSR with fluorescent labeling to establish SSR molecular ID code for cultivars of the Chinese herbaceous peony. *J. Beijing For. Univ.* **2016**, *38*, 101–109.
17. Dutta, S.; Kumawat, G.; Singh, B.P.; Gupta, D.K.; Singh, S.; Dogra, V.; Gaikwad, K.; Sharma, T.R.; Raje, R.S.; Bandhopadhya, T.K. Development of genic-SSR markers by deep transcriptome sequencing in pigeonpea [*Cajanus cajan* (L.) Millspaugh]. *BMC Plant Biol.* **2011**, *11*, 17. [CrossRef]
18. Du, Q.; Gong, C.; Pan, W.; Zhang, D. Development and application of microsatellites in candidate genes related to wood properties in the Chinese white poplar (*Populus tomentosa* Carr.). *DNA Res.* **2012**, *20*, 31–44. [CrossRef]
19. Mishra, R.K.; Gangadhar, B.H.; Nookaraju, A.; Kumar, S.; Park, S.W. Development of EST-derived SSR markers in pea (*Pisum sativum*) and their potential utility for genetic mapping and transferability. *Plant Breed.* **2012**, *131*, 118–124. [CrossRef]
20. Woodhead, M.; McCallum, S.; Smith, K.; Cardle, L.; Mazzitelli, L.; Graham, J. Identification, characterisation and mapping of simple sequence repeat (SSR) markers from raspberry root and bud ESTs. *Mol. Breed.* **2008**, *22*, 555–563. [CrossRef]
21. Wu, J.; Cheng, F.; Cai, C.; Zhong, Y.; Jie, X. Association mapping for floral traits in cultivated *Paeonia rockii* based on SSR markers. *Mol. Genet. Genom.* **2017**, *292*, 187–200. [CrossRef] [PubMed]

22. Guo, Q.; Guo, L.-L.; Zhang, L.; Zhang, L.-X.; Ma, H.-L.; Guo, D.-L.; Hou, X.-G. Construction of a genetic linkage map in tree peony (*Paeonia* Sect. *Moutan*) using simple sequence repeat (SSR) markers. *Sci. Hortic.* **2017**, *219*, 294–301.

23. Li, Y.C.; Korol, A.B.; Fahima, T.; Beiles, A.; Nevo, E. Microsatellites: Genomic distribution, putative functions and mutational mechanisms: A review. *Mol. Ecol.* **2002**, *11*, 2453–2465. [CrossRef] [PubMed]

24. Srivastava, S.; Avvaru, A.K.; Sowpati, D.T.; Mishra, R.K. Patterns of microsatellite distribution across eukaryotic genomes. *BMC Genom.* **2019**, *20*, 153. [CrossRef] [PubMed]

25. Scott, K.D.; Eggler, P.; Seaton, G.; Rossetto, M.; Ablett, E.M.; Lee, L.S.; Henry, R. Analysis of SSRs derived from grape ESTs. *Theor. Appl. Genet.* **2000**, *100*, 723–726. [CrossRef]

26. Wan, Y.; Hong, A.; Zhang, Y.; Liu, Y. Selection and validation of reference genes of *Paeonia lactiflora* in growth development and light stress. *Physiol. Mol. Biol. Plants* **2019**, *25*, 1097. [CrossRef]

27. Wan, Y.; Zhang, M.; Hong, A.; Lan, X.; Yang, H.; Liu, Y. Transcriptome and weighted correlation network analyses provide insights into inflorescence stem straightness in *Paeonia lactiflora*. *Plant Mol. Biol.* **2020**, *102*, 239–252. [CrossRef]

28. Beier, S.; Thiel, T.; Münch, T.; Scholz, U.; Mascher, M. MISA-web: A web server for microsatellite prediction. *Bioinformatics* **2017**, *33*, 2583–2585. [CrossRef]

29. Untergasser, A.; Cutcutache, I.; Koressaar, T.; Ye, J.; Faircloth, B.C.; Remm, M.; Rozen, S.G. Primer3—New capabilities and interfaces. *Nucleic Acids Res.* **2012**, *40*, e115. [CrossRef]

30. Schuelke, M. An economic method for the fluorescent labeling of PCR fragments. *Nat. Biotechnol.* **2000**, *18*, 233. [CrossRef]

31. Lovin, D.D.; Washington, K.O.; deBruyn, B.; Hemme, R.R.; Mori, A.; Epstein, S.R.; Harker, B.W.; Streit, T.G.; Severson, D.W. Genome-based polymorphic microsatellite development and validation in the mosquito *Aedes aegypti* and application to population genetics in Haiti. *BMC Genom.* **2009**, *10*, 590. [CrossRef] [PubMed]

32. Peakall, R.; Smouse, P.E. Genalex 6: Genetic analysis in Excel. Population genetic software for teaching and research. *Mol. Ecol. Notes* **2006**, *6*, 288–295. [CrossRef]

33. Peakall, R.; Smouse, P.E. GenAlEx 6.5: Genetic analysis in Excel. Population genetic software for teaching and research—An update. *Bioinformatics* **2012**, *28*, 2537–2539. [CrossRef] [PubMed]

34. Kalinowski, S.T.; Taper, M.L. Maximum likelihood estimation of the frequency of null alleles at microsatellite loci. *Conserv. Genet.* **2006**, *7*, 991–995. [CrossRef]

35. Kamvar, Z.N.; Tabima, J.F.; Grünwald, N.J. Poppr: An R package for genetic analysis of populations with clonal, partially clonal, and/or sexual reproduction. *PeerJ* **2014**, *2*, e281. [CrossRef]

36. Yu, H.-P.; Cheng, F.-Y.; Zhong, Y.; Cai, C.-F.; Wu, J.; Cui, H.-L. Development of simple sequence repeat (SSR) markers from *Paeonia ostii* to study the genetic relationships among tree peonies (*Paeoniaceae*). *Sci. Hortic.* **2013**, *164*, 58–64. [CrossRef]

37. Ji, L.; da Silva, J.A.T.; Zhang, J.; Tang, Z.; Yu, X. Development and application of 15 novel polymorphic microsatellite markers for sect. *Paeonia* (*Paeonia* L.). *Biochem. Syst. Ecol.* **2014**, *54*, 257–266. [CrossRef]

38. Kantety, R.V.; La Rota, M.; Matthews, D.E.; Sorrells, M.E. Data mining for simple sequence repeats in expressed sequence tags from barley, maize, rice, sorghum and wheat. *Plant Mol. Biol.* **2002**, *48*, 501–510. [CrossRef]

39. Zhang, Y.; Zhang, M.; Hu, Y.; Zhuang, X.; Xu, W.; Li, P.; Wang, Z. Mining and characterization of novel EST-SSR markers of *Parrotia subaequalis* (*Hamamelidaceae*) from the first Illumina-based transcriptome datasets. *PLoS ONE* **2019**, *14*, e0215874. [CrossRef]

40. Park, S.; Son, S.; Shin, M.; Fujii, N.; Hoshino, T.; Park, S. Transcriptome-wide mining, characterization, and development of microsatellite markers in *Lychnis kiusiana* (*Caryophyllaceae*). *BMC Plant Biol.* **2019**, *19*, 12. [CrossRef]

41. Bhandawat, A.; Sharma, V.; Singh, P.; Seth, R.; Nag, A.; Kaur, J.; Sharma, R.K. Discovery and utilization of EST-SSR marker resource for genetic diversity and population structure analyses of a subtropical bamboo, *Dendrocalamus hamiltonii*. *Biochem. Genet.* **2019**, *57*, 652–672. [CrossRef] [PubMed]

42. Wu, J.; Cai, C.; Cheng, F.; Cui, H.; Zhou, H. Characterisation and development of EST-SSR markers in tree peony using transcriptome sequences. *Mol. Breed.* **2014**, *34*, 1853–1866. [CrossRef]

43. Thiel, T.; Michalek, W.; Varshney, R.; Graner, A. Exploiting EST databases for the development and characterization of gene-derived SSR-markers in barley (*Hordeum vulgare* L.). *Theor. Appl. Genet.* **2003**, *106*, 411–422. [CrossRef] [PubMed]

44. Sun, X.; Du, Z.; Ren, J.; Amombo, E.; Hu, T.; Fu, J. Association of SSR markers with functional traits from heat stress in diverse tall fescue accessions. *BMC Plant Biol.* **2015**, *15*, 116. [CrossRef] [PubMed]

45. Xiao, Y.; Zhou, L.; Xia, W.; Mason, A.S.; Yang, Y.; Ma, Z.; Peng, M. Exploiting transcriptome data for the development and characterization of gene-based SSR markers related to cold tolerance in oil palm (*Elaeis guineensis*). *BMC Plant Biol.* **2014**, *14*, 384. [CrossRef] [PubMed]

46. Bosamia, T.C.; Mishra, G.P.; Thankappan, R.; Dobaria, J.R. Novel and stress relevant EST derived SSR markers developed and validated in peanut. *PLoS ONE* **2015**, *10*, e0129127. [CrossRef]

47. Molla, K.A.; Debnath, A.B.; Ganie, S.A.; Mondal, T.K. Identification and analysis of novel salt responsive candidate gene based SSRs (cgSSRs) from rice (*Oryza sativa* L.). *BMC Plant Biol.* **2015**, *15*, 122. [CrossRef]

48. Ruan, X.; Wang, Z.; Su, Y.; Wang, T. Characterization and application of EST-SSR markers developed from the transcriptome of *Amentotaxus argotaenia* (*Taxaceae*), a relict vulnerable conifer. *Front. Genet.* **2019**, *10*, 1014. [CrossRef]

49. Song, Q.; Jia, G.; Zhu, Y.; Grant, D.; Nelson, R.T.; Hwang, E.-Y.; Hyten, D.L.; Cregan, P.B. Abundance of SSR motifs and development of candidate polymorphic SSR markers (BARCSOYSSR_1. 0) in soybean. *Crop Sci.* **2010**, *50*, 1950–1960. [CrossRef]

50. Zhao, H.; Yang, L.; Peng, Z.; Sun, H.; Yue, X.; Lou, Y.; Dong, L.; Wang, L.; Gao, Z. Developing genome-wide microsatellite markers of bamboo and their applications on molecular marker assisted taxonomy for accessions in the genus *Phyllostachys*. *Sci. Rep.* **2015**, *5*, 8018. [CrossRef]

51. Gao, Z.; Wu, J.; Liu, Z.; Wang, L.; Ren, H.; Shu, Q. Rapid microsatellite development for tree peony and its implications. *BMC Genom.* **2013**, *14*, 886. [CrossRef] [PubMed]

52. Botstein, D.; White, R.L.; Skolnick, M.; Davis, R.W. Construction of a genetic linkage map in man using restriction fragment length polymorphisms. *Am. J. Hum. Genet.* **1980**, *32*, 314. [PubMed]

53. Chapuis, M.-P.; Estoup, A. Microsatellite null alleles and estimation of population differentiation. *Mol. Biol. Evol.* **2007**, *24*, 621 631. [CrossRef] [PubMed]

54. Oddou-Muratorio, S.; Vendramin, G.G.; Buiteveld, J.; Fady, B. Population estimators or progeny tests: What is the best method to assess null allele frequencies at SSR loci? *Conserv. Genet.* **2009**, *10*, 1343. [CrossRef]

55. Dakin, E.; Avise, J. Microsatellite null alleles in parentage analysis. *Heredity* **2004**, *93*, 504–509. [CrossRef]

56. Molla, K.A.; Azharudheen, T.P.M.; Ray, S.; Sarkar, S.; Swain, A.; Chakraborti, M.; Vijayan, J.; Singh, O.N.; Baig, M.J.; Mukherjee, A.K. Novel biotic stress responsive candidate gene based SSR (cgSSR) markers from rice. *Euphytica* **2019**, *215*, 17. [CrossRef]

57. Hong, D.; Pan, K. A taxonomic revision of the *Paeonia anomala* complex (*Paeoniaceae*). *Ann. Mo. Bot. Gard.* **2004**, *91*, 87–98

58. Hong, D. *Peonies of the World: Taxonomy and Phytogeography*; Royal Botanic Gardens: Edinburgh, UK, 2010; pp. 33–44.

genes

MDPI

Article

Genetic Analysis of QTL for Resistance to Maize Lethal Necrosis in Multiple Mapping Populations

Luka A. O. Awata [1,2,3], Yoseph Beyene [2], Manje Gowda [2,*], Suresh L. M. [2], McDonald B. Jumbo [2], Pangirayi Tongoona [3], Eric Danquah [3], Beatrice E. Ifie [3], Philip W. Marchelo-Dragga [4], Michael Olsen [2], Veronica Ogugo [2], Stephen Mugo [2] and Boddupalli M. Prasanna [2]

[1] Directorate of Research, Ministry of Agriculture and Food Security, Ministries Complex, Parliament Road, P.O. Box 33, Juba, South Sudan; lawata@wacci.ug.edu.gh
[2] International Maize and Wheat Improvement Center (CIMMYT), World Agroforestry Centre (ICRAF), United Nations Avenue, Gigiri. P.O. Box 1041–00621, Nairobi, Kenya; Y.Beyene@cgiar.org (Y.B.); l.m.suresh@cgiar.org (S.L.M.); b.jumbo@cgiar.org (M.B.J.); M.Olsen@cgiar.org (M.O.); V.Ogugo@cgiar.org (V.O.); s.mugo@cgiar.org (S.M.); b.m.prasanna@cgiar.org (B.M.P.)
[3] West Africa Centre for Crop Improvement (WACCI), College of Basic and Applied Sciences, University of Ghana, PMB 30, Legon, Ghana; ptongoona@wacci.ug.edu.gh (P.T.); edanquah@wacci.ug.edu.gh (E.D.); bifie@wacci.ug.edu.gh (B.E.I.)
[4] Department of Agricultural Sciences, College of Natural Resources and Environmental Studies, University of Juba, P.O. Box 82, Juba, South Sudan; lukatwok11@gmail.com
[*] Correspondence: M.Gowda@cgiar.org; Tel.: +254-727019454

Received: 24 October 2019; Accepted: 24 December 2019; Published: 26 December 2019

Abstract: Maize lethal necrosis (MLN) occurs when maize chlorotic mottle virus (MCMV) and sugarcane mosaic virus (SCMV) co-infect maize plant. Yield loss of up to 100% can be experienced under severe infections. Identification and validation of genomic regions and their flanking markers can facilitate marker assisted breeding for resistance to MLN. To understand the status of previously identified quantitative trait loci (QTL)in diverse genetic background, F_3 progenies derived from seven bi-parental populations were genotyped using 500 selected kompetitive allele specific PCR (KASP) SNPs. The F_3 progenies were evaluated under artificial MLN inoculation for three seasons. Phenotypic analyses revealed significant variability ($P \leq 0.01$) among genotypes for responses to MLN infections, with high heritability estimates (0.62 to 0.82) for MLN disease severity and AUDPC values. Linkage mapping and joint linkage association mapping revealed at least seven major QTL (*qMLN3_130* and *qMLN3_142*, *qMLN5_190* and *qMLN5_202*, *qMLN6_85* and *qMLN6_157 qMLN8_10* and *qMLN9_142*) spread across the 7-biparetal populations, for resistance to MLN infections and were consistent with those reported previously. The seven QTL appeared to be stable across genetic backgrounds and across environments. Therefore, these QTL could be useful for marker assisted breeding for resistance to MLN.

Keywords: multiple population; linkage mapping; JLAM; QTL; validation; genomic prediction; maize lethal necrosis

---

1. Introduction

Maize lethal necrosis (MLN) is a major disease in sub-Saharan Africa (SSA) caused by co-infections of maize chlorotic mottle virus (MCMV) and sugarcane mosaic virus (SCMV) [1]. MCMV can able to interact with any member of the Potyviridae family to cause lethal necrosis in maize [2]. Yield loss due to MLN can reach up to 100% under severe infection and MLN favorable environments [1,3]. Breeding for host resistance to MLN is the most effective means of preventing yield losses in farmer's fields. Application of molecular markers could enhance breeding for resistance to MLN. Although markers are widely used in breeding for crop improvement including maize, the tools are inconsequential unless the linked markers or quantitative trait loci (QTL) are tested for their effectiveness and reproducibility in different genetic backgrounds. QTL validation adds weight to assess the effectiveness of alleles and their linked markers.

QTL mapping approaches are one of the popular genomic tools to dissect the genetic architecture of complex traits [4]. The presence of QTL conferring resistance to several viral diseases in maize has been investigated in numerous linkage and linkage disequilibrium mapping studies [4–8]. QTL mapping or linkage mapping is known for high QTL detection power. Joint linkage association mapping (JLAM) based on different segregating biparental populations is known to provide both the high QTL detection power and high mapping resolution [9,10]. Application of both linkage mapping and JLAM in multiple bi-parental populations is useful to validate earlier findings and to detect possible new sources of resistance for MLN.

Several studies were conducted to validate QTL effects on traits of economic importance in different crops including maize [11–15]. Sukruth et al [16] validated the markers associated with late leaf spot and rust in groundnut using recombinant inbred line (RIL) population and two backcross populations. Zhou et al [17] validated two major QTLs (LEN-3H and LEN-4H) for kernel length in wild barley using a biparental population derived from Fleet (*Hordeum vulgare* L.) and Awcs276. The authors reported significant association between the two QTL and kernel length. Gowda et al [7] employed four biparental maize populations and adopted linkage mapping and joint linkage mapping options to discover and confirm QTL associated with resistance to MLN. They consistently confirmed in two of the populations that three major QTLs were localized on chromosomes 3, 6, and 9. For effective use of trait linked markers and resources, validation of discovered QTL with large expected impact is crucial, as there are too many QTLs to validate and validation population development and assessment is expensive [18–20]. It is a pre-requisite though that for QTL to be effectively used in crop improvement, it should be confirmed that their effects remain consistent across populations and environments [12,16,21].

QTL validation study greatly contributes towards increased resolutions of some of the target QTLs as well as complementarity to previous findings either from genome wide association studies (GWAS) or JLAM or any other approaches. The CIMMYT Global Maize Program has recently identified a number of QTL across maize chromosomes which are associated with resistance to MLN in multiple mapping populations [7,8]. The validation of these QTL using seven different F_3 mapping populations would provide better understanding of QTLs associated with resistance to MLN and justify their application for marker assisted breeding towards improvement of maize lines for resistance to MLN. Further this study also helps to find new source of resistance which might not have been reported in earlier studies. This study was aimed to: (i) evaluate seven different F_3 populations for their responses to MLN under artificial inoculation; (ii) conduct individual population-based QTL mapping; (iii) apply JLAM with three biometric models and compare with linkage mapping results to identify stable and/or unique QTL; and (iv) assess the potential of genomic prediction for MLN resistance within biparental populations with low marker density.

2. Materials and Methods

2.1. Plant Materials

Seven elite maize parental lines with contrasting response to MLN developed by CIMMYT through pedigree and DH breeding schemes were used in this study. Parents CKDHL120918, CML494, CKLTI0227, and CKDHL312 were tolerant to MLN; CML543 and CKDHL221 were moderately tolerant, and CKDHL0089 was susceptible to MLN. These materials were also known for tolerance to various biotic and abiotic stresses with good agronomic performances. The seven bi-parental populations (F_3 pop1–CKDHL120918 × ML494, F_3 pop2–CML543 × CML494, F_3 pop3–CKDHL120918 × CML543, F_3 pop4–CKLTI0227 × CKDHL120918, F_3 pop5–CKDHL0089 × CKDHL120918, F_3 pop6–CKDHL0221 × CKDHL120312 and F_3 pop7–CKDHL0089 × CML494) were used for linkage mapping and JLAM.

To develop F_3 populations, crossing blocks were established in the nursery in Kiboko, Kenya (37°75′ E; 2°15′ S; 975 m a.s.l.; of 530 mm/year of rain fall and temperature ranges from 14.3 to 35.1 °C) during the 2016/2017 cropping season. Seven bi-parental crosses were made among the seven elite parental lines. Single cross (F_1) seeds were grown and F_1 plants were selfed. About 300 to 350 F_2 plants from each population were randomly selfed and F_3 seeds were harvested.

2.2. Phenotypic Evaluation

At least 306 F_3 families from each population with their seven parental lines and six commercial checks were evaluated to determine their response to MLN under artificial inoculation in field. Experiments were conducted for three seasons (April 2017, April 2018, and October 2018) in confined MLN facility in Naivasha (36°26′ E; 0°43′ S; 1896 m a.s.l.; 677 mm/year of rain fall and temperature ranges from 12 to 29 °C), hereafter seasons are referred as environments. Trials were planted using alpha lattice design in a 1-row plot of 3.0 m long, with spacing of 0.75 m between rows and 0.25 m between plants. Two seeds were planted per hill and thinned to one plant per hill 3 weeks after germination, making a total of 13 plants per row. Standard agronomic practices were adopted.

MLN inoculum was obtained from separate, sealed greenhouses maintained in Naivasha for each of SCMV and MCMV [7]. Maintenance of virus stocks in the greenhouses was described earlier [22]. In brief, MLN infected leaf tissues were collected from the field and ground in grinding buffer solution at 1:10 dilution ratio (10 mM potassium-phosphate, pH 7.0) [1,22]. The resulting sap extract was centrifuged at 12,000 rpm for 2 min. The sap was decanted and celite powder was added at the rate of 0.02 g/mL. To propagate the viruses in the greenhouses, a susceptible maize hybrid (H614) was grown and infected by rubbing the sap on the leaves at two to four leaf stages. A separate, sealed greenhouse was the maintained for each virus as stock for further use.

Three weeks before inoculation of the experimental materials, enzyme linked immunosorbent assay (ELISA) test was conducted on leaf samples, randomly collected from infected plants in the greenhouses, to determine the presence and purity of the MCMV and SCMV [7]. Separate extracts from the SCMV and MCMV infected plants was prepared and the two extracts were then mixed to form MLN inoculum at the ratio of 4 parts of SCMV to 1 part MCMV (weight/weight). Two inoculations were applied at the 4th and 5th week after planting. In order to keep uniform disease pressure across experiments, a motorized, backpack mist blower (Solo 423 Mist Blower, 12 L capacity) with an open nozzle (2-inches diameter) was used to inoculate the experimental plants at a high inoculum delivery pressure of 10 kg/cm. Drip irrigation was used to provide water and fertilizer. All other agronomic practices relating to maize production were followed according to standard procedures for field practices. Spreader rows of susceptible maize hybrid (H614) were planted along the experiment to enhance disease spread and intensity [16,23].

MLN Disease severity (MLN-DS) score started 10 days after the 2nd inoculations and was recorded four times at 10 days interval using standardized qualitative scale of 1 to 9, where 1 = resistant, clean, no symptoms; 2 = fine or no chlorotic specks, but vigorous plants; 3 = mild chlorotic streaks on emerging leaves; 4 = moderate chlorotic streaks on emerging new leaves; 5 = chlorotic streaks and mottling throughout plants; 6 = intense chlorotic mottling throughout plants, necrosis on leaf margins; 7 = excessive chlorotic mottling, mosaic and leaf necrosis, at times dead heart symptoms; 8 = excessive chlorotic mottling, leaf necrosis, dead heart and premature death of plants; and 9 = susceptible (complete plant necrosis and dead plants) [7,8]. After analyzing MLN-DS for each time score, we chose the third score (40 days post-inoculation) for further analysis because of its higher heritability and full expression of disease symptoms. The area under the disease progress curve (AUDPC) was calculated for each plot using SAS 9.4 (SAS Institute Inc., 2015, Cary, NC, USA) so as to understand the trend of development of MLN severity across the score intervals.

2.3. Phenotypic Data Analysis

All quantitative genetic parameters were estimated based on the performance of the 2142 F_3 lines. To check the quality of the data, first, test for normality of distribution of error terms for MLN-DS and AUDPC was determined using the Shapiro-Wilk test [24]. Secondly, analyses of description statistics (mean, range, skewness and kurtosis) and correlation among phenotypic traits were performed using META-R [25]. We assumed that each environment was a representative of a replication. Therefore, statistical analysis of phenotypic data was conducted using linear mixed model:

$$Y_{ijk} = \mu + g_i + l_j + b_{kj} + \varepsilon_{ijk,} \tag{1}$$

where Y_{ijk} was the disease severity of the ith genotype at the jth environment in the kth incomplete block, μ was an intercept term, g_i was the genetic effect of the ith genotype, l_j was the effect of the jth environment, b_{kj} was the effect of the kth incomplete block at the jth environment, and ε_{ijk} was the error term confounding with the genotype-by-environment interaction effect. To determine variance components by the restricted maximum likelihood (REML) method, both block and genotype effects were treated as fixed. Significance of variance component estimates was tested by model comparison with likelihood ratio tests where the halved *P*-values were used as approximations [26]. Heritability (H^2) on an entry-mean basis was estimated as the ratio of genotypic to phenotypic variance. The phenotypic variance comprises genotypic variance and the masking GxE interaction variances divided by the number of environments. Further, the mixed linear model (MLN) established in META-R software (http://hdl.handle.net/11529/10201) was adopted and the best linear unbiased predictor (BLUP) and best linear unbiased estimator (BLUE) for each genotype across environments were generated. The BLUPs were used in linkage mapping and joint linkage association mapping analyses whereas BLUEs were used in genomic prediction studies.

2.4. DNA Extraction, Genotyping, Linkage Map Construction and QTL Analysis

In this study, SNP markers were first screened on parental lines and 500 markers, which showed polymorphism between at least two of the seven parents were selected and used to genotype all the populations. Leaf samples were collected from 306 individuals per population 2 to 3 weeks after emergence, based on CIMMYT laboratory protocols [27]. The leaf samples were sent to LGC Genomics (https://www.biosearchtech.com/services/genotyping-services, Herts, UK) and were genotyped using 500 SNPs. Genotypic data obtained from LGC was subjected to quality check and SNPs were called and filtered using TASSEL version 5.0 software [28].

Polymorphic markers for each population were selected. Segregation of each SNP was verified for deviation from classic Mendelian inheritance using x^2-test and SNPs that significantly deviated were discarded [29,30]. Linkage maps were created for the individual populations using the MAP function established in QTL IciMapping v 4.1 software [24,31]. Linkage groups were identified using Group command based on logarithm of odds (LOD) score of 3.0, and recombination rate were converted into centimorgans (cM) using Kosambi mapping function [32]. Ordering of the markers was conducted using the "ordering" instruction with the nnTwoOpt algorithm. Adjustment of the map order was done according to the sum of adjacent recombination frequencies (SARF) and sorted in the "rippling" instruction with a window size of 5 as the amplitude. Instruction generated from the map was used to draw and visualized the map using Map Chart software [33].

The BLUPs obtained from across environments for MLN-DS and AUDPC for each population were subjected to inclusive composite interval mapping (ICIM) analysis so as to determine the QTL linkage. The ICIM is an effective two-step statistical approach that allows separation of co-factor selection from interval mapping process, in order to control the background effects and improve mapping of QTL with additive effects [31]. A LOD threshold of 3.0 with a scanning step of 1 cM were used to declare significant QTL [21,34]. Stepwise regression was adopted to determine the percentages of phenotypic variance explained (R^2) by individual QTL and additive effects at LOD peaks. QTL nomenclature [35] was adopted to nominate QTL conferring MLN resistance, where a two or three letter abbreviation of trait, followed by the chromosome number on which the QTL is found and the marker position to distinguish multiple QTLs were employed. The percentages of phenotypic variance explained (% PVE) by individual QTL, and additive and dominant effects at LOD peaks were generated. Sources of favorable alleles were determined depending on signs of the QTL additive effects [7,36]. For each F_3 population, estimated additive (a) and dominance (d) effects for each QTL were used to calculate the ratio of dominance level ($|d/a|$). This ratio was used to classify the nature of QTL as follow [37]: additive (A; $0 \leq |d/a| \leq 0.2$); partially dominant (PD; $0.2 \leq |d/a| \leq 0.8$); dominant (D; $0.8 < |d/a| \leq 1.2$) and over dominant (OD; $|d/a| > 1.2$).

Based on the SNP markers shared by different populations, an integrated map was built by using IciMapping software [31]. In brief, SNPs overlapped across genetic maps were selected as anchor markers and used to integrate corresponding linkage groups on individual linkage maps. The marker order and marker positions were calculated after calculating the order and the relative position (within each genetic map) of the anchored markers, followed by integrating of all the detected markers into one map. Then all QTL identified from the seven populations were projected onto the integrated map based on their confidence interval.

2.5. Joint Linkage Association Mapping (JLAM)

The JLAM method is a combination of linkage mapping and association mapping approaches [38,39]. Here, BLUPs obtained from across seven bi-parental populations and about 420 SNPs with missing values of <5% and minor allele frequency >0.05 were considered for JLAM study. We employed three biometric models to elucidate the QTL-trait relationships [7]. For each model, first, co-factors were selected using stepwise multiple linear regression based on the Schwarz Bayesian Criterion [40], following the Proc GLM SELECT model of SAS 9.2 [41]. Secondly, *P*-value and *F*-test were determined based on the full model (with QTL effects) and the reduced model (without QTL effects), followed by QTL scan using R software (version 3.5.3). The approach for Model A involved the trait as a function of the co-factor + marker. Model B incorporated population effect as an additional factor to adjust for the population structure so that trait = population + co-factor + marker; and Model C involved nesting of the co-factor and marker within population such that trait = pop + co-factor (pop) + marker (pop) [10,42]. Significance of association between QTL and MLN resistance was determined at $P < 0.05$ following Bonferroni-Holm procedure [43]. The adjusted total phenotypic variance explained (R^2) values for the detected QTL were generated when the significant QTL were simultaneously fitted in linear model. Further principal component analysis (PCA) of all F_3 lines was carried out using TASSEL version 5.2 [28], and the CurlyWhirly version 1.15 software [44] was used to construct PCA biplot for the population structure.

2.6. Genomic Prediction

Due to its advantages over either phenotypic and marker assisted selection, genomic prediction (GP) is becoming a popular approach for breeding, especially for complex traits [45,46]. The technique allows for the selection of superior genotypes without conducting phenotypic evaluation, if a subset of the population has genotypic and phenotypic data. Here, we performed GP analyses using ridge-regression BLUP (RR-BLUP) with five-fold cross-validation. Uniformly distributed, polymorphic, SNPs from each population showing minor allele frequency of less than 5% were used.

Genotypes were sampled from each population and used to constitute both training set and testing set. The sampling of the sets was repeated 100 times [7]. The predictive ability of the GP was computed by dividing the correlation between genomic estimated breeding values (GEBVs) and the observed phenotypic values by the square root of the heritability estimates obtained from the respective populations [46].

3. Results

3.1. Response of Parents and F3 Populations to MLN Infections

The response of F_3 lines for MLN-DS and AUDPC showed continuous distribution, ranging from highly resistant or tolerant to completely susceptible in each individual population as well as across populations (Figure 1).

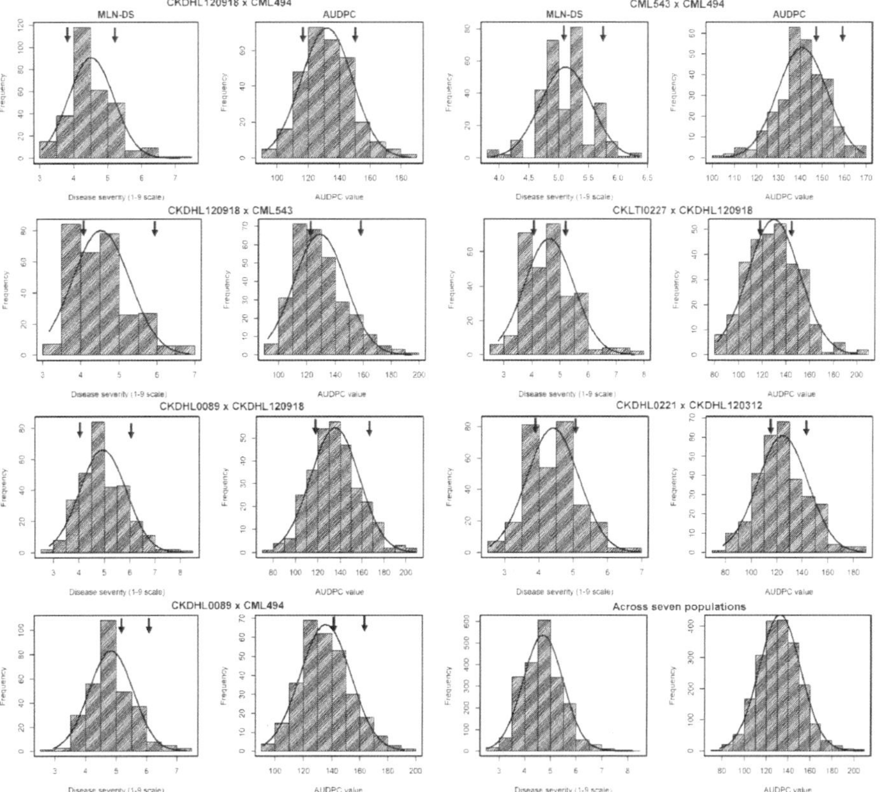

Figure 1. Phenotypic distribution of MLN disease severity (MLN-DS) and the area under disease progress curve (AUDPC) values recorded in seven individual and across F_3 populations. Arrows indicate the performance of parents.

The normality tests indicated that the means were positively and moderately skewed with values ranging from 0.65 to 0.77. Kurtosis values were platykurtotic with maximum of 2.07 for AUDPC. High *W*-test values were observed for all the traits and ranged from 0.95 to 0.97 (data not shown). Individual means across populations ranged from 4.40 to 5.11 with a mean of 4.70 for MLN-DS, and the AUDPC mean values for individual populations ranged from 125.6 to 134.4 with an across population mean of 132.4. The magnitude of genotypic variance for MLN-DS was lowest (0.17) in F_3 pop 2 (CML543 × CML494) and highest (0.69) in F_3 pop 4 (CKLTI0227 × CKDHL120918). Genotypic variances were highly significant ($P < 0.01$) in all seven F_3 populations and across populations for both MLN-DS and AUDPC values. Moderate to high broad–sense heritability estimates of 0.62 to 0.79 were found for MLN-DS and AUDPC values, indicative of high-quality phenotypic data for further genetic analysis (Table 1).

Table 1. Analysis of variance for MLN disease severity (MLN-DS) and area under disease progress curve (AUDPC) values using seven segregating F_3 populations evaluated for three seasons under MLN inoculated plots.

Trait	Mean (Range)	σ^2_G	σ^2_e *	H^2
	CKDHL120918 × CML494 (F_3 pop1)			
MLN-DS	4.52 (3.06–7.12)	0.41 **	0.44	0.74
AUDPC	131.9 (92.9–185.8)	272.20 **	230.73	0.78
	CML543 × CML494 (F_3 pop2)			
MLN-DS	5.11 (3.86–6.35)	0.17 **	0.32	0.62
AUDPC	140.8 (103.8–168.9)	141.05 **	177.68	0.70
	CKDHL120918 × CML543 (F_3 pop3)			
MLN-DS	4.52 (3.17–6.86)	0.46 **	0.50	0.73
AUDPC	126.2 (82.1–201.0)	282.50 **	280.87	0.75
	CKLTI0227 × CKDHL120918 (F_3 pop4)			
MLN-DS	4.60 (2.80–7.65)	0.69 **	0.57	0.79
AUDPC	129.1 (92.6–194.7)	467.78 **	299.86	0.82
	CKDHL0089 × CKDHL120918 (F_3 pop5)			
MLN-DS	4.94 (2.85–8.20)	0.65 **	0.72	0.73
AUDPC	133.9 (79.9–209.3)	411.12 **	379.39	0.76
	CKDHL0221 × CKDHL120312 (F_3 pop6)			
MLN-DS	4.40 (2.56–6.97)	0.50 **	0.49	0.75
AUDPC	125.6 (78.0–187.9)	374.25 **	245.35	0.82
	CKDHL0089 × CML494 (F_3 pop7)			
MLN-DS	4.82 (2.94–7.31)	0.44 **	0.51	0.72
AUDPC	134.4 (91.0–191.8)	300.75 **	260.51	0.78
	Across seven populations			
MLN-DS	4.70 (2.56–8.20)	0.40 **	0.50	0.70
AUDPC	132.4 (78.0–209.3)	265.48 **	267.90	0.75

* and ** indicate significance at $P < 0.05$ and $P < 0.01$, respectively MLN-DS = MLN disease severity 42 dpi; AUDPC = area under disease progress curve; σ^2_G = genotypic variance; σ^2_e* = error variance confounded with GxE variance; and H^2 = broad sense heritability.

Three-dimensional principal component analysis (PCA) of the seven bi-prenatal populations are shown in Figure 2. The results identified five distinct groups including CML543 × CML494, CKDHL120918 × CML543, CKDHL0089 × CKDHL120918, CKDHL0221 × CKDHL120312, and CKDHL0089 × CML494. The distinctiveness of the five populations implies their diversity in their genetic backgrounds. However, populations CKDHL120918 × CML494 and CKLTI0227 × CKDHL120918 were overlapping and CKDHL221 × CKDHO120312 is distinct from other populations. This might indicate their relatedness and distinctness in genetic compositions.

3.2. Molecular Analyses

3.2.1. Linkage Group

Genetic linkage groups consisting of 10 maize chromosomes were constructed for each of the seven F_3 populations. Marker density on each map varied among the seven populations. F_3 pop 4 carried the largest number of polymorphic markers with 298 SNPs and a total length of 1550.07 cM at an average density of 5.20 cM between the markers. F_3 pop 1 had only 112 polymorphic SNPs spanning a total length of 1223.97 cM with a mean spacing of 10.93 cM between markers. Similarly, variation was detected in number of SNPs per linkage group (LG), with marker density per LG ranging from 40 with a mean spacing of 5.71 for LG7 to 153 and an average of 21.86 on LG1. Total linkage map consisted of 389 SNPs with a total length of 2007.85 cM and an average marker interval length of 5.16 cM (Table S1).

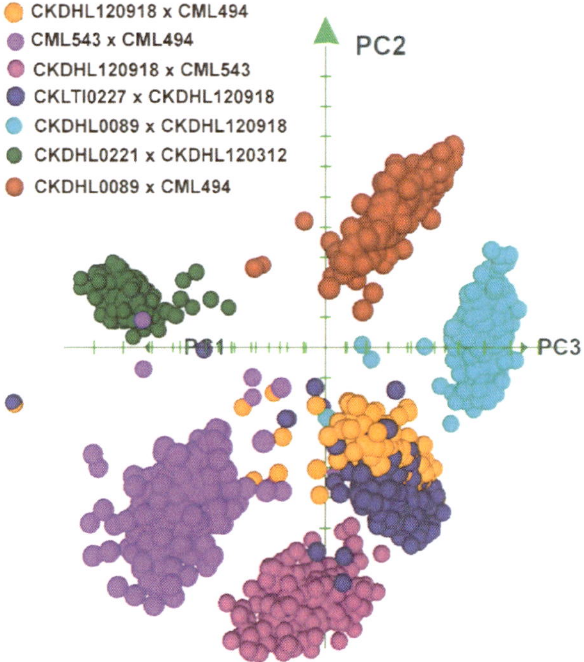

Figure 2. PCA (principal component analysis) biplot showing structures of the seven F_3 populations.

3.2.2. QTL Mapping

QTL analyses across environments revealed 60 and 58 significant QTLs for MLN-DS and AUDPC values for the seven populations, respectively (Table 2), with different sizes of QTL effects. Generally, the study revealed that positive alleles for resistance to MLN were donated either by male or female parent in a cross in each population. Some QTL identified in F_3 pop 6 and 7 showed large additive effects. QTL detected in the other F_3 populations predominantly showed dominant to over-dominant effects. Majority of the QTLs was detected in F_3 pop4 (CKLTI0227 × CKDHL120918), whereby 7% of the QTL showed additive effects for MLN-DS and AUDPC values.

Table 2. Quantitative trait loci (QTL) detected by integrated composite intervals mapping analysis for resistance to MLN in seven F_3 populations evaluated in MLN inoculated plots over three cropping seasons.

Trait Name	QTL Name [a]	Chr	Position (cM)	LOD	PVE (%)	Add	Dom	Nature of QTL	Total PVE (%)	Marker Name		Physical Position (bp)		Fav Parent
										Left M	Right M	Left M	Right M	
								CKDHL120918 × CML494						
MLN-DS	qMLN1_265	1	183	3.99	2.58	−0.46	−0.26	PD		PZB00648_5	d8_3	17,595,139	265,199,938	CKDHL120918
	qMLN2_156	2	3	4.13	2.34	−0.04	1.09	OD		PZA01232_1	PHM3055_9	155,868,024	192,602,324	CKDHL120918
	qMLN2_10	2	122	3.16	0.69	0.07	−0.23	OD		PZA00620_3	PZA00365_2	10,429,405	1,221,385	CML494
	qMLN3_142	3	23	23.92	4.27	0.37	−0.21	PD		PZA00920_1	PHM15449_10	142,821,031	125,077,922	CML494
	qMLN4_150	4	90	4.41	2.53	−0.02	0.70	OD		S4_153520131	S4_149896839	153,520,131	149,896,839	CKDHL120918
	qMLN5_177	5	88	4.24	2.58	0.03	0.83	OD	30.08	PZA01410_1	S5_177634071	172,682,963	177,634,071	CML494
	qMLN6_85	6	78	14.11	3.98	0.34	−0.20	PD		PZB01009_1	PHM8909_12	84,664,840	91,883,155	CML494
	qMLN6_85	6	132	6.25	4.56	0.19	0.76	OD		PZB01009_1	PHM8909_12	84,664,840	918,83,155	CML494
	qMLN7_158	7	87	4.71	2.48	−0.07	1.11	OD		PHM7898_10	S7_157472460	161,993,743	157,472,460	CKDHL120918
	qMLN9_142	9	66	3.64	1.02	−0.11	−0.21	OD		PZB00221_3	PHM229_15	142,271,047	30,003,189	CKDHL120918
	qMLN10_9	10	4	3.82	2.43	0.08	0.84	OD		PZA00866_2	PHM5740_9	124,203,565	87,73,358	CML494
AUDPC	qMLN2_10	2	121	3.03	3.23	1.54	−6.16	OD		PZA00620_3	PZA00365_2	10,429,405	1,221,385	CML494
	qMLN3_142	3	23	12.68	8.92	5.50	−5.57	DO		PZA00920_1	PHM15449_10	142,821,031	125,077,922	CML494
	qMLN4_143	4	140	2.79	1.71	−2.78	1.27	PD	54.72	PHM1505_31	S4_9850443	143,162,745	9,850,443	CKDHL120918
	qMLN6_85	6	78	17.96	21.64	9.80	−5.38	PD		PZB01009_1	PHM8909_12	84,664,840	91,883,155	CML494
	qMLN9_142	9	68	4.79	6.25	−4.08	−5.16	OD		PZB00221_3	PHM229_15	142,271,047	30,003,189	CKDHL120918
								CML543 × CML494						
MLN-DS	qMLN1_47	1	43	3.41	1.10	−0.07	0.04	PD		csu1138_4	S1_46411896	119,018,556	46,411,896	CML494
	qMLN3_146	3	14	11.79	4.08	−0.72	0.25	PD		S3_146966676	S3_146363360	146,966,676	146,363,360	CML494
	qMLN4_30	4	14	2.50	9.94	−0.22	0.52	OD		PZA02457_1	bt2_7	29,031,200	66,290,994	CML494
	qMLN5_160	5	3	3.52	1.22	−0.08	0.04	PD	43.49	S5_42297152	PZA01796_1	42,297,152	160,321,846	CML494
	qMLN6_21	6	4	5.03	1.64	0.09	−0.05	PD		S6_21007530	PZA03063_21	21,007,530	25,335,225	CML543
	qMLN8_10	8	84	3.42	12.37	−0.21	0.53	OD		S8_10001165	PZA02388_1	10,001,165	169,137	CML494
	qMLN9_142	9	3	31.21	12.72	0.70	−0.08	AD		PZA00832_1	PHM7916_4	147,131,097	132,762,904	CML543
AUDPC	qMLN1_47	1	42	6.14	1.64	−2.93	1.62	PD		csu1138_4	S1_46411896	119,018,556	46,411,896	CML494
	qMLN1_265	1	154	2.97	4.17	−7.58	9.26	OD		d8_3	umc128_2	265,199,938	227,602,208	CML494
	qMLN3_146	3	14	15.65	4.18	−24.01	7.57	PD		S3_146966676	S3_146363360	146,966,676	146,363,360	CML494
	qMLN3_142	3	107	3.77	1.17	2.73	0.84	PD		PZA00920_1	PZA02402_1	142,821,031	169,771,952	CML543
	qMLN4_30	4	14	3.88	7.99	−6.91	15.27	OD	49.84	PZA02457_1	bt2_7	29,031,200	66,290,994	CML494
	qMLN5_160	5	3	3.91	1.01	−2.48	0.94	PD		S5_42297152	PZA01796_1	42,297,152	160,321,846	CML494
	qMLN6_21	6	2	8.95	2.38	3.84	−0.71	AD		S6_21007530	PZA03063_21	21,007,530	25,335,225	CML543
	qMLN7_158	7	119	4.81	8.13	5.12	−18.38	OD		PHM7898_10	PHM1912_23	161,993,743	155,970,264	CML543
	qMLN8_10	8	85	3.05	5.28	−6.73	9.00	OD		S8_10001165	PZA02388_1	10,001,165	169,137	CML494
	qMLN9_142	9	3	34.18	10.54	21.95	−4.51	PD		PZA00832_1	PHM7916_4	147,131,097	132,762,904	CML543

Table 2. *Cont.*

Trait Name	QTL Name [a]	Chr	Position (cM)	LOD	PVE (%)	Add	Dom	Nature of QTL	Total PVE (%)	Marker Name Left M	Marker Name Right M	Physical Position (bp) Left M	Physical Position (bp) Right M	Fav Parent
								CKDHL120918 × CML543						
MLN-DS	qMLN1_265	1	14	4.12	2.79	−0.04	−0.21	OD		d8_3	PHM14475_7	265,199,938	256,245,118	CKDHL120918
	qMLN1_252	1	28	3.31	3.08	−0.12	−0.15	OD		PZA02269_4	PHM4942_12	252,721,946	226,461,786	CKDHL120918
	qMLN2_2	2	50	4.30	3.45	−0.05	−0.23	OD		PHM13440_13	PZA00365_2	2,527,344	1,221,385	CKDHL120918
	qMLN3_142	3	73	33.33	28.84	0.46	−0.26	PD	47.88	S3_146966676	S3_68596995	146,966,676	68,596,995	CML543
	qMLN4_143	4	120	4.08	2.87	0.16	−0.04	PD		bt2_7	PHM1505_31	66,290,994	143,162,745	CML543
	qMLN5_202	5	112	2.89	4.43	−0.14	−0.19	OD		PZB00765_1	PHM5484_22	202,174,585	21,449,633	CKDHL120918
	qMLN7_158	7	1	2.62	3.15	−0.39	−0.44	DO		S7_157472460	S7_137455469	157,472,460	137,455,469	CKDHL120918
	qMLN10_9	10	81	6.77	5.10	−0.15	−0.22	OD		PHM5740_9	PZA01642_1	8,773,358	14,703,451	CKDHL120918
AUDPC	qMLN1_265	1	14	4.85	2.50	−1.32	−5.28	OD		d8_3	PHM14475_7	265,199,938	256,245,118	CKDHL120918
	qMLN1_252	1	28	3.44	2.40	−3.01	−3.52	DO		PZA02269_4	PHM4942_12	252,721,946	226,461,786	CKDHL120918
	qMLN2_2	2	50	4.03	2.46	−0.78	−5.45	OD		PHM13440_13	PZA00365_2	2,527,344	1,221,385	CKDHL120918
	qMLN3_154	3	3	2.93	5.23	10.02	−11.50	DO		S3_154250438	S3_150836832	154,250,438	150,836,832	CML543
	qMLN3_142	3	74	17.68	11.09	7.15	−6.24	DO		S3_146966676	S3_68596995	146,966,676	68,596,995	CML543
	qMLN3_207	3	109	2.73	2.02	−3.50	−0.77	PD		PZA00538_15	PZA01931_17	206,889,707	227,682,081	CKDHL120918
	qMLN4_143	4	120	6.03	3.30	4.58	−0.86	AD	52.80	bt2_7	PHM1505_31	66,290,994	143,162,745	CML543
	qMLN5_202	5	112	3.16	3.65	−3.50	−4.49	OD		PZB00765_1	PHM5484_22	202,174,585	21,449,633	CKDHL120918
	qMLN6_157	6	118	3.83	4.79	11.14	−8.68	PD		PZA01618_2	S6_156386857	129,927,781	156,386,857	CML543
	qMLN7_158	7	1	3.85	3.30	−10.90	−11.74	DO		S7_157472460	S7_137455469	157,472,460	137,455,469	CKDHL120918
	qMLN9_109	9	29	2.81	1.63	−3.06	−1.67	PD		PHM229_15	S9_109549230	30,003,189	109,549,230	CKDHL120918
	qMLN10_41	10	2	2.90	3.46	9.86	−10.22	DO		PHM4066_11	PHM1956_90	41,187,565	40,187,565	CML543
	qMLN10_9	10	82	7.45	4.61	−3.83	−5.70	OD		PHM5740_9	PZA01642_1	8,773,358	14,703,451	CKDHL120918
								CKLTI0227 × CKDHL120918						
MLN-DS	qMLN1_282	1	99	3.21	1.55	0.69	−0.60	DO		PHM4752_14	PZA03020_8	298,874,066	282,044,048	CKLTI0227
	qMLN1_265	1	109	3.52	1.56	0.68	−0.63	DO		PZA03020_8	PZA01254_2	282,044,048	106,204,446	CKLTI0227
	qMLN1_47	1	139	4.79	1.38	0.76	−0.82	DO		S1_22744948	S1_87158930	22,744,948	87,158,930	CKLTI0227
	qMLN1_27	1	202	3.54	0.91	0.93	−0.94	DO		S1_15353866	S1_2693968	15,353,866	2,693,968	CKLTI0227
	qMLN3_142	3	37	4.44	1.45	0.73	−0.35	PD		S3_146966676	S3_55444954	146,966,676	55,444,954	CKLTI0227
	qMLN3_130	3	65	5.68	1.48	0.70	−0.82	DO		S3_92864540	S3_136165565	92,864,540	136,165,565	CKLTI0227
	qMLN3_130	3	121	5.32	1.49	−0.68	−1.08	OD		PHM2343_25	S3_154250438	27,981,649	154,250,438	CKDHL120918
	qMLN4_30	4	36	4.36	0.35	0.72	−0.05	AD		PZA02457_1	PZA00726_10	29,031,200	60,768,063	CKLTI0227
	qMLN4_150	4	96	4.29	1.59	0.59	−0.74	OD		S4_155296684	S4_9741874	155,296,684	9,741,874	CKLTI0227
	qMLN5_160	5	130	4.12	1.05	0.90	−0.83	DO	25.13	S5_154350617	S5_198716574	154,350,617	198,716,574	CKLTI0227
	qMLN5_42	5	161	3.49	0.25	−0.08	−1.02	OD		S5_42297152	PZA02164_16	42,297,152	112,179,855	CKDHL120918
	qMLN6_157	6	20	5.87	1.46	0.59	−1.12	OD		S6_156386857	PHM4748_16	156,386,857	158,540,019	CKLTI0227
	qMLN6_85	6	29	7.17	1.68	0.72	−0.34	PD		PHM4748_16	PZB01009_1	158,540,019	84,664,840	CKLTI0227
	qMLN6_85	6	83	8.86	1.79	0.81	−0.43	PD		PZB01009_1	S6_89823772	84,664,840	89,823,772	CKLTI0227
	qMLN6_157	6	116	4.73	1.10	0.07	1.37	OD		PHM2658_129	PZA01884_1	164,999,578	132,316,835	CKLTI0227
	qMLN7_158	7	145	3.56	1.08	0.86	−0.99	DO		S7_167230991	S7_127970539	167,230,991	127,970,539	CKLTI0227
	qMLN8_10	8	100	4.34	1.37	0.73	−0.90	OD		PZA02746_2	PZA02388_1	163,067,200	169,137	CKLTI0227
	qMLN9_109	9	9	3.19	1.64	−0.63	−0.57	DO		PZA00708_3	S9_109549230	147,381,231	109,549,230	CKDHL120918
	qMLN10_114	10	36	5.74	1.61	0.69	−0.83	OD		S10_113832226	PZA01001_2	113,832,226	146,538,889	CKLTI0227

Table 2. *Cont.*

Trait Name	QTL Name [a]	Chr	Position (cM)	LOD	PVE (%)	Add	Dom	Nature of QTL	Total PVE (%)	Marker Name		Physical Position (bp)		Fav Parent
										Left M	Right M	Left M	Right M	
AUDPC	qMLN1_265	1	163	6.85	1.35	24.93	−8.24	PD	28.69	d8_3	PZA02269_4	265,199,938	252,721,946	CKLTI0227
	qMLN2_41	2	2	12.20	2.31	40.41	−11.93	PD		PHM10404_8	PZA02450_1	40,967,991	47,575,949	CKLTI0227
	qMLN3_142	3	34	5.67	3.42	14.61	−8.98	PD		S3_146966676	S3_55444954	146,966,676	55,444,954	CKLTI0227
	qMLN3_146	3	113	10.79	1.98	−40.42	−5.50	AD		S3_146250249	S3_146026612	146,250,249	146,026,612	CKDHL120918
	qMLN3_130	3	125	3.49	3.81	12.36	−18.09	OD		PHM2343_25	S3_154250438	27,981,649	154,250,438	CKLTI0227
	qMLN4_30	4	77	3.29	2.44	19.01	−20.07	DO		S4_6601124	PZA02509_15	6,601,124	3,904,858	CKLTI0227
	qMLN6_157	6	32	8.84	3.77	16.09	−5.31	PD		PHM4748_16	PZB01009_1	158,540,019	84,664,840	CKLTI0227
	qMLN9_147	9	6	3.28	3.17	−15.11	−9.92	PD		S9_146012201	PZA00708_3	146,012,201	147,381,231	CKDHL120918
	qMLN9_109	9	21	3.19	2.66	18.29	−19.22	DO		PZA00708_3	S9_109549230	147,381,231	109,549,230	CKLTI0227
	qMLN10_114	10	36	3.54	3.20	16.73	−18.91	DO		S10_113832226	PZA01001_2	113,832,226	146,538,889	CKLTI0227
							CKDHL0089 × CKDHL120918							
MLN-DS	qMLN1_18	1	4	4.42	7.15	−0.11	0.87	OD	54.34	S1_18838432	PZA00175_2	18,838,432	8,510,027	CKDHL120918
	qMLN2_169	2	168	4.25	6.61	−0.06	1.07	OD		PZA02727_1	PZA00515_10	227,921,381	169,265,278	CKDHL120918
	qMLN3_142	3	73	44.14	24.44	0.62	−0.10	AD		PZA00920_1	S3_133048570	142,821,031	133,048,570	CKDHL0089
	qMLN5_190	5	128	3.75	1.56	−0.14	−0.06	PD		S5_190675983	S5_201226926	190,675,983	201,226,926	CKDHL120918
	qMLN6_157	6	20	3.21	8.20	0.31	−0.41	OD		S6_156386857	PHM3466_69	156,386,857	167,148,384	CKDHL0089
	qMLN6_85	6	126	4.16	1.78	0.16	0.04	PD		PZA00942_2	PHM8909_12	102,566,000	91,883,155	CKDHL0089
	qMLN10_9	10	4	12.06	6.32	−0.28	−0.10	PD		PZA01313_2	PHM5740_9	3,598,262	8,773,358	CKDHL120918
AUDPC	qMLN1_18	1	10	3.69	8.65	−3.67	15.28	OD	57.62	S1_18838432	PZA00175_2	18,838,432	8,510,027	CKDHL120918
	qMLN2_169	2	166	3.18	6.28	−1.81	21.55	OD		PZA02727_1	PZA00515_10	227,921,381	169,265,278	CKDHL120918
	qMLN3_142	3	73	50.02	27.46	16.42	−2.21	AD		PZA00920_1	S3_133048570	142,821,031	133,048,570	CKDHL0089
	qMLN5_190	5	127	5.49	2.11	−4.27	−1.58	PD		PHM7908_25	S5_190675983	191,075,472	190,675,983	CKDHL120918
	qMLN6_157	6	21	2.96	8.12	7.34	−10.37	OD		S6_156386857	PHM3466_69	156,386,857	167,148,384	CKDHL0089
	qMLN6_85	6	133	5.03	1.96	4.23	0.36	AD		PHM8909_12	PZA00427_3	91,883,155	79,815,961	CKDHL0089
	qMLN10_9	10	3	11.23	5.20	−6.46	−2.44	PD		PZA01313_2	PHM5740_9	3,598,262	8,773,358	CKDHL120918
							CKDHL0221 × CKDHL120312							
MLN-DS	qMLN1_47	1	71	5.32	3.74	−0.16	0.07	PD	52.13	PZA00447_8	S1_46411896	9,024,005	46,411,896	CKDHL120312
	qMLN3_130	3	56	44.83	44.51	0.56	−0.11	AD		PZA02402_1	PHM15449_10	169,771,952	125,077,922	CKDHL0221
	qMLN4_150	4	37	2.66	2.74	−0.13	−0.09	PD		PZA01187_1	PHM1505_31	177,666,738	143,162,745	CKDHL120312
	qMLN4_7	4	128	4.63	3.23	0.15	0.01	AD		S4_6544767	PHM16788_6	6,544,767	13,581,955	CKDHL0221
	qMLN8_10	8	45	4.26	3.12	0.15	0.03	AD		PHM5235_8	PZA00368_1	94,414,978	5,632,308	CKDHL0221
AUDPC	qMLN1_47	1	71	4.97	4.38	−4.05	1.41	PD	59.07	PZA00447_8	S1_46411896	9,024,005	46,411,896	CKDHL120312
	qMLN1_47	1	100	2.98	2.49	2.21	−2.90	OD		S1_46411896	PHM12323_17	46,411,896	53,357,797	CKDHL0221
	qMLN3_130	3	56	24.64	26.00	9.70	−2.76	PD		PZA02402_1	PHM15449_10	169,771,952	125,077,922	CKDHL0221
	qMLN3_142	3	63	13.19	12.71	6.76	−1.85	PD		PZA00279_2	PZA00920_1	52,804,070	142,821,031	CKDHL0221
	qMLN4_150	4	39	3.65	5.24	−4.27	−2.01	PD		PZA01187_1	PHM1505_31	177,666,738	143,162,745	CKDHL120312
	qMLN4_7	4	128	3.95	3.44	3.71	0.18	AD		S4_6544767	PHM16788_6	6,544,767	13,581,955	CKDHL0221
	qMLN5_42	5	34	2.63	2.36	−0.23	−4.17	OD		PHM16854_3	PZA00522_12	34,587,029	57,933,548	CKDHL120312
	qMLN8_10	8	45	5.06	4.70	4.16	−0.42	AD		PHM5235_8	PZA00368_1	94,414,978	5,632,308	CKDHL0221
	qMLN10_114	10	43	5.16	4.47	3.72	−2.52	PD		PZA00814_1	PHM1576_25	87,194,491	124,203,168	CKDHL0221

Table 2. *Cont.*

Trait Name	QTL Name [a]	Chr	Position (cM)	LOD	PVE (%)	Add	Dom	Nature of QTL	Total PVE (%)	Marker Name		Physical Position (bp)		Fav Parent
										Left M	Right M	Left M	Right M	
								CKDHL0089 × CML494						
MLN-DS	qMLN5_190	5	89	4.40	5.48	−0.17	0.03	AD		PZA01427_1	PHM7908_25	23,135,578	191,075,472	CML494
	qMLN5_202	5	111	8.94	7.97	−0.20	−0.03	AD	46.74	S5_200938637	PHM563_9	200,938,637	204,993,639	CML494
	qMLN6_157	6	107	3.37	8.42	0.09	0.71	OD		S6_157568432	S6_156386857	157,568,432	156,386,857	CKDHL0089
AUDPC	qMLN3_130	3	1	3.02	6.64	0.75	−4.18	OD		PZA01447_1	S3_133048570	53,549,251	133,048,570	CKDHL0089
	qMLN5_190	5	89	5.14	15.91	−4.78	0.81	AD	50.87	PZA01427_1	PHM7908_25	23,135,578	191,075,472	CML494
	qMLN5_202	5	111	7.62	16.65	−4.81	−0.76	AD		S5_200938637	PHM563_9	20,0938,637	204,993,639	CML494
	qMLN6_157	6	108	3.20	14.95	2.34	16.81	OD		S6_157568432	S6_156386857	157,568,432	156,386,857	CKDHL0089

LOD = logarithm of odds; Add = additive effect; Dom = dominance effect; PVE = phenotypic variance explained; fav parent = parental genotype from where favorable allele for MLN resistance is contributing; [a] QTL name composed by the trait code followed by the chromosome number in which the QTL was mapped and a physical position of the QTL.

In F_3 pop 1, for MLN-DS and AUDPC values, together 15 QTL were detected, which explained 30.08% and 34.72% of the total PVE, respectively. Most of the detected QTL were of small effects except *qMLN6_85*, which had AUDPC values explaining 21.6% of PVE. In F_3 pop 2, seven and 10 QTLs were detected which together explained 43.5 and 49.8% of the total phenotypic variance for MLN-DS and AUDPC values, respectively. Among all the QTLs identified, only three QTLs are specific to AUDPC values. In F_3 pop3, a total of eight QTLs were found for MLN-DS, explaining 47.9% of the total observed phenotypic variability. Among the QTL with major effects, *qMLN3_142* was detected with 28.8% and 11.1% of PVE for MLN-DS and AUDPC values, respectively. There were 19 and 10 QTLs detected which together explained 25.1 and 28.7% of the PVE for MLN-DS and AUDPC values, respectively in F_3 pop4. All the detected QTLs were of small effects. For F_3 pop5, seven QTL each were found for MLN-DS and AUDPC values which together explained 54.3% and 57.6% of the total phenotypic variance, respectively. QTL *qMLN3_142* was found for both MLN-DS and AUDPC values with major effects. F_3 pop6 revealed five QTLs explaining 52.1% of the variation for MLN-DS. Whereas for AUDPC values, nine QTLs were found which explained 59.1% of the total phenotypic variance. QTL *qMLN3_130* consistently explained >25% of the phenotypic variance. F_3 pop7 revealed a total of three and four QTL with 46.7% and 50.8% PVE for MLN-DS and AUDPC values, respectively.

Among several genomic regions identified with stable QTL for MLN resistance, QTL on chromosome 3 was highly consistent. We found two strongly associated SNPs (*PHM15449_10* at 125,077,922 bp and *PZA00920_1* at 142,821,031 bp) within this region and their allelic effects on MLN resistance in different populations were also prominent. The phenotypic values of the different allele classes of these two major-effect SNPs in five F_3 populations and across populations for MLN-DS and AUDPC value were presented in Figure 3.

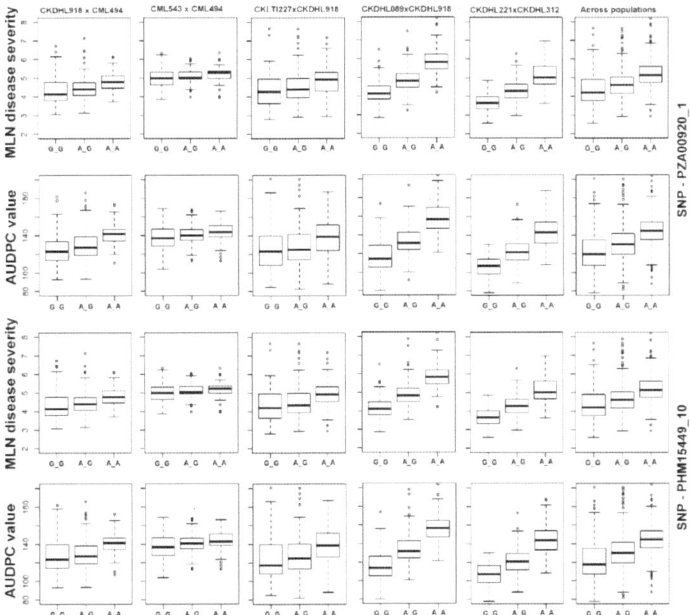

Figure 3. Box plots showing the phenotypic values of the different allele classes of two major-effect SNPs identified in five F_3 populations and across populations for MLN-DS and AUDPC value. The SNP names, alleles and the specific F_3 population where the effect is witnessed are mentioned above. The black horizontal lines in the middle of the boxes are the median values for the MLN-DS and AUDPC value in the respective allele classes. The vertical size of the boxes represents the inter-quantile range. The upper and lower whiskers represent the minimum and maximum values of data.

3.2.3. Consensus Map Construction

Since all populations were genotyped with the same set of SNPs, we integrated the seven maps into a consensus linkage map based on the markers shared by populations (Figure 4). From the total of 118 QTLs (60 for MLN-DS and 58 for AUDPC), similar QTL detected for both traits were treated as single QTL, as a result 77 QTLs were mapped on the consensus map. Among all the QTLs detected from seven populations, 19 QTLs were consistently found in at least two populations; QTL *qMLN3_142* on chromosome 3 was consistently identified in six of the seven populations. Two QTLs (*qMLN1_265* and *qMLN6_157*) were consistently found in four populations, seven QTLs (*qMLN1_47*, *qMLN3_130*, *qMLN4_150*, *qMLN6_85*, *qMLN7_158*, *qMLN8_10*, and *qMLN10_9*) were repeatedly identified in three populations and nine QTLs (*qMLN3_146*, *qMLN4_30*, *qMLN5_42*, *qMLN5_160*, *qMLN5_190*, *qMLN5_202*, *qMLN9_109*, *qMLN9_142*, and *qMLN10_114*) were consistently found in two populations. These clustered QTLs from different populations were placed on the consensus map (Figure 4).

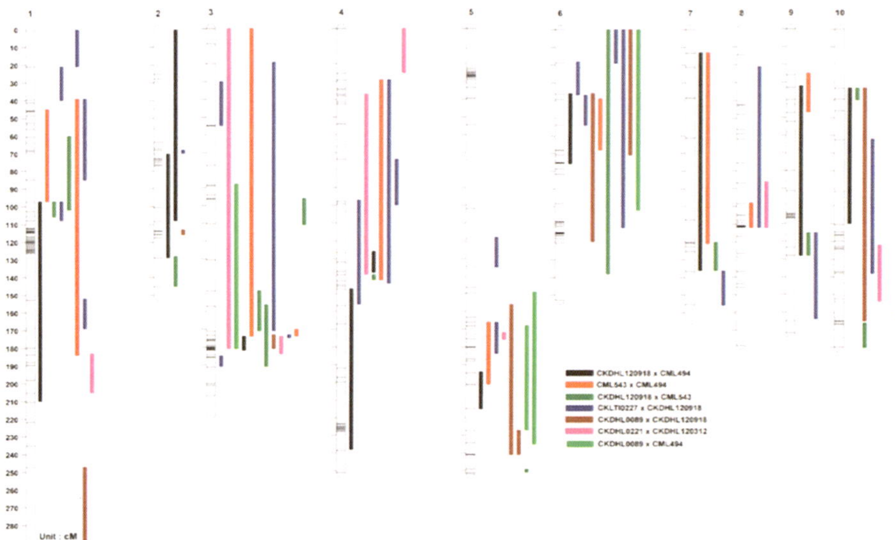

Figure 4. QTL for MLN resistance in the consensus linkage map of seven bi-parental populations. Different colors represent QTL detected from different populations.

3.2.4. Joint Linkage Association Mapping (JLAM)

The JLAM analyses based on three biometric models together found 27 and 28 main effect QTL for MLN-DS and AUDPC values (Table 3). For MLN-DS, 15, 12 and 18 QTLs were detected, which together explained 34.4%, 27.3% and 29.1% of the total variation in model A, B and C, respectively. Among 27 QTLs, five QTLs were consistently detected in all three models for MLN-DS. For AUDPC values, for model A, B and C, we found 14, 12 and 22 QTL which together explained 33.6%, 29% and 39.8% of the total variance. Seven QTLs were consistently detected in all three models. Most of the QTL detected through JLAM showed minor effects except one QTL *qMLN3_148* which was also consistent across models as well as across traits. Genomic prediction (GP) was attempted to predict the genetic values in the seven F$_3$ populations for MLN-DS and AUDPC performances. The mean prediction accuracy ranged from as low as 0.12 in F$_3$ pop1 to 0.75 in F$_3$ pop 7 for MLN-DS (Figure 5). The prediction accuracies were similar for AUDPC values with range of 0.13 in F$_3$ pop1 to 0.75 in F$_3$ pop5.

Table 3. Joint linkage association mapping depicting allele substitution effects and total phenotypic variance explained (PVE) using segregating F_3 progenies derived from seven bi-parental populations evaluated for three seasons under MLN inoculation.

Marker	QTL Name	Chr	Position (Mbp)	Model A		Model B		Model C	
				α-Effect	PVE (%)	α-Effect	PVE (%)	α-Effect	PVE (%)
PZA00447_8	qMLN1_9	1	9.02	−0.08	0.60	–	–	–	–
PHM5622_21	qMLN1_184	1	183.83	–	–	–	–	−0.12	0.40
S3_48493677	qMLN3_48	3	48.49	0.32	7.90	–	–	−0.02	0.40
S3_55444954	qMLN3_55	3	55.44	0.14	0.10	0.10	1.00	–	–
S3_68596995	qMLN3_68	3	68.60	–	–	−0.06	0.20	0.06	0.40
S3_92694873	qMLN3_92	3	92.69	−0.28	0.70	–	–	−0.02	1.80
S3_113429913	qMLN3_113	3	113.43	–	–	−0.38	1.00	–	–
PHM15449_10	qMLN3_125	3	125.08	0.10	3.30	0.11	0.80	0.16	2.20
S3_148291047	qMLN3_148	3	148.29	−0.72	10.20	−0.66	4.70	−0.16	1.40
S3_151342843	qMLN3_151	3	151.34	−0.23	3.20	−0.42	3.30	–	–
PHM2919_23	qMLN3_199	3	199.89	−0.12	0.40	−0.12	0.40	–	–
PZA00726_8	qMLN4_60	4	60.77	–	–	–	–	−0.04	0.70
S4_235381719	qMLN4_235	4	235.38	–	–	–	–	0.03	0.90
PHM565_31	qMLN5_24	5	24.24	–	–	−0.03	0.00	−0.47	0.30
S5_170023563	qMLN5_170	5	170.02	–	–	–	–	0.01	0.10
PHM7908_25	qMLN5_191	5	191.08	–	–	–	–	0.04	0.20
S5_196017729	qMLN5_196	5	196.02	−0.11	0.10	−0.03	0.10	−0.01	0.20
S5_202816906	qMLN5_202	5	202.82	–	–	–	–	−0.10	0.10
PHM563_9	qMLN5_204	5	204.99	–	–	–	–	−0.08	0.20
PZA03167_5	qMLN5_207	5	207.60	–	–	–	–	0.32	0.30
S5_209467974	qMLN5_209	5	209.47	–	–	–	–	−0.07	0.30
S6_13300385	qMLN6_13	6	13.30	0.19	1.90	0.20	1.10	–	–
S6_86475982	qMLN6_86	6	86.48	−0.27	1.00	–	–	–	–
S6_89823772	qMLN6_90	6	89.82	−0.24	2.70	−0.23	0.90	−0.22	2.50
PHM5235_8	qMLN8_94	8	94.41	0.15	0.50	–	–	–	–
PZA01313_2	qMLN10_4	10	3.60	0.11	1.30	0.18	2.90	0.10	0.80
PHM5740_9	qMLN10_9	10	8.77	0.09	0.50	–	–	–	–
Total PVE (%)					34.40		27.30		29.10
PZA00447_8	qMLN1_9	1	9.02	−1.63	0.20	–	–	–	–
PHM5622_21	qMLN1_184	1	183.83	–	–	–	–	−3.54	0.40
S3_48493677	qMLN3_48	3	48.49	6.63	6.80	–	–	–	–
S3_55444954	qMLN3_55	3	55.44	3.53	1.00	1.94	0.60	–	–
S3_68596995	qMLN3_68	3	68.60	–	–	−2.45	0.60	−1.55	2.30
PHM15449_10	qMLN3_125	3	125.08	–	–	3.01	1.00	3.65	3.00
S3_148291047	qMLN3_148	3	148.29	−20.14	10.20	−16.47	4.80	−3.86	1.50
S3_151342843	qMLN3_151	3	151.34	−12.06	4.90	−10.59	3.70	−4.72	1.80
PHM2919_23	qMLN3_199	3	199.89	−4.35	0.50	−4.04	0.80	−1.70	0.80
PZA00726_8	qMLN4_60	4	60.77	–	–	–	–	−1.64	0.70
S4_155378923	qMLN4_155	4	155.38	–	–	−8.03	1.00	–	–
S4_235381719	qMLN4_235	4	235.38	–	–	–	–	−10.74	0.80
S5_170164477	qMLN5_170	5	170.16	5.53	0.50	–	–	10.13	0.50
PHM7908_25	qMLN5_191	5	191.08	–	–	–	–	1.38	0.20
S5_196017729	qMLN5_196	5	196.02	−2.14	0.20	−1.13	0.40	−4.37	0.20
S5_202816906	qMLN5_202	5	202.82	–	–	–	–	−2.53	0.10
PHM563_9	qMLN5_204	5	204.99	–	–	–	–	−5.61	0.30
PZA03167_5	qMLN5_207	5	207.60	–	–	–	–	7.57	0.30
S5_209467974	qMLN5_209	5	209.47	–	–	–	–	2.00	0.40
S6_13300385	qMLN6_13	6	13.30	6.30	4.30	5.69	1.40	0.54	1.00
S6_86475982	qMLN6_86	6	86.48	−6.97	0.90	–	–	–	–
S6_89823772	qMLN6_90	6	89.82	−6.58	0.90	−6.14	1.00	−5.81	2.00
S8_74144408	qMLN8_74	8	74.14	–	–	–	–	3.86	0.30
PHM5235_8	qMLN8_94	8	94.41	4.41	1.50	–	–	–	–
S8_102533570	qMLN8_102	8	102.53	–	–	–	–	0.83	0.30
PZA01313_2	qMLN10_4	10	3.60	3.41	1.20	4.73	3.30	0.66	0.90
PHM5740_9	qMLN10_9	10	8.77	–	–	–	–	−5.51	0.30
PZA00866_2	qMLN10_124	10	124.20	1.71	0.60	1.44	0.40	1.79	0.90
Total PVE (%)					33.60		29.00		39.80

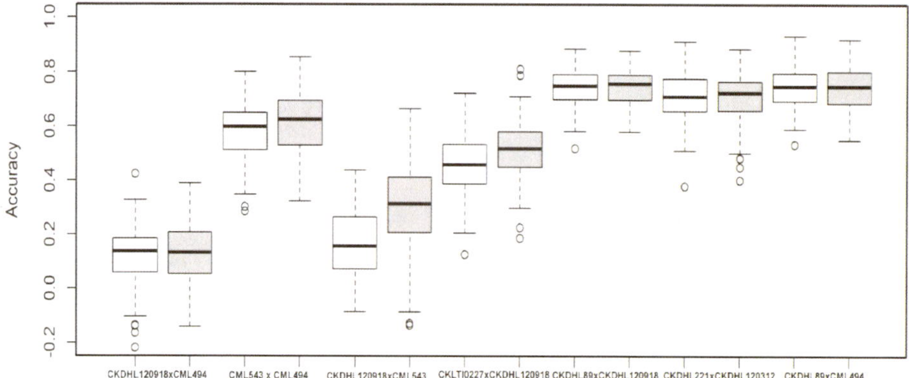

Figure 5. Box-whisker-plots for the accuracy of genomic predictions assessed by five-fold cross-validation within each population for MLN-DS (white box) and AUDPC values (grey box).

4. Discussion

4.1. Response of Parents and F_3 Populations to MLN Infections

A total of 2142 F_3 genotypes and seven parental lines were evaluated for MLN response in an artificially inoculated MLN plots for three seasons (2017–2018). Average mean for the F_3 populations was lower than parental average for MLN-DS. Superiority of the populations over parents could be due to dominance and over-dominance effects resulting from combinations of favorable alleles from different parents and therefore, had better heterosis for resistance to MLN. The concept of heterosis has been widely reported in maize [47–49]. Significant mid-parent and best-parent heterosis for resistance to MLN across locations were reported in Northern Tanzania [50]. The observed heterosis could also be partially explained by higher vigor in the progenies than the inbred parents. The inbred parents are not as vigorous as the heterozygous progenies and less able to cope with stresses of various kinds including pathogens.

The significant variability observed among the populations for resistance to MLN is possibly due to genetic effects and could be an indication of differences in genetic backgrounds of the genotypes for resistance to MLN (Table 1). Further, the study detected high heritability estimates (0.71 to 0.94) for resistance to MLN in all the seven populations. Previous investigations revealed moderate to high heritability estimates (0.34–0.89) for resistance to MLN [7,22,48,51]. The high heritability estimates detected in this study corroborate earlier reports and strongly suggest that resistance to MLN could be largely influenced by several genes with major effects.

4.2. QTL Analyses

In this study, seven major QTL were identified for resistance to MLN, localized on chromosomes 3 (*qMLN3_130* and *qMLN3_142*), 5 (*qMLN5_190* and *qMLN5_202*), 6 (*qMLN6_85* and *qMLN6_157*), 8 (*qMLN8_10*) and 9 (*qMLN9_142*) and spread across the 7-biparetal populations with the percentage of PVE ranging from 10.54% to 44.50% (Table 2). The results indicate that a large portion of the phenotypic variation in MLN resistance can be explained by few QTL with major effects while the remaining portion is due to QTL with minor effects. Similar observations have been made on various quantitative traits [52], including MLN [7,8]. It was observed in the present study that some QTLs showed major effects in some populations and minor effects in other populations. This variability could be attributed to various factors and gene actions including QTL × QTL interactions or QTL × environment interactions which might impact the size of effect of any given QTL in any given biparental population.

To better understand the above interactions and their cause of variability among the genotypes, it is important to consider each component of gene action (additive, dominance and epistasis) separately. Additive effect is realized when each allele is independently contributing to genetic variability. Dominance effect is a variation caused by interaction between alleles at a locus. Epistasis effect is observed when the PVE is due to interactions between alleles at different loci. In this study, it was observed that QTL had varying levels of additive and dominance effects for resistance to MLN (Table 2), suggesting the importance of both modes of gene action in conditioning resistance to the disease. Similarly, some QTL had smaller %PVE values but had larger additive or dominance effects compared to the QTL with larger %PVE. Generally, when epistasis is involved, QTL with moderate and major additive effects are much more affected and they tend to have reduced %PVE. Therefore, epistasis could also be one of the main causes in some QTL having larger additive effects but with reduced %PVE values for resistance to MLN. QTL detected in F_3 pop 1, 2, 3, 4, and 5 predominantly showed dominant to over-dominant effects. The dominance effects observed indicates MLN resistance is possibly due to interactions between individual alleles at specific loci. Other factors may include recombination between QTL and markers, leading to change in number of expected resistant alleles in the progeny hence reduced penetrance in QTL effects [53,54]. For QTL to be useful in a breeding program, it is important to first carry out validation studies to confirm their repeatability in different genetic backgrounds and environments. In line with expectations, the major QTL observed across populations in this study such as qMLN3_130, qMLN3_142, qMLN6_85, qMLN5_190, etc., indicated their stability across different genetic backgrounds.

It was observed that favorable alleles for each QTL were contributed by either resistant or susceptible parent, which is indicated here by sign of the additive effect of the QTL. Negative (−) additive effect was considered for resistant parent being the source of favorable allele for resistance to MLN. Similarly, positive (+) additive effect implied that the susceptible parent was the donor of favorable allele for the observed phenotypic variability. This suggests the importance of major QTL in reduction of MLN infections and their stability across different genetic backgrounds [54]. Earlier studies [7,8] also detected contributions of favorable alleles from both resistant and/or susceptible parents.

At least seven major QTLs localized across chromosomes and across populations with significantly larger %PVE were detected with total %PVE ranging from 25.13 to 59.07. Three of these QTLs (*qMLN3_130*, *qMLN6_85* and *qMLN5_190*) showed effects (%PVE), chromosome positions and confidence intervals (CI) comparable to the previous study [7]. For example, for *qMLN3_130*, the current CI is 125 to 169 Mb, whereas in the previous study CI was 125–130 Mb; for *qMLN6_85* CI is 84–91 Mb and from previous study CI was 85–96 Mb; and for *qMLN5_190* CI is 23–191 Mb and previous CI was 190–191 Mb. The larger CI observed in this study, compared to earlier study, could be due to few SNPs used in this study, however the results confirms the stability of the QTL in specific genomic regions. Another major QTL (*qMLN6_157*) detected on chromosome 6 in the current study was also found in the previous investigations [8]. This observation implies that three QTLs are stable across diverge genetic backgrounds, and hence can be useful in marker assisted breeding for resistance to MLN. Another major effect QTL *qMLN3_142* was detected in six out of the seven populations in this study, which also overlapped with QTL for MLN resistance in DH populations from the earlier study [8]. Two major QTLs (*qMLN5_202* and *qMLN8_10*) mapped across populations in this study, have not been reported previously. This implies the novelty of these genomic regions for MLN resistance. Similarly, it could suggest their specificity to the genetic backgrounds used in the current study.

4.3. Joint Linkage Association Mapping (JLAM)

For JLAM, three biometric models were used to detect the maximum number of QTL linked to genes for MLN resistance. The principle of both models A and B resemble the composite interval mapping approach used in bi-parental populations [55] where co-factors were used to adjust for the population structure and background error within the segregating populations, leading to enhanced capacity to detect QTL. In both MLN-DS and AUDPC, we observed that >65% of QTL detected in model A and B overlapped with slightly higher %PVE in Model A (Table 3). This clearly indicates that control of population stratification by using co-factors alone was moderate, so population effect is also important. The slightly higher number of QTL detected by model A compared to model B could imply that model A was able to more effectively detect the variability among populations, and hence was able to reveal more QTL [10]. Model C with the nested marker effects, exploited the linkage disequilibrium (LD) within segregating populations and assumed population-specific QTL effects [10]. Model C was comparable with Model B in terms of the %PVE by these QTLs. This was well supported by all seven populations with relatively bigger population size ($n = 306$), where the nested model C is able to detect QTL with small to large effect size.

To evaluate the reliability of QTL detected in linkage mapping, we compared the identified 60 and 58 QTLs with 27 and 28 main effect QTLs detected through JLAM for MLN-DS and AUDPC values, respectively. Twenty-three QTLs are common across linkage mapping and JLAM for both the traits (Tables 2 and 3). Comparison of QTL detected through JLAM and linkage mapping revealed five unique QTLs (*qMLN3_199*, *qMLN4_235*, *qMLN5_207*, *qMLN5_209*, and *qMLN6_13*) which were detected only through JLAM. QTLs common across methods helped to pinpoint the specific markers compared to QTLs detected with large CI through linkage mapping. Plausible gene candidates underlie some of the detected QTL regions with significant SNPs found in genes involved in plant defense system, like *PZA00447_8* associated with serine/threonine protein phosphatase, *PZA01313_2* linked to leucine-rich repeat protein and *PZA00726_8* linked to zinc ion binding function which has a role in plant defense responses (Table S2). Among the unique QTL detected in JLAM, QTL on chromosome 6 seems to be overlapping with earlier reported major effect QTL for MLN resistance [7]. Major recessive QTL was reported in this region [56] in F_2 populations. Similarly, with different association panels new QTL detected for resistance to SCMV in the same location on chromosome 6 [57]. Unique QTL on chromosomes 3 and 6 found in both model A and B but not in linkage mapping in this study indicates the variation across populations has helped to detect these QTL, which supports the use of multiple populations to find the novel source of resistance. On the contrary, the other three unique QTLs on chromosomes 4 and 5 were detected only in model C and were QTLs with population specific expression.

We employed the same populations to evaluate GP by sampling both training and testing sets. Moderate to high accuracy of GP was observed for MLN-DS and AUDPC within populations. Low accuracy in F3 pop1 is also due to low number of polymorphic SNPs compared to other populations. Application of GP for selection for improvement of MLN resistance has been reported [7]. The previous investigations detected high prediction accuracy within the populations for both MLN-DS and AUDPC values. Resistance for MLN is complex and requires considerable resources and multiple selection cycles to identify resistant lines using phenotypic selection alone. Therefore, GP could provide an alternative practical approach for breeding for resistance to MLN because it accounts both major and minor effect QTL for accuracy, enhanced breeding gain (shortened breeding cycle) and reduced costs.

5. Conclusions

Seven major QTL were identified for resistance to MLN, localized on chromosomes 3 (*qMLN3_130* and *qMLN3_142*), 5 (*qMLN5_190* and *qMLN5_202*), 6 (*qMLN6_85* and *qMLN6_157*), 8 (*qMLN8_10*) and 9 (*qMLN9_142*) and spread across the 7-biparetal populations with a percentage of PVE ranging from 10.54% to 44.50%. Two SNPs on chromosome 3 (*PHM15449_10* at 125,077,922 bp and *PZA00920_1* at 142,821,031) were strongly associated with MLN resistance and their allelic effects in different populations were also prominent. Several QTL on chromosomes 1, 3, 5, and 6 shared similar CI with those reported previously. These QTL could be adopted for marker assisted breeding for improvement of maize against MLN infection. Additional three major QTL were not reported before further research is warranted to confirm their reproducibility and use in breeding for resistance to MLN. Both linkage mapping and JLAM can be incorporated to confirm earlier reported QTLs and discover new sources of resistance. Incorporation of GP can help to capture both large effect and small effect QTL to improve the level of MLN resistance as a result improved genetic gain. Twenty-three QTL were common across linkage mapping and JLAM however, five unique QTL (*qMLN3_199, qMLN4_235, qMLN5_207, qMLN5_209,* and *qMLN6_13*) were unique to JLAM with the QTL on chromosome 6 appearing to be overlapping with an earlier reported major effect QTL for SCMV resistance. Both linkage mapping and JLAM can be incorporated to confirm earlier reported QTL and discover new source of resistance. The incorporation of genomic selection can help to capture both large effect and small effect QTL to improve the level of MLN resistance.

Supplementary Materials: The following are available online at http://www.mdpi.com/2073-4425/11/1/32/s1, Table S1: genetic linkage groups constructed for resistance to MLN using 500 SNPs in seven F3 populations; Table S2: Twenty-three common QTLs and specific SNP detected in both linkage mapping and joint linkage association mapping across seven F3 populations and their associated candidate genes.

Author Contributions: Conceptualization and methodology, L.A.O.A.; software and data analyses, L.A.O.A. and M.G.; validation, M.G., P.T., E.D. and B.E.I.; investigation, L.A.O.A.; resources, B.M.P. and M.G.; writing—original draft, L.A.O.A.; writing—review and editing, B.E.I., P.T., E.D., P.W.M.-D., V.O., Y.B., S.L.M., M.B.J., M.O., S.M., B.M.P. and M.G.; supervision, B.E.I., P.T., E.D., P.W.M.-D. and M.G.; project administration, P.T., S.L.M. and M.G. All authors have read and agreed to the published version of the manuscript.

Funding: This study was part of a PhD program funded by Intra-ACP Mobility Scheme. The research was supported by the Bill and Melinda Gates Foundation, the Howard G. Buffett Foundation, and the United States Agency for International Development (USAID) through Stress Tolerant Maize for Africa (STMA, Grant # OPP1134248), and the CGIAR Research Program MAIZE. MAIZE receives W1 and W2 support from the Governments of Australia, Belgium, Canada, China, France, India, Japan, Korea, Mexico, Netherlands, New Zealand, Norway, Sweden, Switzerland, U.K., U.S., and the World Bank. We thank CIMMYT technicians for data collection at experimental sites.

Acknowledgments: The study was managed by The Global maize Program, CIMMYT and University of Ghana.

Conflicts of Interest: The authors declare no conflict of interest.

References

1. Mahuku, G.; Lockhart, B.E.; Wanjala, B.; Jones, M.W.; Kimunye, J.N.; Stewart, L.R.; Cassone, B.J.; Subramanian, S.; Nyasani, J.; Kusia, E.; et al. Maize lethal necrosis (MLN), an emerging threat to maize-based food security in sub-Saharan Africa. *Phytopathol.* **2015**, *105*, 956–965. [CrossRef] [PubMed]

2. Wamaitha, M.J.; Nigam, D.; Maina, S.; Stomeo, F.; Wangai, A.; Njuguna, J.N.; Holton, T.A.; Wanjala, B.W.; Wamalwa, M.; Lucas, T.; et al. Metagenomic analysis of viruses associated with maize lethal necrosis in Kenya. *Virol. J.* **2018**, *15*, 90. [CrossRef] [PubMed]

3. Wangai, A.W.; Redinbaugh, M.G.; Kinyua, Z.M.; Miano, D.W.; Leley, P.K.; Kasina, M.; Mahuku, G.; Scheets, K.; Jeffers, D. First Report of Maize chlorotic mottle virus and Maize Lethal Necrosis in Kenya. *Plant Dis.* **2012**, *96*, 1582. [CrossRef] [PubMed]

4. Redinbaugh, M.; Pratt, R. Virus resistance. In *Handbook of Maize: It's Biology*; Bennetzen, J.L., Hake, S.C., Eds.; Springer: New York, NY, USA, 2009; pp. 251–268.

5. Bonamico, N.C.; Di Renzzo, M.; Ibanez, M.; Borghi, M.; Diaz, D.; Salerno, J.; Balzarini, M. QTL analysis of resistance to Mal de Río Cuarto disease in maize using recombinant inbred lines QTL analysis of resistance to Mal de Río Cuarto disease in maize using recombinant inbred lines. *J. Agric. Sci.* **2012**, *150*, 5. [CrossRef]

6. Zambrano, J.L.; Jones, M.W.; Brenner, E.; Francis, D.M.; Tomas, A.; Redinbaugh, M.G. Genetic analysis of resistance to six virus diseases in a multiple virus-resistant maize inbred line. *Theor. Appl. Genet.* **2014**, *127*, 867–880. [CrossRef]

7. Gowda, M.; Beyene, Y.; Makumbi, D.; Semagn, K.; Olsen, M.S.; Bright, J.M.; Das, B.; Mugo, S.; Suresh, L.M.; Prasanna, B.M. Discovery and validation of genomic regions associated with resistance to maize lethal necrosis in four biparental populations. *Mol. Breed.* **2018**, *38*, 66. [CrossRef]

8. Sitonik, C.; Suresh, L.M.; Beyene, Y.; Olsen, M.S.; Makumbi, D.; Oliver, K.; Das, B.; Bright, J.M.; Mugo, S.; Crossa, J.; et al. Genetic architecture of maize chlorotic mottle virus and maize lethal necrosis through GWAS, linkage analysis and genomic prediction in tropical maize germplasm. *Theor. Appl. Genet.* **2019**, *132*, 2381–2399. [CrossRef]

9. Yu, J.; Holland, J.B.; McMullen, M.D.; Buckler, E.S. Genetic Design and Statistical Power of Nested Association Mapping in Maize. *Genet* **2008**, *178*, 539–551. [CrossRef]

10. Wurschum, T.; Liu, W.; Gowda, M.; Maurer, H.P.; Fischer, S.; Schechert, A.; Reif, J.C. Comparison of biometrical models for joint linkage association mapping. *Heredity* **2012**, *108*, 332–340. [CrossRef]

11. Hawkins, L.K.; Warburton, M.L.; Tang, J.D.; Tomashek, J.; Oliveira, D.A.; Ogunola, O.F.; Smith, J.S.; Williams, W.P. Survey of Candidate Genes for Maize Resistance to Infection by Aspergillus flavus and/or Aflatoxin Contamination. *Toxins* **2018**, *10*, 61. [CrossRef]

12. Gelli, M.; Konda, A.R.; Liu, K.; Zhang, C.; Clemente, T.E.; Holding, D.R.; Dweikat, I.M. Validation of QTL mapping and transcriptome profiling for identification of candidate genes associated with nitrogen stress tolerance in sorghum. *BMC Plant Boil.* **2017**, *17*, 123. [CrossRef] [PubMed]

13. Sudheesh, S.; Rodda, M.S.; Davidson, J.; Javid, M.; Stephens, A.; Slater, A.T.; Cogan, N.O.I.; Forster, J.W.; Kaur, S. SNP-Based Linkage Mapping for Validation of QTLs for Resistance to Ascochyta Blight in Lentil. *Front. Plant Sci.* **2016**, *7*, 311. [CrossRef] [PubMed]

14. Gajjar, K.N.; Mishra, G.P.; Radhakrishnan, T.; Dodia, S.M.; Rathnakumar, A.L.; Kumar, N.; Kumar, K.; Dobaria, J.R.; Kumar, A. Validation of SSR markers linked to the rust and late leaf spot diseases resistance in diverse peanut genotypes. *Aust. J. Crop. Sci.* **2014**, *8*, 927–936.

15. Landi, P.; Sanguineti, M.C.; Salvi, S.; Giuliani, S.; Bellotti, M.; Maccaferri, M.; Conti, S.; Tuberosa, R. Validation and characterization of a major QTL affecting leaf ABA concentration in maize. *Mol. Breed.* **2005**, *15*, 291–303. [CrossRef]

16. Sukruth, M.; Paratwagh, S.A.; Sujay, V.; Kumari, V.; Gowda, M.V.C.; Nadaf, H.L.; Motagi, B.N.; Lingaraju, S.; Pandey, M.K.; Varshney, R.K.; et al. Validation of markers linked to late leaf spot and rust resistance, and selection of superior genotypes among diverse recombinant inbred lines and backcross lines in peanut (*Arachis hypogaea* L.). *Euphytica* **2015**, *204*, 343–351. [CrossRef]

17. Zhou, H.; Liu, S.; Liu, Y.; Liu, Y.; You, J.; Deng, M.; Ma, J.; Chen, G.; Wei, Y.; Liu, C.; et al. Mapping and validation of major quantitative trait loci for kernel length in wild barley (*Hordeum vulgare* ssp. spontaneum). *BMC Genet.* **2016**, *17*, 130. [CrossRef]

18. De Leon, T.B.; Linscombe, S.; Subudhi, P.K. Identification and validation of QTLs for seedling salinity tolerance in introgression lines of a salt tolerant rice landrace 'Pokkali'. *PLoS ONE* **2017**, *12*, e0175361. [CrossRef]

19. Ishikawa, G.; Nakamura, K.; Ito, H.; Saito, M.; Sato, M.; Jinno, H.; Yoshimura, Y.; Nishimura, T.; Maejima, H.; Uehara, Y.; et al. Association Mapping and Validation of QTLs for Flour Yield in the Soft Winter Wheat Variety Kitahonami. *PLoS ONE* **2014**, *9*, e111337. [CrossRef]

20. Bernardo, R. Molecular Markers and Selection for Complex Traits in Plants: Learning from the Last 20 Years. *Crop. Sci.* **2008**, *48*, 1649. [CrossRef]

21. Tuberosa, R.; Salvi, S.; Sanguineti, M.C.; Landi, P.; Maccaferri, M.; Conti, S. Mapping QTLs Regulating Morpho-physiological Traits and Yield: Case Studies, Shortcomings and Perspectives in Drought-stressed Maize. *Ann. Bot.* **2002**, *89*, 941–963. [CrossRef]

22. Gowda, M.; Das, B.; Makumbi, D.; Babu, R.; Semagn, K.; Mahuku, G.; Olsen, M.S.; Bright, J.M.; Beyene, Y.; Prasanna, B.M. Genome-wide association and genomic prediction of resistance to maize lethal necrosis disease in tropical maize germplasm. *Theor. Appl. Genet.* **2015**, *128*, 1957–1968. [CrossRef] [PubMed]

23. Tivoli, B.; Baranger, A.; Avila, C.M.; Banniza, S.; Barbetti, M.; Chen, W.; Davidson, J.; Lindeck, K.; Kharrat, M.; Rubiales, D.; et al. Screening techniques and sources of resistance to foliar diseases caused by major necrotrophic fungi in grain legumes. *Euphytica* **2006**, *147*, 223–253. [CrossRef]

24. Wang, J.; Li, H.; Zhang, L.; Meng, L. *Users' Manual of QTL IciMapping, 4.1*; International Maize and Wheat Improvement Center (CIMMYT): Texcoco, Mexico, 2016.

25. Alvarado, G.; López, M.; Vargas, M.; Pacheco, A.; Rodriguez, F.; Burgueno, J.; Crossa, J. META-R (Multi Environment Trail Analysis with R for Windows) Version 5.0—CIMMYT Research Software Dataverse—CIMMYT Dataverse Network. 2015. Available online: http://hdl.handle.net/11529/10201 (accessed on 5 September 2019).

26. Stram, D.O.; Lee, J.W. Variance Components Testing in the Longitudinal Mixed Effects Model. *Biometrics* **1994**, *50*, 1171–1177. [CrossRef] [PubMed]

27. CIMMYT. *Laboratoty Protocols: CIMMYT Applied and Molecular Genetics Laboratoty*, 3rd ed.; CIMMYT: Batan, Mexico, 2005.

28. Bradbury, P.J.; Zhang, Z.; Kroon, D.E.; Casstevens, T.M.; Ramdoss, Y.; Buckler, E.S. TASSEL: Software for association mapping of complex traits in diverse samples. *Bioinformatics* **2007**, *23*, 2633–2635. [CrossRef]

29. Xia, X.; Melchinger, A.E.; Kuntze, L.; Lübberstedt, T. Quantitative Trait Loci Mapping of Resistance to Sugarcane Mosaic Virus in Maize. *Phytopathology* **1999**, *89*, 660–667. [CrossRef]

30. Horn, F.; Habekuss, A.; Stich, B.; Habekuß, A. Linkage mapping of Barley yellow dwarf virus resistance in connected populations of maize. *BMC Plant Boil.* **2015**, *15*, 29. [CrossRef]

31. Meng, L.; Li, H.; Zhang, L.; Wang, J. QTL IciMapping: Integrated software for genetic linkage map construction and quantitative trait locus mapping in biparental populations. *Crop. J.* **2015**, *3*, 269–283. [CrossRef]

32. Kosambi, D. The estimation of map distances from recombination values. *Ann. Eugen.* **1944**, *12*, 172–175. [CrossRef]

33. Voorrips, R.E. MapChart: Software for the graphical presentation of linkage maps and QTLs. *J. Hered.* **2002**, *93*, 77–78. [CrossRef]

34. Li, H.; Ye, G.; Wang, J. A modified algorithm for the improvement of composite interval mapping. *Genetics* **2007**, *175*, 361–374. [CrossRef]

35. Lübberstedt, T.; Melchinger, A.E.; Schön, C.C.; Utz, H.F.; Klein, D. QTL Mapping in Testcrosses of European Flint Lines of Maize: I. Comparison of Different Testers for Forage Yield Traits. *Crop. Sci.* **1997**, *37*, 921. [CrossRef]

36. Liu, Y.; Chen, L.; Fu, N.; Lou, Q.; Mei, H.; Xiong, L.; Li, M.; Xu, X.; Mei, X.; Luo, L. Dissection of additive, epistatic effect and QTL × environment interaction of quantitative trait loci for sheath blight resistance in rice. *Hereditas* **2014**, *151*, 28–37. [CrossRef] [PubMed]

37. Stuber, C.W.; Edwards, M.D.; Wendel, J.F. Molecular Marker-Facilitated Investigations of Quantitative Trait Loci in Maize. II. Factors Influencing Yield and its Component Traits1. *Crop. Sci.* **1987**, *27*, 639. [CrossRef]

38. Carlson, C.H.; Gouker, F.; Crowell, C.R.; Evans, L.; DiFazio, S.P.; Smart, C.D.; Smart, L.B. Joint linkage and association mapping of complex traits in shrub willow (*Salix purpurea* L.). *Ann. Bot.* **2019**, *124*, 701–715. [CrossRef] [PubMed]

39. Mammadov, J.; Sun, X.; Gao, Y.; Ochsenfeld, C.; Bakker, E.; Ren, R.; Flora, J.; Wang, X.; Kumpatla, S.; Meyer, D.; et al. Combining powers of linkage and association mapping for precise dissection of QTL controlling resistance to gray leaf spot disease in maize (*Zea mays* L.). *BMC Genom.* **2015**, *16*, 916. [CrossRef]

40. Schwarz, G. Estimating the Dimension of a Model. *Ann. Stat.* **1978**, *6*, 461–464. [CrossRef]

41. SAS Institute Inc. *SAS® 9.4 Intelligence Platform: Data Administration Guide*, 6th ed.; SAS Institute Inc.: Cary, NC, USA, 2016.

42. Reif, J.C.; Zhao, Y.; Würschum, T.; Gowda, M.; Hahn, V. Genomic prediction of sunflower hybrid performance. *Plant Breed.* **2013**, *132*, 107–114. [CrossRef]

43. Holm, S. A simple sequentially rejective multiple test procedure. *Scand. J. Stat.* **1979**, *6*, 65–70.

44. Chen, J.; Zavala, C.; Ortega, N.; Petroli, C.; Franco, J.; Burgueño, J.; Costich, D.E.; Hearne, S.J. The Development of Quality Control Genotyping Approaches: A Case Study Using Elite Maize Lines. *PLoS ONE* **2016**, *11*, e0157236. [CrossRef]

45. Barría, A.; Christensen, K.A.; Yoshida, G.M.; Correa, K.; Jedlicki, A.; Lhorente, J.P.; Davidson, W.S.; Yanez, J.M. Genomic Predictions and Genome-Wide Association Study of Resistance Against Piscirickettsia salmonis in Coho Salmon (*Oncorhynchus kisutch*) Using ddRAD Sequencing. *G3: Genes|Genomes|Genetics* **2018**, *8*, 1183–1194. [CrossRef]

46. Gao, N.; Teng, J.; Ye, S.; Yuan, X.; Huang, S.; Zhang, H.; Zhang, X.; Li, J.; Zhang, Z. Genomic Prediction of Complex Phenotypes Using Genic Similarity Based Relatedness Matrix. *Front. Genet.* **2018**, *9*, 364. [CrossRef] [PubMed]

47. Hochholdinger, F.; Hoecker, N. Towards the molecular basis of heterosis. *Trends Plant Sci.* **2007**, *12*, 427–432. [CrossRef] [PubMed]

48. Beyene, Y.; Gowda, M.; Suresh, L.M.; Mugo, S.; Olsen, M.; Oikeh, S.O.; Juma, C.; Tarekegne, A.; Prasanna, B.M. Genetic analysis of tropical maize inbred lines for resistance to maize lethal necrosis disease. *Euphytica* **2017**, *213*, 224. [CrossRef]

49. Springer, N.M.; Stupar, R.M. Allelic variation and heterosis in maize: How do two halves make more than a whole? *Genome Res.* **2007**, *17*, 264–275. [CrossRef]

50. Karume, K.; Nazir, N.; Salmin, I. Assessment of heterosis and combining ability in maize (*Zea mays* L.) for maize lethal necrosis disease. *Afr. J. Plant Breed.* **2017**, *4*, 104–200. [CrossRef]

51. Mbega, E.; Ndakidemi, P.; Mamiro, D.; Mushongi, A.; Kitenge, K.; Ndomba, O. Role of Potyviruses in Synergistic Interaction Leading to Maize Lethal Necrotic Disease on Maize. *Int. J. Curr. Microbiol. Appl. Sci.* **2016**, *5*, 85–96. [CrossRef]

52. Mackay, T.F.C. The genetic architecture of quantitative traits: Lessons from Drosophila, Current Opinion in Genetics and Development. *Curr. Opin. Genet. Dev.* **2004**, *14*, 253–257. [CrossRef]

53. Forabosco, P.; Falchi, M.; Devoto, M. Statistical tools for linkage analysis and genetic association studies. *Expert Rev. Mol. Diagn.* **2005**, *5*, 781–796. [CrossRef]

54. Gallois, J.-L.; Moury, B.; German-Retana, S. Role of the Genetic Background in Resistance to Plant Viruses. *Int. J. Mol. Sci.* **2018**, *19*, 2856. [CrossRef]

55. Jansen, R.C.; Stam, P. High Resolution of Quantitative Traits into Multiple Loci via Interval Mapping. *Genet.* **1994**, *136*, 1447–1455.

56. Jones, M.W.; Penning, B.W.; Jamann, T.M.; Glaubitz, J.C.; Romay, C.; Buckler, E.S.; Redinbaugh, M.G. Diverse chromosomal locations of quantitative trait loci for tolerance to maize chlorotic mottle virus in five maize populations. *Phytopathology* **2018**, *108*, 748–758. [CrossRef] [PubMed]

57. Gustafson, T.J.; de Leon, N.; Kaeppler, S.M.; Tracy, W.F. Genetic analysis of sugarcane mosaic virus resistance in the Wisconsin diversity panel of maize. *Crop. Sci.* **2018**, *58*, 1853–1865. [CrossRef]

 *genes*

Article

Natural Variation and Domestication Selection of *ZmPGP1* Affects Plant Architecture and Yield-Related Traits in Maize

Pengcheng Li, Jie Wei, Houmiao Wang, Yuan Fang, Shuangyi Yin, Yang Xu, Jun Liu, Zefeng Yang and Chenwu Xu *

Jiangsu Key Laboratory of Crop Genetics and Physiology/Key Laboratory of Plant Functional Genomics of the Ministry of Education/Jiangsu Key Laboratory of Crop Genomics and Molecular Breeding/Jiangsu Co-Innovation Center for Modern Production Technology of Grain Crops, Agricultural College of Yangzhou University, Yangzhou 225009, China
* Correspondence: qtls@yzu.edu.cn; Tel.: +86-0514-87979358

Received: 3 July 2019; Accepted: 28 August 2019; Published: 30 August 2019

Abstract: *ZmPGP1*, involved in the polar auxin transport, has been shown to be associated with plant height, leaf angle, yield traits, and root development in maize. To explore natural variation and domestication selection of *ZmPGP1*, we re-sequenced the *ZmPGP1* gene in 349 inbred lines, 68 landraces, and 32 teosintes. Sequence polymorphisms, nucleotide diversity, and neutral tests revealed that *ZmPGP1* might be selected during domestication and improvement processes. Marker–trait association analysis in inbred lines identified 11 variants significantly associated with 4 plant architecture and 5 ear traits. SNP1473 was the most significant variant for kernel length and ear grain weight. The frequency of an increased allele T was 40.6% in teosintes, and it was enriched to 60.3% and 89.1% during maize domestication and improvement. This result revealed that *ZmPGP1* may be selected in the domestication and improvement process, and significant variants could be used to develop functional markers to improve plant architecture and ear traits in maize.

Keywords: natural variation; maize; nucleotide diversity; domestication selection; *ZmPGP1* gene

---

1. Introduction

Maize (*Zea mays* L.) is one of the most widely grown and important cereal crops, which plays a critical role in ensuring food security. Maize was domesticated from the wild grass teosinte more than 8700 years ago [1]. The domestication of maize went through two stages: domestication selection and subsequent genetic improvement (post-domestication selection) [2]. Strong directional selection had profound effects on the morphological structure of maize, and genetic improvement affected its productivity [3]. For example, from 2000 to 2014, the total maize production in the United States and China increased by 31% and 49% respectively, of which half could be attributed to genetic advances [4,5]. Human selection has profound effects on the genetic diversity for the genomic region under selection and target genes [3]. Genetic consequences during the domestication and breeding history will enable us to understand its important role on yield increase in the modern maize breeding.

Grain yield (GY) is a complicated quantitative trait and is mainly determined by three yield components: effective ear number, kernel number, and kernel weight [6]. Maize kernel and ear morphological traits are the most important factors determining grain yield. Kernel weight is mainly affected by kernel size, which is usually measured by kernel length (KL), kernel width (KW), and kernel thickness (KT). Ear length (EL), ear diameter (ED), and kernel row number (KRN) are important traits determining the kernel number [7]. Planting density is a major factor in determining the effective ear number. The increased maize productivity is predominantly due to higher planting density, resulting from the domestication and improvement of plant shoot architecture [4]. Plant architecture is influenced by aboveground phenotypes, such as plant height (PH), ear height (EH), and leaf number (LN). From 1930 to 2001 in the United States, maize ear height was reduced by 3 cm per decade, leaf angle became more upright, tassel branch numbers became averaged 2.5 fewer branches per decade, and leaf number increased from 12.2 in the 1930s to 13.8 in the 1970s [8]. Identification of genes associated with grain yield and plant architecture traits will be helpful for maize yield improvement.

Most plant and ear traits are quantitative traits, which are controlled by a large number of small effect quantitative trait locus (QTLs). Many QTLs related to yield components and plant architecture traits have been identified in several maize linkage populations. A total of 163 QTLs were detected for four ear traits in 10 different RIL populations, accounting for 55.4–82% of phenotypic variation [9]. In the same panel, approximately 800 QTLs with major and minor effects were identified for 10 plant architecture-related traits [10]. Martinez et al. [11] assembled a yield QTLome database, and 808 QTLs for GY and seven additional GY components of common interest in maize breeding from 32 mapping populations were used for meta-QTL analysis. A total of 84 meta-QTLs were projected on the 10 maize chromosomes [11]. A number of genes that affect plant and ear traits have been identified, such as *fea2*, *fea3*, the ramose genes and *KRN4* for kernel row number, *df3*, *df8*, *df9*, and *br2* for plant height, and *td1*, *bif2*, *ba1*, and *tsh4* for tassel morphology [12–21]. Numerous kernel and morphological traits have changed during maize domestication and improvement, and some key genes have been cloned. *Tb1* has been shown to be associated with maize branching [22], the teosinte allele *gt1* confers multiple ears per branch [23], and *tga1* was associated with kernel structure [24]. In addition, genome-wide selection signals during maize domestication and improvement were assessed, 484 domestication and 695 improvement selective sweeps were detected, and a number of genes with stronger signals for selection underlie major morphological changes [3].

Indole-3-acetic acid (IAA), an active form of auxin, is a key regulator of plant growth and development. *ZmPGP1* (*ABCB1* or *br2*) was firstly cloned using a Mu element, the mutant was characterized by compact lower stalk internodes and the plants showed semi-dwarf stalks [16]. *ZmPGP1* was an adenosine triphosphate (ATP) binding cassette (ABC) transporter which belonged to the multidrug resistant (MDR) class of P-glycoproteins (PGPs), and functioned as an efflux carrier in polar auxin transport. The protein had two transmembrane domains that provide the translocation pathway of auxin and two cytoplasmic nucleotide-binding domains that hydrolyse ATP and drive the transport reaction [25,26]. Different alleles of *ZmPGP1* have been shown to be associated with plant height, ear height, leaf angle, ear length, yield traits, and root development under aluminum stress [16,27–31]. Although several mutations of *ZmPGP1* have also been identified, the sequence polymorphism and natural variations of the gene have not been investigated. It is also unclear whether *ZmPGP1* exists as a signal of selection during maize domestication and improvement. In the present study, we re-sequenced *ZmPGP1* in 349 inbred lines, 68 landraces, and 32 teosintes, and aimed to: (1) examine the *ZmPGP1* nucleotide diversity between maize inbred lines, landraces, and teosintes, (2) identify natural variations in candidate genes associated with grain yield and plant architecture traits, and (3) examine the significant associations for their involvement in maize domestication and improvement.

2. Materials and Methods

2.1. Plant Materials and the Phenotypic Evaluation

A total of 349 inbred lines, 68 landraces, and 32 wild relatives were selected in this study [32]. The inbred lines were grown in the field in a randomized block design with two replicates in 2016, 2017, and 2018 in Sanya, Hainan Province (18°23′ N, 109°44′ E). Each inbred line was grown in a single row with 13 plants, 3 m in length, and 0.5 m between adjacent rows. Fifteen days after pollination, 6 plants in the middle of each row were selected to measure leaf number above the topmost ear (LNAE), plant height (PH), tassel branch number (TBN), and tassel main axis length (TMAL), the first leaf above the ear position leaf was selected to measure leaf angle (LA, the angle between the horizontal and the midrib of the leaf) and leaf width (LW). The measure method of plant architecture traits referred to are as described in Pan et al. (2017) [10]. After harvesting and drying, 3 well-developed ears were selected to measure ear traits, including ear grain weight (EGW), 100-kernel weight (HKW), ear diameter (ED), ear weight (EW), ear length (EL), kernel length (KL), kernel width (KW), kernel thickness (KT), and kernel number per row (KNR). The root and shoot traits at the seedling stage were determined by Li et al. [32] in a hydroponic system.

2.2. DNA Isolation and ZmPGP1 Resequencing

A modified cetyl trimethylammonium bromide (CTAB) method was used to exact genomic DNA from young leaves of each line at the seedling stage. The sequences of the *ZmPGP1* gene were sequenced by BGI (Beijing Genomics Institute) Life Tech Co. China using targeted sequence capture technology on the NimbleGen platform [33]. The genomic sequence of *ZmPGP1* (GRMZM2G315375) of the B73 inbred line was used as a reference for target sequence capture.

2.3. Analysis of Sequence Data

Multiple sequence alignment of the maize *ZmPGP1* gene was performed using MAFFT software and was further edited manually [34]. Using DNASP5.0 software [35], we analyzed single nucleotide polymorphisms (SNP) and allelic diversities across all tested lines. Two parameters, π and θ, were used to estimate the degree of polymorphism within the tested population. Tajima's D [36], Fu and Li's D*, as well as Fu and Li's F* [37] statistical tests were used to test for neutral evolution within each group and each defined region. The sequence data and markers were shown in Dataset 1–2.

2.4. Marker–Trait Association Analysis in Inbred Lines

Genotyping-by-sequencing (GBS) was used to identify the genotypes of 349 inbred lines [32]. A total of 163,931 SNPs were obtained by filtering out markers with more than 20% of missing data and below 1% minor allele frequency. Three models were used to conduct marker–trait associations: (1) the K model, controlling for kinship, (2) the PCA + K model, controlling for both population structure (principal component, PC) and kinship, and (3) the Q + K model, controlling for both population structure (Q) and kinship. Principal component analysis (PCA) and kinship were calculated using Tessel5.0, and Q was calculated by admixture. A total of 499 *ZmPGP1*-based markers with minor allele frequency (MAF) $\geq$0.05 were selected for association analysis in 349 inbred lines, and the *p* value threshold was set at 2.00×10^{-3} (0.5/499).

3. Results

3.1. Sequence Polymorphisms of ZmPGP1

The *ZmPGP1* sequence alignment of 349 inbred lines, 68 landraces, and 32 teosintes spanned 9710 bp, which covered a 1762 bp upstream region, a 182 bp 5′UTR region, a 6821 bp coding region containing five exons and four introns, a 400 bp 3′UTR region, and a 545 bp downstream region (Table 1). Sequence polymorphisms, including SNPs and InDels, at *ZmPGP1* were identified, and 1070 variations were detected, including 878 SNPs and 192 InDels. On average, SNPs and InDels were found every 11.06 bp and 50.57 bp, respectively. The highest frequency of SNPs and InDels were found in the 3′UTR (5.86 bp) and 5′UTR (14 bp). The overall nucleotide diversity (π) of the *ZmPGP1* locus was 0.007. Among five regions of the *ZmPGP1*, and when the coding regions were less diverse than other regions (0.006), the downstream and 3′UTR showed high nucleotide diversity (0.016 and 0.015, respectively).

Table 1. Summary of parameters for the analysis of nucleotide polymorphisms of the maize genes *ZmPGP1*.

Parameters	Upstream	5′UTR	Coding Region	3′UTR	Downstream	Entire Region
Total length of amplicons (bp)	1762	182	6821	400	545	9710
Number of all of the sequence variants	69	32	779	86	104	1070
Frequency of all of the sequence variants	0.039	0.176	0.114	0.215	0.191	0.110
Number of nucleotide substitutions (bp)	43	19	663	67	86	878
Frequency of polymorphic sites per bp	0.024	0.104	0.097	0.168	0.158	0.090
Number of indels	26	13	116	19	18	192
Number of indels sites	34	42	445	36	73	640
Average indel length	1.308	3.231	3.836	1.895	4.056	3.333
Frequency of indels per bp	0.015	0.071	0.017	0.048	0.033	0.020
$\pi \times 1000$	12.890	7.990	6.020	15.270	15.660	7.110
$\theta \times 1000$	40.870	22.720	19.110	36.120	53.510	22.080
Tajima's D	−1.905 *	−1.620	−2.089 *	−1.659	−2.065 *	−2.071 *
Fu and Li's D	−5.458 **	−2.149	−8.874 **	−8.232 **	−7.498 **	−9.242 **
Fu and Li's F	−4.605 **	−2.342 *	−5.917 **	−6.037 **	−5.699 **	−6.057 **

* indicates a statistical significance at $p < 0.05$ level, ** indicates a statistical significance at $p < 0.01$ level. "UTR" indicated untranslated region.

3.2. Nucleotide Diversity and Selection of ZmPGP1 in Inbred Lines, Landraces and Teosinte

To investigate the genetic diversity of *ZmPGP1* in inbred lines, landrace, and teosinte, the sequence conservation (C) and nucleotide diversity (π) were analyzed and compared. For all test lines, the values of C and $\pi \times 1000$ were 0.793 and 7.110, respectively (Figure 1a). Compared with teosintes, landraces and inbred lines showed higher conservation ($C_T = 0.845$, $C_L = 0.920$ and $C_I = 0.923$) and lower diversity ($\pi \times 1000_T = 20.724$, $\pi \times 1000_L = 9.970$ and $\pi \times 1000_I = 6.558$). The highest divergence between inbred lines and teosintes was observed in the upstream and downstream regions (Figure 1b). A divergence peak was found in the fourth intron by comparing landraces to inbred lines. To investigate the involvement in maize domestication and improvement of *ZmPGP1*, the entire sequence was tested by the neutral test, including Tajima's D and the D* and F* of Fu and Li. The values for Tajima's D and the D* and F* of Fu and Li of *ZmPGP1* were significantly less than 0, indicating this gene maybe selected in the domestication and improvement process (Figure 1a).

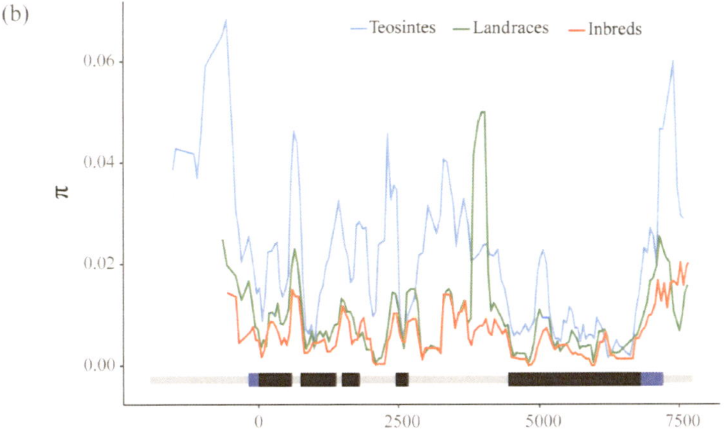

(a)

Populations	*Hd*	Dens.	C	$\pi \times 1000$	$\theta \times 1000$	Tajima's D	D	F
Teosintes	1.000	97.631	0.845	20.724	33.650	-1.488	-2.181	-2.116
Landraces	1.000	38.414	0.920	9.970	11.230	-0.394	-0.966	-0.810
Inbreds	0.969	33.059	0.923	6.558	7.170	-0.262	-2.556*	-1.578
All	0.980	90.422	0.793	7.110	22.080	-2.071*	-9.242**	-6.057**

(b)

Figure 1. Nucleotide diversity in inbred lines, landraces, and teosinte. (**a**) Summary of nucleotide polymorphisms and neutrality test of ZmPGP1, Hd represents haplotype diversity, Dens denotes number of single nucleotide polymorphisms (SNP) per 1000 bp, C represents sequence conservation, and D* and F* represent Fu and Li's D*and F*. (**b**) Nucleotide diversity (π) of inbred lines, landraces, and teosinte. π was calculated using the sliding windows method with a window size of 100 bp and a step length of 25 bp. * indicates a statistical significance at $p < 0.05$ level, ** indicates a statistical significance at $p < 0.01$ level.

3.3. Association Analysis of Phenotypic Traits with ZmPGP1

To identify significant variants associated with phenotypic traits, association analysis was performed using 499 variants, including 269 SNPs and 230 InDels with minor allele frequency (MAF) $\geq$0.05 in 349 inbred lines. Three mixed linear models (MLM), MLM + K, MLM + Q + K, and MLM + PCA + K, were employed to perform marker-traits association analysis. Comparing the Quantile-Quantile (QQ) plots generated for these models, we selected MLM + PCA + K to minimize both false positives and false negatives (Figure 2a).

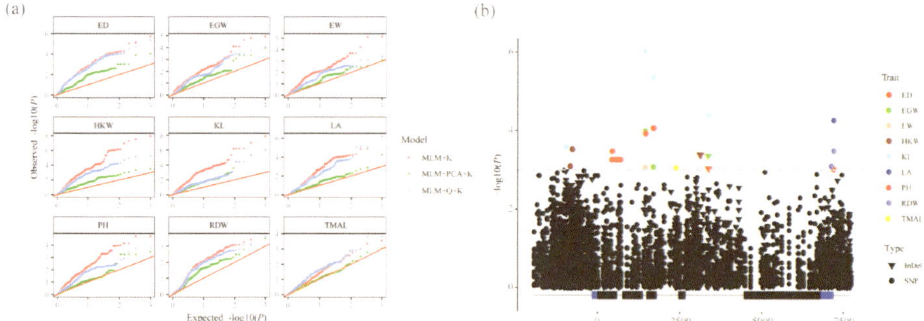

Figure 2. *ZmPGP1*-based association mapping. (**a**) QQ plot for the association analysis under three models, red, green and blue dots denote MLM + K, MLM + PCA + K, and MLM + Q + K, respectively. (**b**) Manhattan plot by using the MLM + PCA + K model. Triangles and dots represent InDels and SNPs, respectively. Abbreviations for traits are as follows: ED, ear diameter; EGW, ear grain weight; EW, ear weight; HKW, 100-kernel weight; KL, kernel length; LA, leaf angle; PH, plant height; RDW, root dry weight; TMAL, tassel main axis length.

Using a Bonferroni correction based on 499 *ZmPGP1*-based markers, the *P*-value thresholds were set at 0.001 (0.5/499). A total of 24 significant marker–trait associations involved 15 variants (12 SNPs and 3 InDels) were identified for 9 traits using the MLM + PCA + K model (Table S1). Among these 24 associations, 9 and 15 sites were associated with 4 plant architecture (PH, LA, TMAL, and RDW [root dry weight]) and 5 ear traits (ED, EGW, EW, HKW and KL), respectively (Figure 2b; Table S2). A total of 3, 4, 6, and 2 variants were distributed in the upstream, exon, intron, and 3′UTR regions, respectively. The SNP at site 1708 in exon 3, which was associated with ED, EGW, and KL, caused synonymous changes. SNPs at sites 438, 453, and 555 in exon 1 caused non-synonymous changes in the amino acid sequence. Three high LD SNPs 438, 453, and 555 were associated with PH (Table S1). All significant variants could explain 2.98–6.91% of the phenotypic variation. Most of the associations were small effect variants and could explain less than 4% of the phenotypic variation. SNP1473, associated with KL ($p = 9.34 \times 10^{-7}$), explained the most phenotypic variation, up to 6.93%. In addition, 4 pleiotropic sites, including SNP1473, SNP1708, SNP7213, and InDel3387, were significantly associated with ED, EGW, KL, EGW, LA, and RDW (Figure 3). SNP1473 in intron 2 was associated with four ear traits (ED, EGW, KL, and EW), SNP1708 was associated with three ear traits (ED, EGW, and KL), SNP7213 in the 3′UTR was associated with ED, LA, and RDW, and the InDel at site 3387 in intron 4 was associated with ED, EGW, and KL (Table 2).

Linkage disequilibrium (LD) analysis found that SNP438, SNP453, SNP555, SNP628 and SNP706 showed strong LD ($r^2 > 0.95$) with each other in inbred lines. After the clumping of variants, 11 significant sites were identified. Six major haplotypes which contained more than 10 lines emerged from the 11 significant sites across inbred lines, and a significant phenotypic difference was observed between haplotypes in 8 traits, except for TMAL (Table S2). Four significant variants were significantly associated with KL, including InDel-970, SNP1473, SNP1708, and InDel3387. Three major haplotypes, which contained more than 20 lines, emerged from the 4 significant sites across 349 inbred lines (Figure 4c). The phenotypic differences in KL between the three major haplotypes were compared, and a significant difference was detected by ANOVA ($p = 6 \times 10^{-10}$) between haplotypes. Hap1, carrying all increased alleles, had the longest kernel length, followed by Hap2, which included the majority of tested inbred lines. Hap3, carrying all decreased alleles, had the shortest kernel length. SNP1473 was the most significant sites, the allele T group had a significantly longer KL than the allele C group ($p = 6.9 \times 10^{-8}$, Figure 4d). Further, we analyzed the allele frequencies among the three populations. The results showed that the frequency of the SNP1473T in teosintes was 40.6%, and in landraces and inbred lines, the frequency increased to 60.3% and 89.1%, respectively

(Figure 4e). These results suggested that SNP1473 might have been selected during domestication and improvement of maize. Three variants at sites 1473, 1708, and 3387 were significantly associated with EGW, which could divide the tested inbred lines into 2 major haplotypes (Figure S1). A significant difference between haplotypes was observed for EGW ($p = 1.3 \times 10^{-4}$). The SNP at site 1473 also had the most significant association with EGW. Three variants were identified for HKW that divided the inbred lines into four groups (Figure S2). The HKW of Hap1 was higher than the other three haplotypes ($p = 8.3 \times 10^{-9}$). The most significant site was SNP-769, and the frequency of the increased allele, SNP-769T, increased from 8.3% in teosintes to 33.3% in inbred lines. Five SNPs with high LD ($r^2 > 0.95$) were significantly associated with PH. The plant height in the inbred lines carrying allele SNP453G was higher than those containing the C allele (Figure S3). The frequency of the G allele decreased from 50.0% in teosintes to 16.1% in inbred lines. Two SNPs significantly associated with RDW divided the tested inbred lines into 3 major haplotypes (Figure S4). The frequency of increased allele SNP7137C increased from 0 in teosintes to 74.1% in inbred lines.

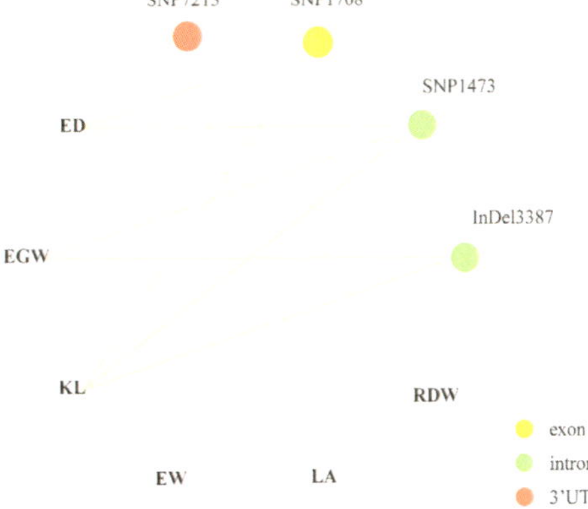

Figure 3. The network between pleiotropic site and associated traits. Yellow, green, and orange circle indicated the variations were in exon, intron, and 3'UTR, and the lines represent *p* value.

Table 2. Significant markers associated with phenotypic traits.

Trait	Marker	Allele	*p* Value	−lg (*P*)	R^2	Region	Position [a]
ED	SNP1473	T/C	1.20×10^{-4}	3.92	4.74%	intron2	1473
ED	SNP1708	G/C	8.75×10^{-5}	4.06	4.93%	exon3	1708
ED	InDel3387	-/G	9.34×10^{-4}	3.03	3.49%	intron4	3387
ED	SNP7213	T/A	9.52×10^{-4}	3.02	3.47%	3'UTR	7213
EGW	SNP1473	T/C	1.04×10^{-4}	3.98	4.06%	intron2	1473
EGW	SNP1708	G/C	8.53×10^{-4}	3.07	2.98%	exon3	1708
EGW	InDel3387	-/G	4.47×10^{-4}	3.35	3.31%	intron4	3387
EW	SNP1473	T/C	8.64×10^{-4}	3.06	3.28%	intron2	1473
HKW	SNP-769	C/T	3.01×10^{-4}	3.52	4.12%	upstream	−769
HKW	SNP-836	C/A	8.11×10^{-4}	3.09	3.19%	upstream	−836
HKW	InDel3129	T/-	4.17×10^{-4}	3.38	3.42%	intron4	3129
KL	InDel-970	GACAG/——	2.58×10^{-4}	3.59	3.78%	upstream	−970
KL	SNP1473	T/C	9.34×10^{-7}	6.03	6.91%	intron2	1473
KL	SNP1708	G/C	4.42×10^{-6}	5.36	6.03%	exon3	1708
KL	InDel3387	-/G	4.06×10^{-5}	4.39	4.79%	intron4	3387
LA	SNP7213	T/A	5.44×10^{-5}	4.26	3.94%	3'UTR	7213
PH	SNP453	C/G	3.38×10^{-4}	3.47	3.21%	exon1	453
RDW	SNP7137	C/G	8.07×10^{-4}	3.09	4.21%	3'UTR	7137
RDW	SNP7213	T/A	3.30×10^{-4}	3.48	4.85%	3'UTR	7213
TMAL	SNP2414	G/A	8.94×10^{-4}	3.05	4.36%	intron3	2414

[a] The position of the start codon (ATG) is labelled as "0".

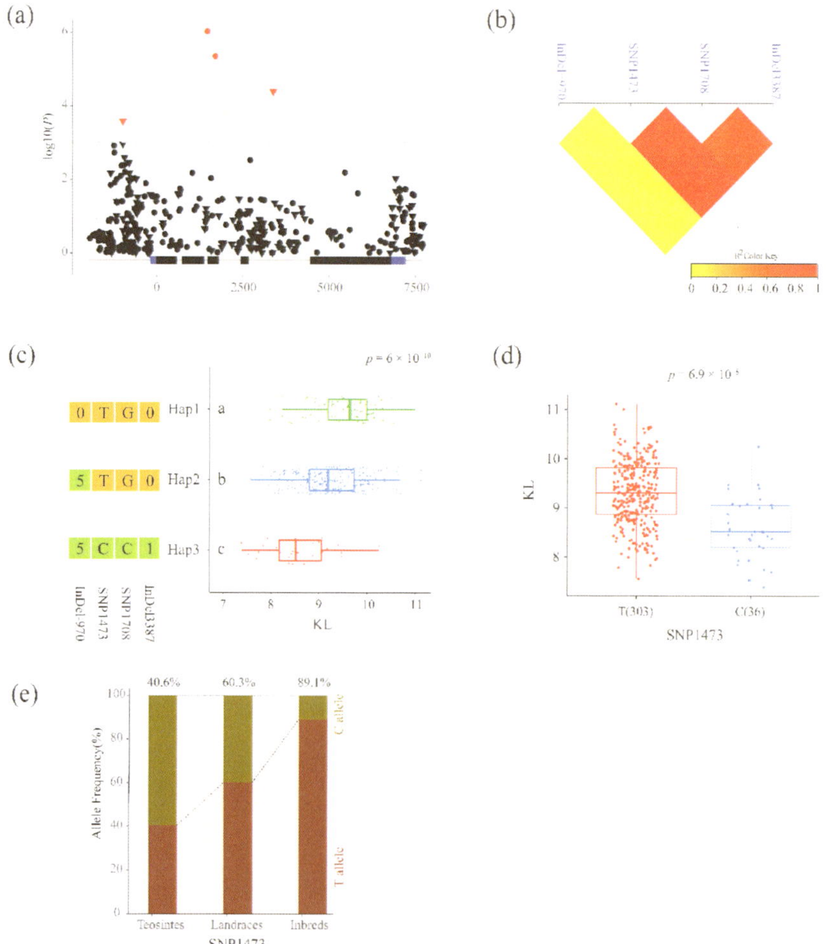

Figure 4. Natural variations in *ZmPGP1* were significantly associated with KL. (**a**) *ZmPGP1*-based association mapping for kernel length (KL). (**b**) Linkage disequilibrium (LD) heatmap for six significant variants associated with KL. (**c**) Haplotypes of *ZmPGP1* among natural variations in inbred lines. (**d**) Comparison of kernel length between different alleles of SNP1473. (**e**) The allele frequency of SNP1473 in teosinte, landraces, and inbred lines.

4. Discussion

The process of maize domestication and improvement has been studied with population genetics–genomics [3], QTL mapping [38], and gene expression assays [39]. During domestication and improvement, the plant morphology and productivity of maize have changed dramatically. Maize plants typically have one or two short branches and only two ears, each with several hundred kernels [38]. These changes involved artificial selection of specific genes controlling key morphological and agronomic traits [40], resulting in reduced genetic diversity. Previous studies have identified several genes underlying maize evolution: 484 domestication and 695 improvement regions were identified from population genetics analyses [3]. It is estimated that approximately 2–4% of genes have been selected during maize domestication and improvement [40]. Here, we examined DNA sequence variation in *ZmPGP1*, which is involved in the polar movement of indole-3-acetic acid (IAA).

Plant hormones, such as auxins, play a key role in plant growth, development, defenses, and stress tolerance [41]. A previous study reported that an auxin response factor might contribute to the morphological difference between maize and teosinte [40]. We found that the level of nucleotide diversity ($\pi \times 1000$) in teosintes is 20.724, decreased to 9.970 in landraces and 6.558 in maize inbred lines (Figure 1a), suggesting that approximately half of the genetic diversity has been lost during domestication process. Similar results were observed in several plants, such as soybeans and cucumbers [42,43]. Many previous studies employed only a limited number of teosinte, landraces, and maize to identify the domestication signals. For example; a total of 14 inbred lines, 16 landraces, and 16 teosinte accessions were chosen to artificial selection of 1095 genes. 28 inbred lines, 16 landraces, and 16 teosinte accessions were used to investigate the involvement of 32 MADS-box genes during maize domestication and improvement [44,45]. In this study, a larger population including 349 inbred lines, 68 landraces and 32 wild relatives were used to re-sequence *ZmPGP1* with high sequencing depth (more than 100×), which could help us to identify the selection signals with larger effective and high accurate.

Plant architecture and kernel and ear traits, the key factors affecting grain yield, were the main traits targeted of maize breeding. The identification of the natural variations in these traits could help to improve the efficiency of breeding selection. Although hundreds of QTLs related to these traits have been identified [10,11], few genes have been cloned from the natural germplasm. *ZmPGP1* (ABCB1 or br2), involved in auxin polar transport, has been shown to be associated with plant height, stalk diameter, leaf length and leaf angle [28]. Three Mu insertions were detected in the exon and intron of *ZmPGP1* [16]. These mutations dramatically affected height reduction but were rare variations in natural accessions. Natural germplasm with a broad genetic base could be a potential resource for improving yield [46]. Natural variations of *ZmPGP1* have also been identified [16,27–31], and some alleles have great potential in maize improvement. One rare SNP variant in the exon could reduce plant height without affecting yield [47]. A new 241-bp deletion in the last exon of PGP1 also had no negative effect on yield, but significantly reduced plant height and ear height and increased stalk diameter and erected leaves. The deletion was a rare allele that could be detected in only one line of 311 diverse maize accessions [28]. The result revealed that *ZmPGP1* has good potential to reshape plant architecture without the loss of yield in maize breeding. Candidate gene association analysis can identify the elite variation and the best haplotype for target traits. The elite variations of more than 30 genes involved in flowering time, kernel composition, drought tolerance, and root development were detected by candidate gene association analysis [48]. For example, *crtRB1* was proved to be associated with β carotene concentration and conversion in maize kernels, and the most favorable alleles were developed to inexpensive markers to use for crop provitamin A biofortification [49]. In this study, to identify the natural variations and favorable haplotypes of *ZmPGP1*, 1070 variations were detected from 9710 bp re-sequenced genomic region of *ZmPGP1*. In total, 11 variants were identified for 5 yield-related traits and 4 plant architecture (Figure 2; Table 2). However, two previously rare variations [28–31,47] were not found in our study. SNP1473 was the most significant variant for KL and EGW. The frequency of the increased allele T was 40.6% in teosintes and was enriched to 60.3% and 89.1% during maize domestication and improvement (Figure 4; Figure S1). The selection patterns were similar with the 1.2-Kb presence-absence variant of *KRN4*, which is likely responsible for increased kernel row number in maize [15]. In conclusion, we re-sequenced the *ZmPGP1* gene in 349 inbred lines, 68 landraces, and 32 teosintes, sequence polymorphisms, nucleotide diversity and neutral tests revealed that *ZmPGP1* might be selected during domestication and improvement processes. A total of 11 variants significantly associated with 4 plant architecture and 5 ear traits were identified by marker–trait association analysis in inbred lines. The significant variants could be used to develop new markers to improve plant architecture and ear traits in maize.

Supplementary Materials: The following are available online at http://www.mdpi.com/2073-4425/10/9/664/s1, Table S1: All significant markers associated with phenotypic traits. Table S2: Phenotypic differences among different haplotypes. Figure S1: Natural variations in *ZmPGP1* were significantly associated with EGW. Figure S2:

Natural variations in *ZmPGP1* were significantly associated with HKW. Figure S3: Natural variations in *ZmPGP1* were significantly associated with PH. Figure S4: Natural variations in *ZmPGP1* were significantly associated with RDW. Dataset 1: The sequence of ZmHKT1.fa. Dataset 2: The variaton of ZmPGP1.fa.

Author Contributions: P.L., J.W., Z.Y. and C.X. conceived and designed research. J.W., Y.F., S.Y., and J.L. conducted experiments. P.L., J.W. and Y.X. analyzed data. P.L., H.W., Z.Y. and C.X. wrote the manuscript. All authors read and approved the manuscript.

Funding: This research was funded by This work was supported by grants from the National Key Technology Research and Development Program of MOST (2016YFD0100303), the Natural Science Foundations of Jiangsu Province (BK20180920), the National Natural Science Foundations (31601810, 31801028), the Innovative Research Team of Universities in Jiangsu Province, a project funded by the Priority Academic Program Development of Jiangsu Higher Education Institutions (PAPD) and Research and innovation program for graduate students of Yangzhou university (XKYCX18_082).

Conflicts of Interest: The authors declare no conflict of interest.

References

1. Matsuoka, Y.; Vigouroux, Y.; Goodman, M.M.; Sanchez, G.J.; Buckler, E.; Doebley, J. A single domestication for maize shown by multilocus microsatellite genotyping. *Proc. Natl. Acad. Sci. USA* **2002**, *99*, 6080–6084. [CrossRef]

2. Fan, L.J.; Bao, J.D.; Wang, Y.; Yao, J.Q.; Gui, Y.J.; Hu, W.M.; Zhu, J.Q.; Zeng, M.Q.; Li, Y.; Xu, Y.B. Post-Domestication Selection in the Maize Starch Pathway. *PLoS ONE* **2009**, *4*, e7612. [CrossRef] [PubMed]

3. Hufford, M.B.; Xu, X.; van Heerwaarden, J.; Pyhajarvi, T.; Chia, J.M.; Cartwright, R.A.; Elshire, R.J.; Glaubitz, J.C.; Guill, K.E.; Kaeppler, S.M.; et al. Comparative population genomics of maize domestication and improvement. *Nat. Genet.* **2012**, *44*, 808. [CrossRef]

4. Haarhoff, S.J.; Swanepoel, P.A. Plant Population and Maize Grain Yield: A Global Systematic Review of Rainfed Trials. *Crop. Sci.* **2018**, *58*, 1819–1829. [CrossRef]

5. Duvick, D.N. Genetic Progress in Yield of United States Maize (*Zea mays* L.). *Maydica* **2005**, *50*, 193–202.

6. Messmer, R.; Fracheboud, Y.; Banziger, M.; Vargas, M.; Stamp, P.; Ribaut, J.M. Drought stress and tropical maize: QTL-by-environment interactions and stability of QTLs across environments for yield components and secondary traits. *Theor. Appl. Genet.* **2009**, *119*, 913–930. [CrossRef]

7. Zhang, C.S.; Zhou, Z.Q.; Yong, H.J.; Zhang, X.C.; Hao, Z.F.; Zhang, F.J.; Li, M.S.; Zhang, D.G.; Li, X.H.; Wang, Z.H.; et al. Analysis of the genetic architecture of maize ear and grain morphological traits by combined linkage and association mapping. *Theor. Appl. Genet.* **2017**, *130*, 1011–1029. [PubMed]

8. Duvick, D.N. The contribution of breeding to yield advances in maize (*Zea mays* L.). *Adv. Agron.* **2005**, *86*, 83–145.

9. Xiao, Y.J.; Tong, H.; Yang, X.H.; Xu, S.Z.; Pan, Q.C.; Qiao, F.; Raihan, M.S.; Luo, Y.; Liu, H.J.; Zhang, X.H.; et al. Genome-wide dissection of the maize ear genetic architecture using multiple populations. *New Phytol.* **2016**, *210*, 1095–1106. [CrossRef]

10. Pan, Q.C.; Xu, Y.C.; Li, K.; Peng, Y.; Zhan, W.; Li, W.Q.; Li, L.; Yan, J.B. The Genetic Basis of Plant Architecture in 10 Maize Recombinant Inbred Line Populations. *Plant Physiol.* **2017**, *175*, 858–873. [CrossRef]

11. Martinez, A.K.; Soriano, J.M.; Tuberosa, R.; Koumproglou, R.; Jahrmann, T.; Salvi, S. Yield QTLome distribution correlates with gene density in maize. *Plant Sci.* **2016**, *242*, 300–309. [CrossRef] [PubMed]

12. Bommert, P.; Je, B.I.; Goldshmidt, A.; Jackson, D. The maize G α gene COMPACT PLANT2 functions in CLAVATA signalling to control shoot meristem size. *Nature* **2013**, *502*, 555. [CrossRef] [PubMed]

13. Je, B.I.; Gruel, J.; Lee, Y.K.; Bommert, P.; Arevalo, E.D.; Eveland, A.L.; Wu, Q.Y.; Goldshmidt, A.; Meeley, R.; Bartlett, M.; et al. Signaling from maize organ primordia via FASCIATED EAR3 regulates stem cell proliferation and yield traits. *Nat. Genet.* **2016**, *48*, 785. [CrossRef]

14. McSteen, P. Branching out: The ramosa pathway and the evolution of grass inflorescence morphology. *Plant Cell* **2006**, *18*, 518–522. [CrossRef] [PubMed]

15. Liu, L.; Du, Y.F.; Shen, X.M.; Li, M.F.; Sun, W.; Huang, J.; Liu, Z.J.; Tao, Y.S.; Zheng, Y.L.; Yan, J.B.; et al. KRN4 Controls Quantitative Variation in Maize Kernel Row Number. *PLoS Genet.* **2015**, *11*, e1005670. [CrossRef]

16. Multani, D.S.; Briggs, S.P.; Chamberlin, M.A.; Blakeslee, J.J.; Murphy, A.S.; Johal, G.S. Loss of an MDR transporter in compact stalks of maize br2 and sorghum dw3 mutants. *Science* **2003**, *302*, 81–84. [CrossRef] [PubMed]

17. Lawit, S.J.; Wych, H.M.; Xu, D.P.; Kundu, S.; Tomes, D.T. Maize DELLA Proteins dwarf plant8 and dwarf plant9 as Modulators of Plant Development. *Plant Cell Physiol.* **2010**, *51*, 1854–1868. [CrossRef]

18. Bommert, P.; Lunde, C.; Nardmann, J.; Vollbrecht, E.; Running, M.; Jackson, D.; Hake, S.; Werr, W. thick tassel dwarf1 encodes a putative maize ortholog of the Arabidopsis CLAVATA1 leucine-rich repeat receptor-like kinase. *Development* **2005**, *132*, 1235–1245. [CrossRef] [PubMed]

19. McSteen, P.; Hake, S. Barren inflorescence2 regulates axillary meristem development in the maize inflorescence. *Development* **2001**, *128*, 2881–2891.

20. Gallavotti, A.; Zhao, Q.; Kyozuka, J.; Meeley, R.B.; Ritter, M.; Doebley, J.F.; Pe, M.E.; Schmidt, R.J. The role of barren stalk1 in the architecture of maize. *Nature* **2004**, *432*, 630–635. [CrossRef]

21. Chuck, G.; Whipple, C.; Jackson, D.; Hake, S. The maize SBP-box transcription factor encoded by tasselsheath4 regulates bract development and the establishment of meristem boundaries. *Development* **2010**, *137*, 1243–1250. [CrossRef]

22. Doebley, J.; Stec, A.; Hubbard, L. The evolution of apical dominance in maize. *Nature* **1997**, *386*, 485. [CrossRef]

23. Wang, H.; Nussbaum-Wagler, T.; Li, B.L.; Zhao, Q.; Vigouroux, Y.; Faller, M.; Bomblies, K.; Lukens, L.; Doebley, J.F. The origin of the naked grains of maize. *Nature* **2005**, *436*, 714–719. [CrossRef] [PubMed]

24. Wills, D.M.; Whipple, C.J.; Takuno, S.; Kursel, L.E.; Shannon, L.M.; Ross-Ibarra, J.; Doebley, J.F. From Many, One: Genetic Control of Prolificacy during Maize Domestication. *PLoS Genet.* **2013**, *9*, e1003604. [CrossRef] [PubMed]

25. Chang, G.; Roth, C.B. Structure of MsbA from E. coli: A homolog of the multidrug resistance ATP binding cassette (ABC) transporters. *Science* **2001**, *293*, 1793–1800. [CrossRef] [PubMed]

26. Aller, S.G.; Yu, J.; Ward, A.; Weng, Y.; Chittaboina, S.; Zhuo, R.P.; Harrell, P.M.; Trinh, Y.T.; Zhang, Q.H.; Urbatsch, I.L.; et al. Structure of P-glycoprotein reveals a molecular basis for poly-specific drug binding. *Science* **2009**, *323*, 1718–1722. [CrossRef]

27. Knoller, A.S.; Blakeslee, J.J.; Richards, E.L.; Peer, W.A.; Murphy, A.S. Brachytic2/ZmABCB1 functions in IAA export from intercalary meristems. *J. Exp. Bot.* **2010**, *61*, 3689–3696. [CrossRef] [PubMed]

28. Wei, L.; Zhang, X.; Zhang, Z.H.; Liu, H.H.; Lin, Z.W. A new allele of the Brachytic2 gene in maize can efficiently modify plant architecture. *Heredity* **2018**, *121*, 75–86. [CrossRef] [PubMed]

29. Balzan, S.; Carraro, N.; Salleres, B.; Dal Cortivo, C.; Tuinstra, M.R.; Johal, G.; Varotto, S. Genetic and phenotypic characterization of a novel brachytic2 allele of maize. *Plant Growth Regul.* **2018**, *86*, 81–92. [CrossRef]

30. Pilu, R.; Cassani, E.; Villa, D.; Curiale, S.; Panzeri, D.; Cerino Badone, F.; Landoni, M. Isolation and characterization of a new mutant allele of brachytic 2 maize gene. *Mol. Breeding* **2007**, *20*, 83–91. [CrossRef]

31. Zhang, M.L.; Lu, X.D.; Li, C.L.; Zhang, B.; Zhang, C.Y.; Zhang, X.S.; Ding, Z.J. Auxin efflux carrier ZmPGP1 mediates root growth inhibition under aluminum stress. *Plant Physiol.* **2018**, *177*, 819–832. [CrossRef] [PubMed]

32. Li, P.C.; Pan, T.; Wang, H.M.; Wei, J.; Chen, M.J.; Hu, X.H.; Zhao, Y.; Yang, X.Y.; Yin, S.Y.; Xu, Y.; et al. Natural variation of ZmHKT1 affects root morphology in maize at the seedling stage. *Planta* **2019**, *249*, 879–889. [CrossRef] [PubMed]

33. Choi, M.; Scholl, U.I.; Ji, W.Z.; Liu, T.W.; Tikhonova, I.R.; Zumbo, P.; Nayir, A.; Bakkaloglu, A.; Ozen, S.; Sanjad, S.; et al. Genetic diagnosis by whole exome capture and massively parallel DNA sequencing. *Proc. Natl. Acad. Sci. USA* **2009**, *106*, 19096–19101. [CrossRef]

34. Katoh, K.; Standley, D.M. MAFFT Multiple Sequence Alignment Software Version 7: Improvements in Performance and Usability. *Mol. Biol. Evol.* **2013**, *30*, 772–780. [CrossRef] [PubMed]

35. Librado, P.; Rozas, J. DnaSP v5: A software for comprehensive analysis of DNA polymorphism data. *Bioinformatics* **2009**, *25*, 1451–1452. [CrossRef] [PubMed]

36. Tajima, F. Statistical method for testing the neutral mutation hypothesis by DNA polymorphism. *Genetics* **1989**, *123*, 585–595. [PubMed]

37. Fu, Y.X.; Li, W.H. Statistical tests of neutrality of mutations. *Genetics* **1993**, *133*, 693–709.

38. Yang, C.J.; Samayoa, L.F.; Bradbury, P.J.; Olukolu, B.A.; Xue, W.; York, A.M.; Tuholski, M.R.; Wang, W.D.; Daskalska, L.L.; Neumeyer, M.A.; et al. The genetic architecture of teosinte catalyzed and constrained maize domestication. *Proc. Natl. Acad. Sci. USA* **2019**, *116*, 5643–5652. [CrossRef]

39. Swanson-Wagner, R.; Briskine, R.; Schaefer, R.; Hufford, M.B.; Ross-Ibarra, J.; Myers, C.L.; Tiffin, P.; Springer, N.M. Reshaping of the maize transcriptome by domestication. *Proc. Natl. Acad. Sci. USA* **2012**, *109*, 11878–11883. [CrossRef]

40. Yamasaki, M.; Wright, S.I.; Mcmullen, M.D. Genomic screening for artificial selection during domestication and improvement in maize. *Ann. Bot.* **2007**, *100*, 967–973. [CrossRef]

41. Wolters, H.; Jurgens, G. Survival of the flexible: Hormonal growth control and adaptation in plant development. *Nat. Rev. Genet.* **2009**, *10*, 305–317. [PubMed]

42. Qi, J.J.; Liu, X.; Shen, D.; Miao, H.; Xie, H.Y.; Li, X.X.; Zeng, P.; Wang, S.H.; Shang, Y.; Gu, X.F.; et al. A genomic variation map provides insights into the genetic basis of cucumber domestication and diversity. *Nat. Genet.* **2013**, *45*, 1510.

43. Zhou, Z.K.; Jiang, Y.; Wang, Z.; Gou, Z.H.; Lyu, J.; Li, W.Y.; Yu, Y.J.; Shu, L.P.; Zhao, Y.J.; Ma, Y.M.; et al. Resequencing 302 wild and cultivated accessions identifies genes related to domestication and improvement in soybean. *Nat. Biotechnol.* **2015**, *33*, 408. [CrossRef] [PubMed]

44. Zhao, Q.; Weber, A.L.; McMullen, M.D.; Guill, K.; Doebley, J. MADS-box genes of maize: Frequent targets of selection during domestication. *Genet. Res.* **2011**, *93*, 65–75. [CrossRef] [PubMed]

45. Yamasaki, M.; Tenaillon, M.I.; Bi, I.V.; Schroeder, S.G.; Villeda, H.; Doebley, J.; McMullen, M.D. A large-scale screen for artificial selection in maize identifies candidate agronomic loci for domestication and crop improvement. *Plant Cell* **2005**, *17*, 2859–2872. [CrossRef]

46. Wang, H.W.; Li, K.; Hu, X.J.; Liu, Z.F.; Wu, Y.J.; Huang, C.L. Genome-wide association analysis of forage quality in maize mature stalk. *BMC Plant Biol.* **2016**, *16*, 227.

47. Xing, A.Q.; Gao, Y.F.; Ye, L.F.; Zhang, W.P.; Cai, L.C.; Ching, A.; Llaca, V.; Johnson, B.; Liu, L.; Yang, X.H.; et al. A rare SNP mutation in Brachytic2 moderately reduces plant height and increases yield potential in maize. *J. Exp. Bot.* **2015**, *66*, 3791–3802. [CrossRef]

48. Yan, J.B.; Warburton, M.; Crouch, J. Association Mapping for Enhancing Maize (*Zea mays* L.) Genetic Improvement. *Crop Sci.* **2011**, *51*, 433–449. [CrossRef]

49. Yan, J.B.; Kandianis, C.B.; Harjes, C.E.; Bai, L.; Kim, E.H.; Yang, X.H.; Skinner, D.J.; Fu, Z.Y.; Mitchell, S.; Li, Q.; et al. Rare genetic variation at Zea mays crtRB1 increases β-carotene in maize grain. *Nat. Genet.* **2010**, *42*, 322. [CrossRef] [PubMed]

Article

Transferability and Polymorphism of SSR Markers Located in Flavonoid Pathway Genes in *Fragaria* and *Rubus* Species

Vadim G. Lebedev [1,2], Natalya M. Subbotina [1,2], Oleg P. Maluchenko [3], Tatyana N. Lebedeva [4], Konstantin V. Krutovsky [5,6,7,8,9,*] and Konstantin A. Shestibratov [2]

[1] Pushchino State Institute of Natural Sciences, Prospekt Nauki 3, 142290 Pushchino, Russia; vglebedev@mail.ru (V.G.L.); natysubbotina@rambler.ru (N.M.S.)

[2] Branch of the Shemyakin-Ovchinnikov Institute of Bioorganic Chemistry, Russian Academy of Sciences, Prospekt Nauki 6, 142290 Pushchino, Russia; schestibratov.k@yandex.ru

[3] All-Russian Research Institute of Agricultural Biotechnology, Timiriazevskaya Str. 42, 127550 Moscow, Russia; oleg.maluchenko@mail.ru

[4] Institute of Physicochemical and Biological Problems of Soil Science, Russian Academy of Sciences, Institutskaya Str. 2, 142290 Pushchino, Russia; tanyaniko@mail.ru

[5] Department of Forest Genetics and Forest Tree Breeding, Georg-August University of Göttingen, Büsgenweg 2, 37077 Göttingen, Germany

[6] Center for Integrated Breeding Research, Georg-August University of Göttingen, Albrecht-Thaer-Weg 3, 37075 Göttingen, Germany

[7] Laboratory of Population Genetics, N. I. Vavilov Institute of General Genetics, Russian Academy of Sciences, Gubkin Str. 3, 119333 Moscow, Russia

[8] Laboratory of Forest Genomics, Genome Research and Education Center, Institute of Fundamental Biology and Biotechnology, Siberian Federal University, 660036 Krasnoyarsk, Russia

[9] Department of Ecosystem Science and Management, Texas A&M University, 2138 TAMU, College Station, TX 77843-2138, USA

* Correspondence: kkrutov@gwdg.de; Tel.: +49-551-393-35-37

Received: 30 November 2019; Accepted: 19 December 2019; Published: 21 December 2019

Abstract: Strawberry (*Fragaria*) and raspberry (*Rubus*) are very popular crops, and improving their nutritional quality and disease resistance are important tasks in their breeding programs that are becoming increasingly based on use of functional DNA markers. We identified 118 microsatellite (simple sequence repeat—SSR) loci in the nucleotide sequences of flavonoid biosynthesis and pathogenesis-related genes and developed 24 SSR markers representing some of these structural and regulatory genes. These markers were used to assess the genetic diversity of 48 *Fragaria* and *Rubus* specimens, including wild species and rare cultivars, which differ in berry color, ploidy, and origin. We have demonstrated that a high proportion of the developed markers are transferable within and between *Fragaria* and *Rubus* genera and are polymorphic. Transferability and polymorphism of the SSR markers depended on location of their polymerase chain reaction (PCR) primer annealing sites and microsatellite loci in genes, respectively. High polymorphism of the SSR markers in regulatory flavonoid biosynthesis genes suggests their allelic variability that can be potentially associated with differences in flavonoid accumulation and composition. This set of SSR markers may be a useful molecular tool in strawberry and raspberry breeding programs for improvement anthocyanin related traits.

Keywords: *Fragaria*; *Rubus*; microsatellites; transferability; polymorphism; introns; exons; flavonoid biosynthesis pathway; transcription factor genes; chitinase

1. Introduction

The Rosaceae family comprises approximately 3000 species and includes very important fruit, berry, and ornamental plants. This family has been relatively recently reorganized into three subfamilies: Dryadoideae, Spiraeoideae, and Rosoideae. The latter one includes cultivated berries in the genera *Fragaria* (strawberry) and *Rubus* (raspberry and blackberry) [1]. Strawberry and raspberry are in especially high demand among consumers due to their appearance, taste, and aroma [2,3]. They are also rich in antioxidants and other bioactive compounds beneficial for human health. The strawberry is the most consumed berry worldwide—more than 9 million tons were harvested in 2017, while the production of raspberry and blackberry increased by 50% for the period 2010–2017 and exceeded 800,000 tons [4]. It is suggested that the consistent demand for healthy and delicious berryfruit observed in the current decade will increase in the nearest future [2].

Recently, the high interest in berry crops has led not only to increased production levels, but also to the expansion of *Rubus* and *Fragaria* breeding programs; several dozens of them are now known [3,5]. For a long period of time, the main directions in breeding were crop yield and disease resistance, but in the past years, fruit sensorial and nutritional qualities have become major objectives [6]. However, *Rubus* and *Fragaria* breeding is complicated because of several genetic problems including polyploidy and the highly heterozygous nature of the germplasm [3,5]. The genus *Rubus* consisted of several hundred species, and the ploidy level can widely vary among them. Raspberries are mainly diploids ($2n = 14$), while blackberries may vary from diploids to dodecaploids ($2n = 84$), whereas the hybrids between them can be hexaploid (loganberry) or septaploid (boysenberry) [7]. A total of 22 wild species of *Fragaria* have been described. Almost half of them are diploids ($2n = 14$), while tetra-, hexa-, octo-, and decaploid species are also known [8]. The main cultivated strawberry crop, *F.* × *ananassa*, is a hybrid between *F. chiloensis* and *F. virginiana*, but the origin of its octoploid genome from four diploid ancestors has long been unknown. In addition to *F. vesca* and *F. iinumae*, contribution of different species was assumed, but only in 2019 the phylogenetic analyses of Edger et al. [9] provided a strong genome-wide support that *F. iinumae*, *F. nipponica*, *F. viridis*, and *F. vesca* are diploid progenitor species. *F. vesca* (the wild strawberry) and *F. viridis* (the green strawberry) can be used in strawberry breeding as donors of abiotic and biotic stress resistance and fruit aroma [10] and firm flesh, remontant flowering habit, and an acidic apple-like aroma [11], respectively.

Developing a new cultivar by traditional methods is a very long process that can take up to 15 years for raspberry [5] or 10 years for strawberry [12]. Molecular markers, however, can be used at any stage of plant growth and can increase the speed and accuracy of germplasm assessment. A good choice for breeding purposes is a simple sequence repeat (SSR) or microsatellite markers consisting of tandem repeats (1–6 nucleotides). Due to their codominant inheritance, high level of polymorphism, and abundance in genome, they play an important role in identifying genomic regions associated with the traits of economic importance [13]. SSR markers for the *Rubus* and *Fragaria* species were first developed in the early 2000s [14,15] and in subsequent years, a number of studies were carried out, including evaluation of marker transferability. SSR transferability depends on genetic distance between individual specimens. SSRs are more transferable, overall, within the species of the same genus or among related genera within families than between remote genera and different families [16]. The *Fragaria* and *Rubus* species from Rosoideae subfamily have both the same basic number of chromosomes ($x = 7$) and close phylogenetic relationships based on their chloroplast and nuclear DNA markers, as well as similar morphology. These facts suggest collinearity between *Fragaria* and *Rubus* genomes [17]. In raspberry breeding, interspecific hybridization is widely used, and development of molecular genetic markers that can be transferred between different species, especially with different ploidy, becomes an important task. However, very few molecular genetic markers are known for *Rubus*, and fewer are transferable between species [13]. All strawberry cultivars now available at the market have been produced using traditional breeding methods [3]. Among raspberry cultivars, there are currently only two productive cultivars with root rot resistance that have been produced using the marker-assisted selection (MAS), but they are still in the commercial trial stage [18]. The development of new molecular genetic markers that could be used for molecular genetic characterization of both wild species and germplasm would accelerate the breeding of new cultivars [2]. The transferability is very important for their wide use.

The SSR markers can be developed using either genomic or expressed sequence tag (EST) nucleotide sequences. The EST-based markers are more transferable and more useful in MAS as they are linked with expressed genes and could be associated with important agronomical traits [19]. The strawberry studies have shown that EST-SSR markers were more transferable compared to random genomic nongenic noncoding SSR (ncSSR) markers [20,21]. At the same time, it is known that ESTSSR markers are generally less polymorphic than ncSSRs that mostly represent noncoding genomic regions because of a greater DNA sequence conservation of transcribed regions [22]. There are no markers that would be universal or ideal for all practical applications and tasks. For some tasks such as analyses of population structure, genetic drift, migration, gene flow, mating system, and individual identification of clones, cultivars, etc. selectively neutral ncSSRs would be the most appropriate markers, but for functional analysis of adaptive variation, MAS, QTL mapping, etc., ESTSSRs that represent functional alleles and haplotypes and link to important adaptive and breeding traits would be more appropriate and useful markers. Among SSR markers, those that are located in introns and untranslated regions (UTRs) are more polymorphic and potentially can combine advantages of both ncSSR and EST-SSR markers, while those that are located in exons are less polymorphic, but more likely to be under direct selection or represent selectively different alleles, and therefore are more useful for MAS because they might better represent functional traits important for breeding. We studied diploid *Rubus* species and confirmed that SSR markers located in introns were more variable in comparison to EST-SSR located in exons [23]. However, development of such markers complete nucleotide sequences of important adaptive and breeding trait related genes.

Many MAS studies are aimed at developing markers representing key genes, but only few studies have focused on the developing markers representing structural and regulatory genes of metabolic pathways in cultivars with contrasting phenotypic traits of interest [24]. We are especially interested in developing SSR markers representing flavonoid pathway genes because the main polyphenols in fruits are flavonoids (Figure 1). Complete nucleotide sequences are already available for many of these genes. Thus, breeding of new berry cultivars with the improved nutritional value using MAS and SSR markers representing flavonoid biosynthesis-related genes seems to be a highly promising approach. The combinatorial interactions of the regulatory genes with structural genes that act to control the flux of various branches of the pathway ultimately determines the flavonoid composition [25]. However, as far as we know, SSR markers representing *Fragaria* and *Rubus* transcription factors (TFs) were not developed before our study. Main aims of our study were to: (1) develop SSR markers using coding (CDS) and non-coding (NCDS) sequences of structural and regulatory genes of flavonoid biosynthesis pathways, (2) evaluate them in *Rubus* and *Fragaria* species of different ploidy, (3) test cross-species transferability within and among *Rubus* and *Fragaria* genera, (4) assess the relationship of transferability and polymorphism of SSR markers with location of primer binding sites for polymerase chain reaction (PCR) and SSR loci in respective genes. Some of these genes could control anthocyanin content that correlates strongly with color. Therefore, strawberry and red raspberry cultivars with a wide range of berry colors were used in our study. In addition to nutritional value, disease resistance is also considered to be a valuable trait for the berry crops. The key factor for developing pathogen resistance in new cultivars is the identification and introgression of genes from cultivated varieties or their wild relatives [26]. Strawberry and raspberry production suffers from a number of agriculturally important diseases, and, therefore, we included in our study developing of SSR markers in strawberry genes encoding pathogenesis-related (PR) proteins – β-1,3-glucanases (PR-2 family) and chitinases (PR-3, 4, 8, 11 families). Since the sequences for raspberry chitinase genes were absent in the NCBI GenBank database, we sequenced the fragments of chitinase III genes in raspberry cultivars.

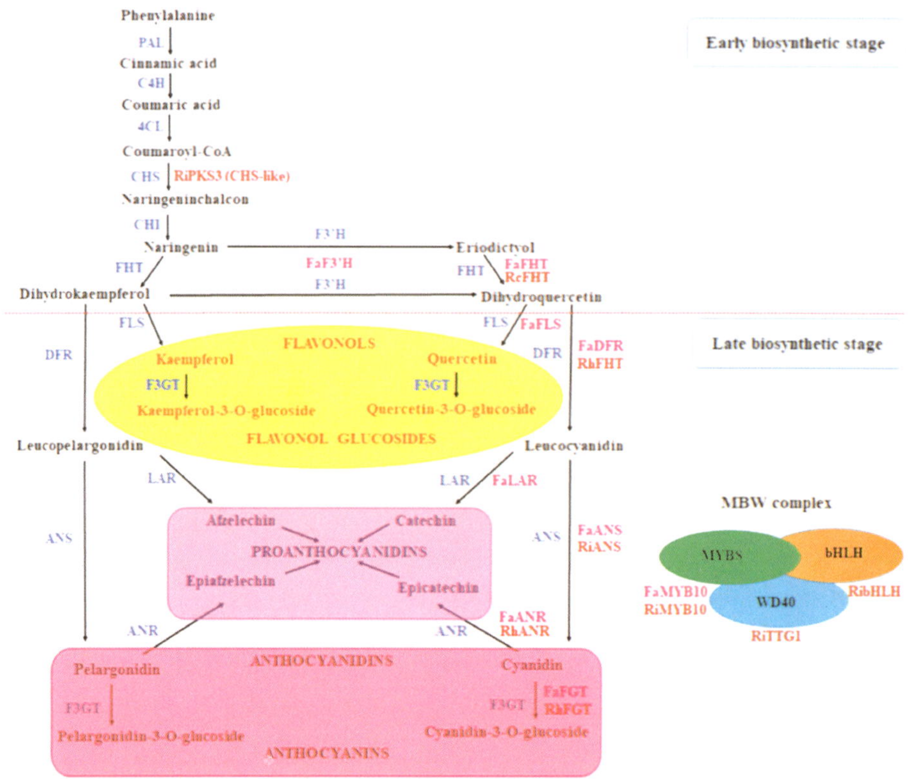

Figure 1. Schematic representation of flavonoid biosynthesis in strawberry and raspberry. PAL—phenylalanine lyase; C4H—cinnamate 4-hydroxylase; 4CL—4-coumarate CoA ligase; CHS—chalcone synthase; CHI—chalcone isomerase; F3'H—flavonoid 3'-hydroxylase; FHT—flavonone 3-hydroxylase; FLS—flavonol synthase; DFR—dihydroflavonol 4-reductase; LAR—leuco-anthocyanidin reductase; ANS—anthocyanidin synthase; ANR—anthocyanidin reductase; F3GT—flavonoid 3-O-glycosyl transferase. Pink color indicates strawberry genes and dark-red color indicates *Rubus* genes used in this study.

2. Materials and Methods

2.1. Plant Materials

Sixteen *Fragaria* specimens, including *F. × ananassa*, *F. vesca*, *F. viridis* and (*F. × ananassa*) × *Comarum palustre*, and 32 *Rubus* cultivars, including red raspberry (*R. idaeus*; Idaeobatus subgenus), black raspberry (*R. occidentalis*; Idaeobatus subgenus), blackberry (*Rubus* subgenus), cloudberry (*R. chamaemorus*; Chamaemorus subgenus), arctic bramble (*R. arcticus*; Cyclastis subgenus), and hybrid arctic bramble (*R. × stellarcticus*; Cyclastis subgenus) were chosen to genotype newly developed SSR loci located in the flavonoid biosynthesis and pathogenesis-related genes. These cultivars have a wide range of fruit color, ploidy and various geographic and genetic origins, but mostly of Russian origin (Table 1). The plants used in this study were kindly provided by I.A. Pozdniakov (OOO Microklon, Pushchino, Russia). Each cultivar represented a microclonally vegetatively propagated line containing genetically identical plants. Therefore, a single specimen per culture was used for further DNA isolation and genotyping.

Table 1. List of 48 *Fragaria* and *Rubus* specimens genotyped in the study.

Species	Specimen	Pedigree	Ploidy	Origin
F. × ananassa	White D	*F. virginiana × F. chiloensis*	8x	Sweden
	Vechnaya Vesna	Grenader-V2 × Rannyaya Plotnaya	8x	Russia
	Girlyanda	Elsanta × Korona	8x	Russia
	Zolushka	Festivalnaya × Senga Sengana	8x	Russia
	Lakomka	Krasavitsa × Korona	8x	Russia
	Lyubava	Solovushka × Geneva	8x	Russia
	Solovushka	Syurpriz Olimpiade × Festivalnaya Romashka	8x	Russia
	Tsaritsa	Venta × Red Gauntlet	8x	Russia
	Honeoye	Vibrant × Holiday	8x	USA
	Korona	Tamella × Induka	8x	Netherlands
	Red Gauntlet	New Jersey 105 × Climax	8x	UK
	Senga Sengana	Sieger × Markee	8x	Germany
	Black Prince	unknown	8x	unknown
F. × Comarum	Lipstick	Hybrid (*F. × ananassa*) × *Comarum palustre*	7x	Netherlands
F. vesca	-		2x	Russia
F. viridis	-		2x	Russia
R. chamaemorus	NyBy	wild strain	8x	Finland
R. arcticus	Elpee		2x	Finland
	Mespi		2x	Finland
R. × stellarcticus	Anna		2x	Sweden
	Beata		2x	Sweden
	Linda	Hybrid arctic bramble *R. arcticus* ssp. *stellatus* × *R. arcticus* ssp.	2x	Sweden
	Sofia	*arcticus*	2x	Sweden
	Astra		2x	Finland
	Aura		2x	Finland
R. idaeus	Babye Leto II	Autumn Bliss × Babye Leto	2x	Russia
	Oranzhevoe Chudo	Shapka Monomaha (open pollination)	2x	Russia
	Zheltyj Gigant	Marosejka × Ivanovskaya	2x	Russia
	Zolotaya Osen	13-39-11 (open pollination)	2x	Russia
	Patritsiya	Marosejka × M102	2x	Russia
	Gusar	Canby × pollen mix	2x	Russia
	Fenomen	Stolichnaya × Odarka	2x	Ukraine
	Joan J	Terri-Louise × Joan Squire	2x	UK
	Marosejka	7324/50 × 7331/3	2x	Russia
	Pingvin	interspecific hybrid	2x	Russia
	Fall Gold	NH 56-1 × (Taylor × *R. pungens* var. oldhamii) F2 (open pollination)	2x	USA
	Himbo Top	Autumn Bliss × Rafzeter	2x	Switzerland
	Polana	Heritage × Zeva Herbsterne	2x	Poland
	Zhar-Ptitsa	7-43-2 (open pollination)	2x	Russia
R. occidentalis	Cumberland	Gregg selfed	2x	USA
	Jewel	(Bristol × Dundee) × Dundee	2x	USA
Blackberry	Brzezina	90,402 × 89,403	4x	Poland
	Natchez	Ark. 2005 × Ark. 1857	4x	USA
	Ebony King	unknown	4x	USA
Hybrid	Boysenberry	complex hybrid	7x	USA
	Loganberry	*R. ursinus × R. idaeus*	6x	USA
	Tayberry	Aurora × SCRI 626/67	6x	UK
	Buckingham Tayberry	chimeral spineless sport of Tayberry	6x	UK

2.2. Simple Sequence Repeat (SSR) Marker and Polymerase Chain Reaction (PCR) Primer Development

The WebSat software [27] was used to detect SSR loci in the nucleotide sequences of *F. × ananassa* and *Rubus* genes available in the NCBI GenBank database (http://www.ncbi.nlm.nih.gov) (Table S1). To search for SSRs, the following threshold criteria were used: ten for mononucleotide repeats, five for dinucleotide motifs, four for tri-, three for tetra-, and two for penta-, and hexanucleotide repeats. The Primer 3 software (http://primer3.org) was used to design appropriate (PCR) primers based on the sequences flanking the SSR loci.

Primers were designed using the following criteria: primer length of 18–27 bp (optimally 22 bp), GC content of 40–80%, annealing temperature of 57–68 °C (optimally 60 °C), and expected amplified product size of 100–400 bp. Primers for the *RiG001* locus were as in [28]. Primers were synthesized by Syntol Comp. (Moscow, Russia) and are presented in Table S1.

2.3. DNA Isolation, PCR Amplification and Fragment Analysis

A single DNA sample per each specimen was produced from young expanding leaves representing a single plant per each sample. Total genomic DNA was extracted using the STAB method [29]. The quality and quantity of extracted DNA were determined by the NanoDrop 2000 spectrophotometer (Thermo Fisher Scientific Inc., Waltham, MA, USA). The final concentration of each DNA sample was adjusted to 50 ng/μL in TE buffer before the PCR amplification.

For genotyping, PCR was performed separately for each primer pair using a forward primer labeled with the fluorescent dye 6-FAM and an unlabeled reverse primer (Syntol Comp., Moscow, Russia). The PCR amplification was performed in a total volume of 20 μL consisted of 50 ng of genomic DNA, 10 pmol of the labeled forward primer, 10 pmol of an unlabeled reverse primer, and PCR Mixture Screenmix (Evrogen JSC, Moscow, Russia). After an initial denaturation at 95 °C for 3 min, DNA was amplified during 33 cycles in a gradient thermal cycler (Bio-Rad Laboratories, Inc., Hercules, CA, USA) programmed for a 30 s denaturation step at 95 °C, a 20 s annealing step at the optimal annealing temperature of the primer pair, and a 35 s extension step at 72 °C. A final extension step was done at 72 °C for 5 min.

The PCR generating clear, stable, and specific DNA fragments within an expected length (200–400 bp) were considered as successful PCR amplifications. If a primer pair failed three times to amplify template DNA that was amplified with other primers, then it was scored as a null genotype.

Separation of amplified DNA fragments was performed in an ABI 3130xl Genetic Analyzer using S450 LIZ size standard (Syntol Comp.). Peak identification and fragment sizing were done using the Gene Mapper v4.0 software (Applied Biosystems, Foster, CA, USA).

2.4. Genetic Data Analysis

Genetic statistics were calculated for each polymorphic microsatellite marker. The number of alleles, observed (H_o) and expected (H_e) heterozygosities, and polymorphic information content (PIC) for 32 diploid *Rubus* cultivars were calculated using the PowerMarker 3.25 software [30]. Analogous parameters for 13 octoploid *Fragaria* cultivars were calculated using the GenoDive 3.0 software [31]. Principal component analysis (PCA) and construction of the box plots were performed with the PCORD 5 software [32].

2.5. Chitinase Gene Sequencing and Sequence Alignment

Based on the expected homology between *Fragaria* and *Rubus* species, the following two primers were used for PCR amplification of a 528 bp long fragment homologous to the strawberry chitinase III gene in three raspberry cultivars: Ch-Up1 5′-GAAGATGCCCGCCAAGTTG and Ch-Low2S 5′-TTGATGGAGGAGCTGTATC. The amplification reaction mixture (25 μL) contained ~0.15 μg of genomic DNA, ScreenMix-HS buffer (Evrogen JSC, Moscow, Russia), 0.8 mM of each primer, and Milli-Q water. The PCR protocol included an initial denaturation step at 95 °C for 5 min followed by 31 cycles consisting of 45 s at 95 °C, 30 s at 59 °C, and 60 s at 72 °C each. A final step of 10 min at 72 °C ended the cycles followed by a hold at 4 °C. The PCR products were purified and sequenced by Evrogen JSC. The chitinase III sequences were aligned, visualized and manually inspected using the MView 1.63 software (www.ebi.ac.uk/Tools/msa/mview).

3. Results

3.1. Development and Characterization of SSR Markers

A total of 118 SSR loci were detected in 21 gene sequences (45.6 Kb). The number of SSRs ranged from one to 13 per gene (5.6 on average). One SSR was found per every 387 bp on average; less frequent in exons with one SSR per every 628 bp, but more frequent in introns with one SSR per every 263 bp on average. In our SSR analysis, loci with pentanucleotide motifs were detected at the highest frequency (45%), followed by loci with hexa-(25%) and dinucleotide (13%) motifs. Loci with tetra-, mono- and trinucleotide motifs were relatively less frequent—8%, 5%, and 4%, respectively. All loci with mononucleotide repeats consisted of only T nucleotide and contained from 10 to 12 T nucleotides. Among 17 loci with dinucleotide repeats, the AT/TA motif was the most frequent (47%), followed by CT/GA (24%) and GT/CA (18%). On average, number of repeats were 9.7 for loci with dinucleotide motifs and 7.5 for loci with trinucleotide motifs ranging from 5 to 32 and 4 to 14 motifs per locus, respectively. Loci with tetranucleotides motifs contained only three repeats, and loci with penta- and hexanucleotide motifs contained only two repeats.

Location of microsatellite loci in CDS (exons) and NCDS (introns, 5′ and 3′UTRs, and upstreams— upfront regions further than 500 bp from the first exon) was determined, with majority of them (35%) being located in introns, 28% in exons, 14% in 5′UTRs and upstreams, and 9% in 3′-UTRs (Table 2). Loci with pentanucleotide repeats prevailed everywhere (41–55%), except in upstreams (29%), where the proportion of hexanucleotide SSRs was higher (41%). A relatively high proportion of hexanucleotide SSRs was also in exons (33%). The nucleotide distribution was approximately equal in the exons (21, 24, 26, and 28% for T, G, A, and C, respectively), while in NCDS T (46%) and A (34%) prevailed over C (12%) and G (8%).

Table 2. Number of microsatellite loci with different nucleotide repeat motifs and their location in gene.

Location	SSR Motif						Total
	Mono	Di	Tri	Tetra	Penta	Hexa	
Upstream	0	2	2	1	5	7	17
5′UTR	0	3	0	2	7	4	16
Exon	0	1	3	0	18	11	33
Intron	2	9	0	6	17	7	41
3′UTR	4	0	0	0	6	1	11
Total	6	15	5	9	53	30	118

However, not all microsatellite loci could be developed into useful SSR markers. For example, some microsatellite loci were located at the end of the sequenced DNA fragment. Microsatellite loci mononucleotide repeats were also not used for developing SSR markers. We also tried to use various combinations of the location of SSR loci and annealing sites for PCR primer pairs. This analysis resulted to selection of 24 sequences ranging from 122 to 400 bp long and harboring 43 microsatellites. Finally, 24 primer pairs were successfully designed (Table S1). In addition to the newly developed markers, we used the *RiG001* marker from *R. idaeus* [28]. The developed SSR loci were in all gene regions except 3′UTRs: ten were in introns, eight in exons, two in upstreams, and one was in 5′UTR (Table S1). In addition, four loci were located at the junction of CDS and NCDS (two loci at the junction of 5′UTR and exon, and intron and exon each). Eleven markers contained more than one SSR locus. The *FaFS01*, *FaAR01*, *RhUF01*, *FaMY02*, and *FaFG01* markers contained two SSR loci, the *FaCH01*, *FaCH02*, *FaF3H01*, *FaBG01*, and *RiMY01* markers contained three SSR loci, and the marker *FaMY01* contained five SSR loci. Among the new SSR markers developed, 15 were developed using *F.* × *ananassa* sequences whereas nine were developed using *Rubus* species sequences.

3.2. Cross-Specific Transferability of SSR Markers

The 24 genic SSRs developed in this study and one published SSR from *R. idaeus* (Table S1) were evaluated for cross-amplification in two important genera of the subfamily Rosoideae, *Fragaria* and *Rubus*. These 25 SSR loci represented 18 structural and regulatory flavonoid biosynthesis genes and three PR protein genes. A total of 48 specimens belonging to 11 species and hybrids with a wide range of ploidy (di-, tetra-, hexa-, hepta- and octoploids) were used for marker validation (Table 1). The collection of selected specimens included samples from different breeding programs worldwide, including specimens from Finland, Germany, The Netherlands, Poland, Russia, Sweden, Switzerland, UK, Ukraine, and USA. Cross-amplification results and allele sizes are presented in Table S1. Twenty two of the 25 primer pairs amplified a PCR product or products of approximately the size expected for a homologous gene. Only primer pairs for the *RiMY01* locus generated multiple bands of approximately the expected size in *R.* × *stellarcticus* hybrids. In total, 10 primer pairs, representing nine out of 21 genes, amplified a product of the expected size in all two genera, indicating that primer binding sites were conserved across two rosaceous genera screened.

All primer pairs (15) from *F.* × *ananassa* amplified a PCR product in each of the 14 strawberry specimens including *Fragaria* × *Comarum* hybrid that showed 100% transferability despite the various genetic origin. More than two (up to eight) fragments were amplified in the *F.* × *ananassa* specimens, which was expected because of their polyploid nature. Ten SSR markers (67.7%) revealed genetic differentiation among strawberry specimens, while polymorphism was not detected in five SSR markers. Among them, two loci (*FaAR01* and *FaCH01*) each had two amplified fragments, but they were identical in all tested specimens. In order to evaluate the transferability of SSR markers in the diploid *Fragaria* species, the developed PCR primer pairs were used to genotype *F. vesca* and *F. viridis*.

Transferability of SSRs within *Fragaria* was high. Eleven of these primers (73.3%) amplified fragments in *F. vesca* and *F. viridis*. However, four strawberry primer pairs for the *FaDR01*, *FaLR01*, *FaMY01*, and *FaMY02* loci failed to amplify in two *Fragaria* species, suggesting that either these sequences have diverged between octoploid and diploid species or are not present in *F. vesca* and *F. viridis*, but are present in two other ancestors of the octoploid genome of *F.* × *ananassa*.

The transferability of *Rubus* markers within the genus was lower than that of *Fragaria*—50–70% for 10 markers, depending on the species. The maximum transfer was in red raspberry, the minimum—in hybrid arctic bramble: four markers worked in all six hybrid arctic bramble cultivars, but *RcFH01*—only in two of them. All of the five *Rubus idaeus*-derived SSR markers successfully amplified fragments in all red raspberry cultivars. In addition, a marker from *R. coreanus* and only one markers from blackberry (*RiMY01*) were amplified in red raspberry cultivars. Six *Rubus* SSRs amplified PCR products in all *Rubus* species and five in all six hybrid arctic bramble cultivars. In addition, the *RiG001* marker was amplified only in red raspberry. Typically, markers are amplified within the same species, but three markers *RhDR01*, *RhDR02*, and *RhAR01* developed originally in a blackberry (cultivar Arapoho, NCBI GenBank accession number JF764809) did not produce a product in our blackberry cultivars and hybrids. Moreover, these three markers were not amplified in any tested specimen.

Transferability from *F.* × *ananassa* to *Rubus* species was demonstrated for 5–7 out of the 15 primer pairs (33.3–46.7%). Thus, the transferability of strawberry markers decreased as cultivars become less related: all 15 markers were amplified in *F.* × *ananassa*, 11 markers in two diploid *Fragaria* species, and 5–7 in *Rubus* species. Six out of the 15 *F.* × *ananassa* primer pairs (40%) amplified fragment of the expected size in red raspberry and hybrid cultivars such as Loganberry, Tayberry, and Boysenberry. Successful cross amplification in other *Rubus* species ranged from 33.3% in black raspberry and blackberry to 46.7% in Nordic species (*R. chamaemorus*, *R. arcticus*, and *R.* × *stellarcticus*), where the *FaLR01* marker was also amplified, while it did not produce any PCR product in other *Rubus* species. The majority of *Rubus* cultivars had one or two amplified fragments per primer pair, however, for some polyploidy cultivars, as well as the blackberry and hybrids, there were more than two fragments (up to four) amplified by some primer pairs. The octoploid species *R. chamaemorus* almost always showed the presence of only one or two fragments.

Transferability of *Rubus* SSRs to the *Fragaria* species was relatively low: only three out of 10 markers (all from *R. idaeus*) were amplified, and there were no differences among species. In total, four out of 14 SSR markers had amplified fragments in *F. vesca* and *F. viridis* in the same range of size as that in *F.* × *ananassa*, and the rest can be used to separate diploid species and octoploid strawberry. Six SSRs had *F.* × *ananassa* and unique fragments amplified, and four markers only unique fragments: the same for two diploid species (*FaFH01*) and different (*FaCH01*, *FaBG01*, and *RiMY01*). Only the *FaFS01* marker amplified two fragments unique to *Fragaria* × *Comarum* hybrid.

Twelve out of 18 SSR markers (66.7%) were found to be polymorphic in the 13 strawberry specimens. In addition, three markers (*FaAR01*, *RiAS01*, and *FaCH01*) had the same two fragments amplified in all specimens. Thus, these markers were monomorphic in these specimens, but may be polymorphic in a wider collection. The majority of the polymorphic SSR markers in *Fragaria* genes (7 out of 10) had more than two fragments amplified in strawberry specimens, probably, originating from four genomes of octoploid strawberry. In total, 11 out of 14 markers (78.6%) were polymorphic in the genus *Rubus*. However, only one of them (*RhUF01*) was polymorphic in all species. There are two reasons for this, firstly, not all markers were amplified in all species. Secondly, a small number of specimens were used for most species. Most of the markers were polymorphic in species with a large number of tested specimens and/or having a hybrid origin. In red raspberry (14 cultivars), hybrid Arctic bramble (six specimens), and *Rubus* hybrid (four specimens) the proportion of polymorphic markers was 61.5%, 58.3% and 66.7%, respectively. This approximately corresponds to the proportion of polymorphic markers in strawberry. In black raspberry and arctic bramble, each having two specimens tested, 27.3% and 30.8% markers were polymorphic, respectively.

Monomorphic markers of strawberry *FaFS02*, *FaTG01* and *FaCH01* produced alleles of the same size in *Fragaria* and *Rubus* species—271, 324, and 324 + 329 bp, respectively. In addition, the polymorphic strawberry markers *FaFS01*, *FaLR01*, and *FaFH01* had one allele amplified in northern *Rubus* species, which was almost the same (different by only one or two nucleotides) as one of the main strawberry alleles. Interestingly, the alleles of *Rubus* species for seven markers from *Fragaria* had almost the same size as expected or were different by no more than 10–20 nucleotides in both directions, while the alleles in strawberry for *Rubus* markers (*RiAS01*, *RiMY01*, and *RiHL01*) were always less than the expected size by about 40–100 nucleotides. The most multi-allelic SSR markers (17–19 alleles; *FaFS01*, *FaMY02*, and *RiMY01*) contained dinucleotide repeats in introns and presented series of consecutive alleles in 2-bp steps. In addition, the *RiAS01* marker with exon-located locus showed a number of alleles with a step of 24: 261, 285, 309, 333, and 357 in *Rubus* species; 261, 285, and 333 in *Fragaria* species.

3.3. Allelic Polymorphism and Genetic Diversity

Number of alleles, expected (H_e) and observed (H_o) heterozygosities, and polymorphism information content (PIC) were calculated for 12 polymorphic SSR markers in 13 octoploid strawberry specimens (Table 3) and in 24 diploid *Rubus* cultivars (Table 4). The number of alleles in strawberry specimens varied widely among these markers ranging from two in *RiMY01* to 14 in *FaFS01*, with 6.7 on average, respectively. The H_o and H_e values ranged from 0.37 to 0.90 and 0.35 to 0.90, with 0.63 and 0.66 on averages, respectively. The *FaCH02* locus, located in exon, demonstrated the lowest heterozygosity. The PIC ranged from 0.34 to 0.89 with an average of 0.66. The most markers (nine out of 12) had PIC values higher than 0.5, suggesting that these markers can efficiently measure genetic diversity in strawberry.

Table 3. Diversity of 12 polymorphic SSR markers in 13 *F.* × *ananassa* specimens.

Locus	Location	Number of Alleles	H_o	H_e	PIC
FaF3H01	5′UTR	4	0.42	0.52	0.52
FaFH01	intron	5	0.78	0.72	0.79
FaFS01	intron/exon	14	0.90	0.90	0.89
FaDR01	upstream	9	0.58	0.82	0.81
FaLR01	intron	4	0.77	0.71	0.70
FaFG01	5′UTR/exon	8	0.74	0.72	0.71
FaMY01	intron	4	0.42	0.45	0.45
FaMY02	intron	13	0.75	0.84	0.83
RiMY01	intron	2	0.48	0.50	0.50
RiHL01	intron	4	0.60	0.60	0.59
FaCH02	exon	4	0.37	0.35	0.34
FaBG01	upstream	9	0.68	0.82	0.81
Mean		6.7	0.63	0.66	0.66

Table 4. Diversity of 10 polymorphic SSR markers in 24 diploid *Rubus* cultivars.

Marker	Location	Number of Alleles	H_o	H_e	PIC
RiG001	intron	3	0.08	0.53	0.43
FaFH01	intron	5	0.00	0.57	0.51
RcFH01	intron	4	0.21	0.43	0.40
FaFS01	intron/exon	5	0.29	0.70	0.65
FaLR01	intron	3	0.04	0.34	0.29
RiAS01	exon	4	0.33	0.62	0.56
RhUF01	exon	4	0.33	0.38	0.34
RiMY01	intron	13	0.42	0.84	0.82
RiHL01	intron	2	0.04	0.04	0.04
FaCH01	5′UTR/exon	3	0.50	0.64	0.57
Mean		4.60	0.23	0.51	0.46

From two alleles in the *RiHL01* marker to 13 alleles in the *RiMY01* marker were amplified in 24 diploid *Rubus* cultivars, with a mean number of 4.6 alleles per locus for 10 polymorphic markers (Table 4). Observed (H_o) and expected (H_e) heterozygosity ranged from 0 to 0.5 and 0.04 to 0.84 with mean values of 0.23 and 0.51, respectively.

Unexpectedly, the lowest heterozygosity was observed in the intron-located *RiHL01* marker. The PIC ranged from 0.04 to 0.82 with a mean of 0.46, which was noticeably lower than in strawberries (0.66), and only six markers had PIC values higher than 0.50 suggesting their high potential to measure high genetic diversity. The other four markers had *PIC* ranging from 0.25 to 0.50 showing their rather moderate potential to measure genetic diversity, and only the *RiHL01* marker was slightly informative (<0.25).

Principal Component Analysis (PCA) was used to reveal genetic relations among *Fragaria* (Figure 2) and *Rubus* (Figure 3) species and cultivars based on SSR markers representing the flavonoid biosynthesis pathway genes. The first three PCs explained 22.5%, 16.3%, and 9.8% of the total variance in *Fragaria*, respectively (Figure 2). All cultivars of *F. × ananassa* formed a compact group and were completely separated from diploid *Fragaria* species and (*F. × ananassa*) × *C. palustre* hybrid. *F. vesca* and *F. viridis* were grouped along the PC1, whereas *F. virids* and *F. × ananassa* along the PC2. Interestingly, *Fragaria × Comarum* hybrid (*F. × ananassa*) × *C. palustre* was very distant from the rest *Fragaria* cultivars and species, which is in agreement with a pink color of flowers in this hybrid, which makes it also very different from other cultivars. The PC3 did not contribute much in delineation of cultivars, therefore plots with PC3 are not presented here.

The PCA in Figure 3 represents the relationships between 32 individual *Rubus* cultivars and species. The first three PCs explained 20.3%, 11.4%, and 9.7% of the total variance, respectively. The PC3 did not contribute much in delineation of cultivars, therefore plots with PC3 are not presented here. In general, the grouping was as expected, and a good discrimination was observed between four *Rubus* subgenera—all of them were well-separated along PC1 and PC2. Unexpectedly, arctic bramble (*R. arcticus*; Cyclastis subgenus) was very distant from other *Rubus* species. There was a clear overlap between two *R. arcticus* and *R. × stellarcticus* clusters. Despite belonging to different subgenera black raspberry (*R. occidentalis*; Idaeobatus subgenus) cultivars were closer to blackberry (*Rubus* subgenus), which is in agreement with having also a common black color of their berries. The *Rubus* hybrids cluster coincided with the red raspberry (*R. idaeus*; Idaeobatus subgenus) cluster and was clearly distinguished from blackberry group, which suggests closer relationship of hybrids with red raspberry. All these hybrids have berries in different shades of red color.

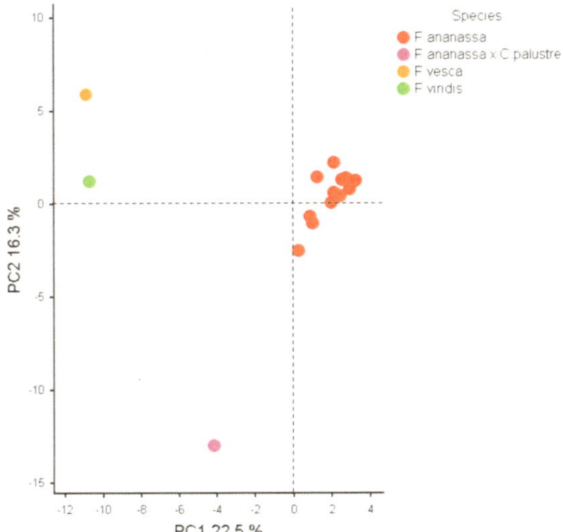

Figure 2. PCA of genotyped strawberry (*Fragaria*) species and specimens based on 13 polymorphic SSR markers.

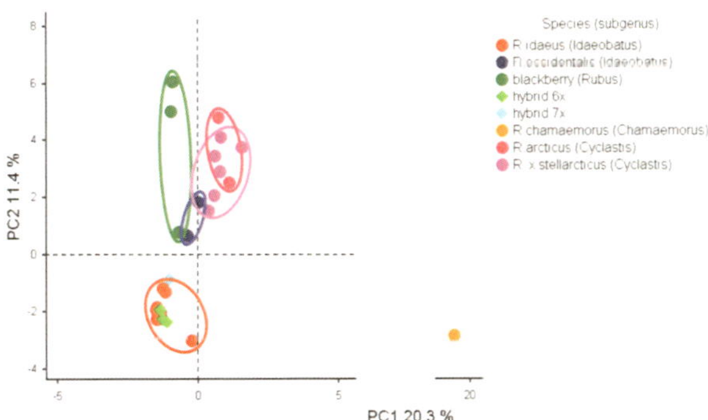

Figure 3. PCA of *Rubus* cultivars based on 10 polymorphic SSR markers.

3.4. Genetic Data Analysis

To determine the relationship between transferability and polymorphism of SSR markers and the location of loci and of primer pairs, the data were grouped in the Table 5. The identification of the location of the primer binding sites and the separation of the SSR markers into four groups based on this trait showed their clear connection with transferability level. When both primers are located in the conserved exons, the complete transferability is observed both within *Fragaria* and *Rubus* species and the cross-amplificatons between the *Fragaria* and *Rubus* genera. In the opposite direction (from *Rubus* to *Fragaria*), only one out of three markers (*RiMY01*) was amplified.

Table 5. Relationships between transferability and polymorphism of SSR markers and location of primer pairs and SSR loci in gene.

Genus	Locus	Primer Binding Site Location [1]	Amplification		Locus Location	*F.* × *ananassa*		*Rubus* [3]	
			Fragaria	*Rubus* [2]		Number of Alleles	Polymorphism	Number of Alleles	Polymorphism
					Both Binding Sites Located in Exons				
Fragaria	*FaFH01*	ex1–ex2	+	+	in1	5	+	8	–/+/+/+/–/+
	FaFS01	ex2–ex3	+	+	in2 ex3	14	+	8	+/–/+/+/–/+
	FaFS02	ex1–ex1	+	+	ex1	1	–	1	–
	FaAR01	ex4–ex5	+	+	ex4 in4	2	–	1	–
	FaTG01	ex2–ex2	+	+	ex2	1	–	3	–/–/–/+/–/–
Rubus	*RiAS01*	ex2–ex2	+	+	ex2	2	–	5	+/–/+/+/+/+
	RhUF01	ex2–ex2	–	+	ex2	–	n	5	+
	RiTT01	ex–ex	–	+	exon	–	n	1	–
					One of Binding Sites Located in Exon another in Intron or 5′UTR				
Fragaria	*FaMY02*	in2–ex3	+/–	–	in2	14	+	–	n
	FaLR01	ex2–in2	+/–	–/–/–/+/+	in2	4	+	2	n/n/n/n/+/+
	FaFG01	5′UTR–ex1	+	–	5′UTR ex1	8	+	–	n
	FaAS01	in–ex2	+	–	ex2	1	–	–	n
	FaCH01	5′UTR–ex1	+	+/–/+/+/+	5′UTR ex1	2	–	3	+/n/n/+/–/+
	FaCH02	in–ex2	+	–	ex2	4	+	–	n
Rubus	*RiHL01*	ex1–in	+	+	in	4	+	2	+/–/–/–/–/–
	RcFH01	in2–ex3	–	+	in2	–	n	5	+/+/–/+/–/+
	RiG001	in–ex2	–	+/–/–/–/–	in	–	n	2	+/n/n/n/n/n
					Both Binding Sites Located in Intron, 5′UTR or Upstream				
Fragaria	*FaMY01*	in2–in2	+/–	–	in2	4	+	–	n
	FaDR01	up–5′UTR	+/–	–	up	9	+	–	n
	FaF3H01	5′UTR–5′UTR	+	–	5′UTR	4	+	–	n
	FaBG01	up–up	+	–	up	9	+	–	n
					One of Binding Sites Located across Exon/Intron Junction				
Rubus	*RiMY01*	ex1/in1–ex2	+	+/+/+/+/–	in1	2	+	17	+/–/+/+/+/n
	RhDR01	in2–in2/ex3	–	–	in2	–	n	–	n
	RhDR02	ex1–in1/ex2	–	–	in1	–	n	–	n
	RhAR01	ex5–ex5/in5	–	–	e5	–	n	–	n

[1] in—intron, ex—exon; up—upstream; the number is not provided in case, if gene has only a single or no intron.
[2] *R. idaeus/R. occidentalis* and blackberry/hybrids/*R. chamaemorus* and *R. arcticus/R.* × *stellarcticus*. [3] *R. idaeus/R. occidentalis*/blackberry/hybrids/*R. arcticus/R.* × *stellarcticus*. n—no amplification.

When one of the primer binding site is located in more variable NCDS (introns or 5′UTR), the transferability level decreases. With intrageneric transferability, two of the *F.* × *ananassa* markers, *FaMY01* and *FaLR01*, were no amplified product in the diploid *F. vesca* and *F. viridis*, while out of the three *Rubus* markers, only the *RiHL01* (representing TF) marker was amplified in all *Rubus* cultivars. The *RiG001* marker was amplified only in red raspberry, and *RcF3H* was transferred in Anna and Beata, but not in Linda, Sofia, Astra, and Aura hybrid arctic bramble cultivars. The transferability between the genera was also much lower. Out of the five *Fragaria* markers, only *FaLR01* was amplified in the Nordic species: cloudberry, Arctic bramble, and hybrid Arctic bramble, and *FaCH01*—in some cultivars of red raspberry, hybrids and Nordic species, but was not amplified in black raspberry and blackberry at all. A similar situation was observed when both primer binding sites were located in NCDS: the amplification of some markers failed in *Fragaria* diploid species, and none of the *Fragaria* markers was amplified in *Rubus*.

The location of one of the binding sites across intron-exon junction had the worst effect on transferability. Three out of four markers were not amplified in any specimen, even when the second primer site was located in exon and in the same species (blackberry). Only the *RiMY01* marker amplified some fragments, but it was inconsistent; multiple fragments were generated in Anna and Beata, but no fragments in Linda, Sofia, Astra, and Aura hybrid Arctic bramble cultivars.

We also observed a clear relationship between the location of loci and polymorphism level. When loci were located in introns or 5′UTR, the larger number of alleles, up to 14 alleles in *F.* × *ananassa* and 17 alleles in *Rubus*, was observed, and all markers were polymorphic. On the other hand, the allele number in exon-located loci did not exceed four in *F.* × *ananassa* (*FaCH02*) and five in *Rubus* species (*RiAS01, RhUF01*) and many markers were monomorphic. For example, all three markers that amplified one fragment in strawberry specimens were located in exons. Markers that contained more than one microsatellite locus located both in CDS and NCDS demonstrated high polymorphism (*FaFS01, FaFG01*) as well as monomorphism (*FaCH01*). In total, nine out of 12 polymorphic markers in strawberry were in NCDS, two in exons + introns, and only one in exon (*FaCH02*). More than half (six out of 11) of polymorphic markers in *Rubus* were also located in introns. With regard to the relationship between polymorphism and SSR motif type, the highest allelic variation was revealed in markers with dinucleotide motifs and large number of repeats, such as in the *FaMY02, FaFS01,* and *RiMY01* markers. It should be noted that two of these markers represented the *MYB10* TF genes in raspberry and strawberry.

3.5. Sequence Analysis of Chitinase III Genes

Based on the *F.* × *ananassa* chitinase III (chi3) sequence (GenBank accession number AF134347), we designed a pair of primers and amplified cDNA fragments from three raspberry cultivars with yellow-, orange- and red-colored berries (Zolotaya Osen, Oranzhevoe Chudo, and Babye Leto II, respectively). Sequencing confirmed that these fragments were composed of 528 nucleotides within the full length of open reading frame (ORF) and encoded 176 amino acids. The red-fruited Babye Leto II cultivar differed from the other two cultivars by two synonymous nucleotide substitutions. The nucleotide sequences of three chitinase III gene fragments were deposited in the NCBI GenBank (accession numbers MK333194, MK333195, and MK333196, respectively). The translated amino acid sequences of the raspberry chitinase III were aligned with published sequences of strawberry (cv. Chandler) and raspberry of unknown origin (Figure S1) [33]. The identity between amino acid sequences of the amplified chitinase III gene was 93.1% for all three Russian raspberry cultivars vs. unknown raspberry and 86.9% vs. strawberry. Twenty amino acid substitutions were the same for all three Russian and one unknown raspberry cultivars compared to strawberry sequence. In addition, eight substitutions were unique only for the unknown raspberry cultivar, and two substitutions were unique for our three cultivars.

4. Discussion

Modern plant breeding, including also berry crop breeding, seems to be almost impossible without modern genomic methods. They are needed to develop DNA based molecular genetic functional markers, such as single nucleotide polymorphisms (SNPs) and SSRs in adaptive and breeding trait related genes. The SSRs are highly variable in length due to insertion-deletion of the entire repeat units or motifs, mainly as a result of recombination errors or DNA polymerase slippage [34]. Genic SSRs could be even more informative than SNPs because unlike biallelic SNPs they usually have multiple alleles that can mark multiple alleles and haplotypes in these genes. It makes them very suitable for the MAS [35]. In comparison to random noncoding genomic SSRs the genic and EST-SSR markers are more transferable [21,36] and are better amplified [37] because of more conservative primer annealing sites. In addition, in silico development of genic and EST-SSR markers can now be relatively easily done using publicly available nucleotide sequence databases. Genic SSRs in functional genes can be used as "functional genetic markers", and they have a much higher transferability across different taxa than random genomic SSRs [38]. Sargent et al. [39] used primer pairs based on the binding sites in the *Fragaria* exons flanking polymorphic introns and found that their transferability was significantly higher compared to the random genomic SSRs. However, genic SSRs are usually less polymorphic than random genomic SSRs, which can limit their use in MAS [34]. Thus, the most optimal marker could be those that have a SSR locus in a variable gene region such as intron and the primer annealing sites in the conserved exons flanking this intron. In case of long introns, it would be important to have at least one annealing site in an exon. However, information on the exon-intron structure is not available in the

EST databases, more genome sequence data become available allowing to design reliable, consistent, polymorphic, functional, and transferable genic SSRs.

4.1. Choice of Genes and Genic SSR Marker Development

The flavonoid pathway is initiated by chalcone synthase (CHS) and involves more than 10 enzymes that act at early and late stages leading to the biosynthesis of different compounds such as flavonols, condensed tannins (proanthocyanidins) and anthocyanins (Figure 1). It is well known that pelargonidin-3-glucoside is a major anthocyanin in strawberry [40], whereas cyanidin-3-sophoroside is a major anthocyanin in red raspberry, followed by other cyanidin glycosides [41]. The late structural genes are regulated at transcriptional level by a ternary protein complex named MBW, which is formed by R2R3-MYB TFs, basic helix-loop-helix (bHLH) proteins, and WD40-repeat proteins [42].

For development of new SSR markers we used 13 structural and four regulatory flavonoid biosynthesis genes from GenBank (9 *F.* × *ananassa* and 8 *Rubus* species genes) (Table S1). Particular attention was paid to the flavonol synthase (*FLS*) and dihydroflavonol 4-reductase (*DFR*) genes, for which two SSR markers were developed. These enzymes competed for common substrates, dihydroflavonols (Figure 1), in order to direct the biosynthesis to colorless flavonols or colored anthocyanins, respectively, and may determine color phenotype [43]. For comparison, we also used a pair of primers designed for the *RiG001* locus from the *R. idaeus* aromatic polyketide synthase (*RiPKS3*) gene [28]. Unlike the RiPKS1, the typical naringenin chalcone synthase (CHS), the *RiPKS3* produced predominantly p-coumaryltriacetic acid lactone [44], but the sequences of both genes amplified by their PCR primer pairs are almost identical [23]. Among the *MYB TFs* genes we used the *MYB10* orthologs involved in the anthocyanin biosynthesis during ripening in more than 20 commercially important *Rosaceae* species [45]. Assuming its important role in flavonoid pathway regulation, we developed two markers for this gene. The *bHLH* gene from red raspberry is very similar to the *MdbHLH33* gene that is closely associated with anthocyanin production in apple [46]. In addition, we included in the study genes encoding PR proteins—chitinase (*FaChi2-1*) and ß-1,3-glucanases (*FaBG2-2* and *ToyoGluIII*). Previous studies demonstrated that expression of the *FaChi2-1* and *FaBG2-2* genes in strawberry are induced in response to pathogen inoculation [47].

A total of 118 SSR loci were identifies in 21 genes, and pentanucleotide motifs were the most abundant. Our result is in contrast to previous findings identifying trinucleotide [21,37] as the most frequent genic repeats in Rosaceae plant species, unlike dinucleotide motifs that were the most frequent among non-genic random genomic repeats [36,48,49]. The difference can be also explained, at least partly, by less stringent search parameters that were used in our study compared to those that are typically used in searches for random genomic SSRs, but they were similar with those that are usually used for the search of SSRs in coding regions. For example, Park et al. [37] found that the most profound allelic variation was revealed by the primer pairs flanking the penta-repeats (91%), whereas authors noted no significant difference among di-, tri-, and tetra-repeat (61–66%) motifs.

Identified SSRs were categorized by location in exons, introns, 5′UTRs, 3′UTRs, and upstreams. SSRs were located mainly in NCDS in genes: although exons occupied 45% of the gene length, but they contained only 28% of the SSR loci. It is known that genic SSRs located mainly in variable NCDS, but not in conserved exons. For example, development of genome-wide SSR markers in such different species as papaya and chickpea showed a similar distribution of SSRs across genome: 78–87% were in the intergenic regions, 9–10% in introns, and 2–3% in exons [34,50]. It has been repeatedly reported, that tri- and hexanucleotide motifs were more abundant in exons since they do not lead to frame shift and do not effect protein function and property as much as mutation of other repeats that could be under mutation pressure. Meanwhile, di-, tetra-, and pentanucleotide motifs are abundant in NCDS [35,36,50]. Such selection also reduces the variability of SSRs in exons. This is not consistent with our data, where pentanucleotide SSRs were the most common in the exons. However, apparently selection nevertheless occurred, and we did not find SSRs with mono-, di-, and tetranucleotide motifs in exons in our study, although they made up a significant part of all SSRs in introns and UTRs (Table 2).

We found also that A and T nucleotides prevailed both in general in the studied genes (72%), and to a greater extent in NCDS (80%). It is known that some motifs, such as AT/TA, showed a greater abundance in most species [51].

When developing markers, we took into account the location of loci in genes (so that they were in different gene regions), as well as the location of several loci in the same marker. In addition, preference was given to dinucleotide motifs and a greater number of motif repeats, because such SSRs are more polymorphic [28,51]. As a result, we designed 15 and 9 primer pairs for SSR markers based on nucleotide sequences of genes with known function in *F. × ananassa* and *Rubus* species, respectively. These markers included all flavonoid biosynthesis genes available for *Fragaria* and *Rubus* in the NCBI GenBank database. Earlier, sets of SSRs for poplar genes involved in wood formation [52] or stress related genes in peanut [53] were developed, but not on flavonoid biosynthesis genes. We also developed SSR markers for the TF regulatory genes, which were not previously reported for *Fragaria* and *Rubus*.

4.2. Choice of Cultivars and Transferability of SSR Markers

It was shown that total anthocyanin content correlated with color of berry in both strawberry [54] and raspberry [55], and cultivars were selected primarily to have a broad variety of berry colors. In addition, we took into account their commercial value and application in the breeding programs. Other *Rubus* and *Fragaria* cultivars were chosen due to their diverse ploidy (black raspberry, blackberry, loganberry, and boysenberry) and as wild potential donors of traits of interest (*F. vesca* and *F. viridis*). The rare cultivated species of *Rubus* were also included in the study, such as cloudberry (*R. chamaemorus*), arctic bramble (*R. arcticus*), and hybrid arctic bramble (*R. × stellarcticus*). Only a few rare reports on the SSR markers are available for these boreal species [56,57] that are rich in ellagic acid and are regionally extremely important and valuable crops. Moreover, *R. arcticus* is used to develop new cultivars. The hybrids of the octoploid *F. × ananassa* and the hexaploid *C. palustre*, which unlike the white-flowered strawberry have red and pink flowers and are usually grown for ornamental purposes, are also of high interest for the flavonoids biosynthesis research.

The amplification of the SSR markers in the strawberry specimens using primer pairs based on the *F. × ananassa* sequences was 100% successful, and this is in agreement with data obtained by many other researchers. The transferability of the strawberry SSR markers to *F. vesca* and *F. viridis* was noticeably lower (73.3%): *FaMY01*, *FaMY02*, *FaDR01*, and *FaLR01* were not amplified using primer pairs developed from *F. × ananassa* genes. It is known that rapid genomic changes can occur after polyploidization, and loss of homologous copies of many duplicated genes was often observed [21]. Hirakawa et al. [58] calculated that the size of the octoploid *Fragaria* genome (698 Mb) approximately equalled 80% of the combined genomes of its four diploid wild relatives, *F. iinumae*, *F. nipponica*, *F. nubicola*, and *F. orientalis* (~200 Mb each). It is also possible that primer binding sites in *F. vesca* and *F. viridis* for these SSR markers have mismatches with sites that were used to design primers. It cannot be also excluded that the prolonged selection (with the purpose to change the quality of strawberries) contributed to the discrepancy between binding sites. The alleles unique for some diploid species were identified at a number of loci. The marker transferability to (*F. × ananassa*) × *C. palustre* hybrid is not different from the strawberry specimens. This happened due to the fact that *C. palustre* is a close relative of *Fragaria*—they are members of the same subtribe Fragariinae [1]. Twelve out 15 markers (80%) were polymorphic in *F. × ananassa* (produced two and more fragments). This is consistent with other authors who reported 81% [59] and 91% [21] of polymorphic loci.

The transference among the *Rubus* species was not as successful as among *Fragaria* species. The transferability of loci from the *Rubus* flavonoid pathway genes was 70% for red raspberry, 50% for hybrid arctic bramble, and 60% for the other *Rubus* species. This fact suggests a close genetic relationship among the studied *Rubus* species and a high conservativeness of the flavonoid biosynthesis genes. Mnejja et al. [16] showed that transferability negatively correlated with genetic distance between the genera in the Rosaceae family and between species within the *Prunus* genus. However, many authors reported similar and high (80–100%) cross-species amplification of raspberry loci (*Idaeobatus* subgenus) in other species, such as black raspberry (*Idaeobatus* subgenus) [60,61], blackberry (*Rubus* subgenus) [28,60,61], and Arctic bramble (*Cyclastis* subgenus) [57]. The *RiG001* marker was amplified only in red raspberry. Castillo et al. [28] found that this marker was not amplified in blackberry and hybrids. We demonstrated it also in black raspberry earlier [23] and in cloudberry, arctic bramble, and hybrid Arctic bramble in this study. This marker appears to be a good identifier for red raspberry among cultivated *Rubus* species.

The transferability of *F. × ananassa* loci to *Rubus* species was moderate: 46.7% in Nordic species, 40% in red raspberry and hybrids, and 33.3% in black raspberry and blackberry (33.3%). Other authors [20,62] reported even a lower transferability from *F. × ananassa* to diploid red and black raspberries (*Idaeobatus* subgenus, 8–23%) than to tetraploid blackberry (*Rubus* subgenus, 26–36%). Only two of the 15 strawberry markers were polymorphic in raspberries (*FaFS01* and *FaCH01*) and five markers if all *Rubus* species were taken into account. This is consistent with other data that demonstrated the difficulty of identifying polymorphic loci in *Rubus*—only 10% of the markers from *Fragaria* amplified a polymorphic product in *Rubus* [17].

There are no published data on transferability of markers from *Rubus* to *Fragaria*. In our study, only three of 10 *Rubus* primer pairs amplified in all *Fragaria* specimens. All of them were from *R. idaeus* (the *ANS* locus and two TF genes—*MYB10* and *bHLH*). This level is slightly higher than the transferability from *Fragaria* to *Rubus*. It should be noted that the primers for the similar genes in *Fragaria*, *ANS* and *MYB10*, did not amplified in *Rubus*. The amplification of *RiMY01* showed the presence of unique alleles in *F. vesca* and *F. viridis*.

The moderate transferability level between *Fragaria* and *Rubus* suggests a remote relationship between these two genera. Potter et al. [1] reported that the phylogenetic relationship between *Fragaria* and *Rosa* is closer than between *Fragaria* and *Rubus*. Qi et al. [63] evaluated the SSR primer pairs using in silico PCR and demonstrated that strawberry is the closest to the rose followed by the raspberry. We found an interesting regularity in the size of alleles during cross-species amplification: alleles in *Rubus* species amplified using *Fragaria* primers had sizes similar to the expected, while alleles in *Fragaria* species amplified using *Rubus* primers were significantly smaller than expected. This was not previously reported, since there was no work on the transferability of *Rubus* markers to *Fragaria*. Perhaps this is due to the fact that the genus *Rubus* is evolutionary older than the *Fragaria* genus [1,64].

The SSR transferability to the poorly studied northern *Rubus* species did not differ from raspberries and blackberries, except for the amplification of the *FaLR01* marker in them. Kostamo et al. [57] reported amplification of all seven markers in arctic bramble cultivars using primer pairs developed for raspberry SSR loci, but the annealing temperature was lowered to 50 °C from the original 60 °C. It is known that lowering the annealing temperature may increase transferability, but nonspecific amplification can occur [21]. We have repeatedly observed null-alleles among hybrid arctic bramble cultivars for the *FaFS01*, *RcFH01*, *FaLR01*, *FaFH01*, and *FaCH01* markers and only once for cultivars of other *Rubus* species (for FaCH01 marker). This is probably due to the high heterozygosity in primer binding sites that results in mismatch between primers and binding sites in the *R. × stellarcticus* cultivars. Closer similarity of allele sizes to strawberry markers in northern *Rubus* species compared to raspberries and blackberries, as well as the amplification of the *FaLR01* marker in them, suggests their closer relationship with *Fragaria* compared to other tested *Rubus* species.

4.3. Genetic Diversity of Fragaria and Rubus

The number of alleles varied among 12 SSR markers genotyped in 13 strawberry specimens and ranged from two to 14 alleles (6.67 alleles on average) (Table 3). This is slightly higher than previously reported for *F. × ananassa* (5.6) [65]. The low number of alleles per locus can reflects a poor choice of microsatellites or low levels of genetic diversity. Hilmarsson et al. [10] found a mean number of alleles in *F. vesca* was only 4.5 and the authors believed that this was due to low levels of genetic diversity in the species, but the most polymorphic marker had 16 alleles. Within the strawberry collection, the mean H_0 and H_e were 0.63 and 0.66, respectively, and the mean polymorphism information content (PIC) was 0.66. Our H_e value is similar to 0.66 obtained by Yoon et al. [65], but H_0 and PIC significantly exceed their values of 0.51 and 0.45, respectively. The mean H_0 and H_e vary significantly across different *Fragaria* species: from 0.08 and 0.17 in *F. vesca* [10] up to 0.75 and 0.86 in *F. virginiana* [66], respectively. Nine informative markers having high polymorphism information content (PIC) values (0.52–0.89) can be used for efficient evaluation of large collections of strawberry samples.

The parameters of polymorphism were also calculated for 24 diploid *Rubus* cultivars based on 10 SSR markers (Table 4). From two to 13 (mean 4.6) alleles were observed for polymorphic markers in diploid *Rubus* cultivars. These values were rather similar with published results obtained in other diploid *Rubus* species. From two to 5 (mean 3) alleles per polymorphic locus were observed in 21 *R. occidentalis* cultivars [67] and 2–15 (7.5) in 24 *R. idaeus* cultivars [28]. In our study, the mean values of H_0, H_e, and PIC were 0.23, 0.51, and 0.46, respectively. These values were lower than those observed in *R. idaeus* [28] and *R. coreanus* [49]. This may be because the gene sequences, from which our SSR markers are derived, are more conserved compared to random genomic SSRs used in these published reports. However, average H_e and PIC in our study were higher than in the published study in *R. occidentalis* [67], while H_0 was lower. The most polymorphic SSR loci based on high H_0, H_e, and PIC were *RiMY01* and *FaFS01*. *FaFS01* was also the most polymorphic marker for strawberry. These loci contained long dinucleotide repeats, which usually have higher levels of polymorphism compared to other repeats [21]. The *RiG001* marker demonstrated the PIC value, that was very similar with results, reported by Castillo et al. [28]—0.43 vs. 0.46, respectively. Our results demonstrated that the SSR markers developed in this study might be useful for the genetic assessment of *Fragaria* and *Rubus* species.

PCA of 16 *Fragaria* specimens genotyped with 13 SSR markers demonstrated clear separation between *F.* × *ananassa*, *F. vesca*, *F. viridis*, and *Fragaria* × *Comarum* hybrid (Figure 2). Biswas et al. [68] also demonstrated a clear separation of 26 *F.* × *ananassa* and 7 *F. vesca* specimens by SSR. Vallarino et al. [69] showed a clear separation between domesticated (*F.* × *ananassa*) and wild (*F. moschata*, *F. vesca*, and *F. chiloensis*) specimens using profiles of primary and secondary metabolites, but wild species formed a common cluster. Unlike our study, Sanchez-Sevilla et al. [70] showed no differences between (*F.* × *ananassa*) × *C. palustre* hybrid (Pink Panda cultivar) and strawberry cultivars using PCA based on DArT markers, although hybrid and cultivars were most diverse according to the phylogenetic dendrogram.

PCA of 32 *Rubus* cultivars genotyped with 10 SSR markers demonstrated, in general, a good separation of the four *Rubus* subgenera, except *R. occidentalis*, which clustered separately from *R. idaeus* despite belonging to the same *Idaeobatus* subgenus (Figure 3). Earlier, PCA of 50 *Rubus* cultivars based on genomic SSRs showed that black raspberry is the most distant from other *Rubus* species [60]. Interestingly, in report of Simlat et al. [71], Jewel cultivar was not separated from red raspberry using PCA based on SSRs. According to our data, *Rubus* hybrids clustered together with red raspberry, while in Graham et al. [60] they were located approximately in the middle between red raspberry and blackberry cultivars.

4.4. Relationship between Properties of SSR Markers and Their Location

The more transferable SSR markers are characterized by a lower frequency of null alleles. The most likely reason for null alleles are mutations in one or two primer binding sites creating mismatch between primer and binding site sequences [22]. The chances for mismatch depend on location of these sites in gene, and, therefore, location of binding sites is important for amplification. We showed a clear relationship between the location of primer binding sites and transferability of SSR markers within and between *Fragaria* and *Rubus* genera (Table 5). The most transferability demonstrated markers that were amplified by primer with binding sites located in exons. The location of binding sites in NCDS resulted in a significant decrease of transferability. It should be noted that location only one of two binding sites in variable sequences was sufficient to drastically reduce transferability, especially between genera. It is known, that chances of mutations in binding sites and their mismatch with primers will increase in taxa that are phylogenetically distant [48]. In our study, all null alleles in some *R. × stellarcticus* hybrids were observed only for loci with primer binding sites located in NCDS.

The lowest transferability was found also when one of the two primer binding sites were located across the intron/exon junction. Three out of four markers with such binding sites were not amplified in any of the 48 tested specimens, even in the blackberry which sequences were used to develop these markers. Thus, we experimentally confirmed the assumption of Vidal et al. [34] that PCR amplification failures can occur in the case of designing primers with binding sites across exon-intron junctions. It should be noted that for two SSR loci detected in the *RhDFR* gene and one SSR in the *RhANR* gene all computer programs used in this study to design primers selected the intron-exon junctions for binding sites, and these two markers were not amplified. Thus, either parameters for designing primers for these markers should be changed or primers should be designed manually to develop successful SSR markers for these genes.

The polymorphism of SSR markers in transcribed regions can affect transcription, translation, and/or gene function. SSR polymorphisms within exons can result in amino acid change that can lead to a gain or loss of function, in the 5′UTR they can regulate gene expression by affecting transcription and translation, in the 3′UTR they can be responsible for gene silencing or transcription slippage, and in introns they can affect gene transcription, mRNA splicing, or export to cytoplasm [35,52]. Ultimately, all these polymorphisms can affect phenotypes. However, the likelihood of polymorphism in different gene regions may vary. Our studies have shown a clear association between polymorphism of SSRs and their location in CDS and NCDS (Table 5). Highly polymorphic loci were located in introns, 5′UTRs, and upstreams, whereas loci with moderate polymorphism and monomorphic ones were located in exons. This relationship has been observed for both *Fragaria* and *Rubus*. For example, the *FaFS01* marker representing SSR in the second intron of the flavonol synthase gene was one of the most polymorphic in strawberries (14 alleles) and *Rubus* (eight alleles), while the *FaFS02* marker representing SSR in the first exon of the same gene was monomorphic in both genera. In general, SSRs located in introns had 5.5 and 6.0 alleles per locus on average in *Fragaria* and *Rubus*, respectively, whereas in SSRs located in exons—only 1.8 and 3.0 alleles per locus, respectively. Our results confirm the data of Du et al. [52] that intronic SSRs of *Populus tomentosa* were more variable (3.7 alleles/locus) than exonic SSRs (2.4 alleles/locus), likely due to higher selection pressure on CDS than on NCDS.

4.5. SSR Markers Representing Transcription Factor (TF) Genes

The TFs of MYB and bHLH families are widespread in plants. Using the lily transcriptome Biswas et al. [72] developed 71 SSRs in genes of 31 different TF families and most of them represented the bHLH TF (10 SSRs) and MYB (nine SSRs) families. The SSR markers in the TF gene in plants were first developed by Kujur et al. [73], which showed the influence of the SSR polymorphism on secondary structure of proteins and on such traits as seed weight and number of pods and seeds. Since then, there have been few similar studies in Rosaceae where similar markers have been developed for Japanese plum [24], but no such markers have been reported for *Rubus* and *Fragaria* before our first study, where we developed a SSR marker representing the *RiMYB10* TF gene, an activator of transcription of flavonoid biosynthesis [23]. The association between the *MYB10* gene and fruit color has been demonstrated for apple [45] and peach [74]. Gonzalez et al. [24] found three allele variants in the EST–SSRs designed for the *PsMYB10* TF gene. In our study of 21 varieties of red and black raspberries presented here, we showed that the *RiMY01* marker representing the *R. idaeus MYB10* gene had a significantly greater polymorphism compared to six markers representing structural genes of flavonoid biosynthesis—9 vs. 2–4 alleles per locus and *PIC* of 0.82 vs. 0.05–0.35 [23]. This is consistent with high PCR amplification efficiency and high polymorphism found for SSR markers representing TF genes in a number of crops [72].

In this study, we developed five SSRs using sequences of all publicly available TF genes of *Rubus* and *Fragaria*, which are part of the MBW transcriptional complex. As a result, two of the three most multiallelic markers were the *RiMY01* and *FaMY02* from TF genes. We assume that their high variability is caused by a combination of several factors: arrangement in introns, long dinucleotide repeats, and a noticeable AT enrichment. It is interesting that according to Biswas et al. [72] the SSRs in most TF genes were GC-rich, but genes in the bHLH and MYB families have higher AT content than other GC-rich genes. It is possible that this is a feature of these TF families, causing their increased variability. High level polymorphism of these SSR markers suggests association of alleles in flavonoid biosynthesis TF genes with variation in flavonoid accumulation and composition. Thus, we believe that SSR markers developed in this study and representing the TF genes can serve as useful tools for MAS of *Fragaria* and *Rubus* species.

4.6. Chitinase III as Allergen in Raspberry

Chitinases are able to degrade the chitin, a major component of fungal cell walls, and they play key roles in plant defense system from fungal pathogens. Plant chitinases are also a well-known group of food allergens. It is a relatively small group, but it is present in highly consumed fruits [75]. Chitinases can be divided into five classes (I-V), and plant class III chitinases (PR-8 proteins) have an additional lysozyme activity, which is not found in other classes of chitinases [76]. Marzban et al. [33] identified four allergen proteins including class III chitinase in raspberry fruits. They showed high sequence identity of raspberry proteins to different PR protein families in Rosaceous species (especially to Fra a strawberry proteins) and suggest that the consumption of raspberries might be responsible for adverse reactions in sensibilized individuals. We sequenced fragments of chitinase III genes in three Russian raspberry cultivars that have yellow-, orange- and red-colored berries and compared them with the chitinase III genes in Chandler strawberry (GenBank AF134347) and unknown raspberry [33]. These sequences matched both raspberry and strawberry sequences, but there were nonsynonymous substitutions in the raspberry sequences. Two nonsynonymous substitutions differed both strawberry and previous published raspberry chitinase III sequences. The amino acid sequences of all three raspberry cultivars with different berry colors were identical.

It is known that fruit color correlates with allergen content in strawberry. Hjerno et al. [77] demonstrated that the strawberry Fra a 1 allergen (a homolog of the major birch pollen allergen Bet v 1) is synthesized in red ripe fruits of *F.* × *ananassa*, but not in white (colorless) cultivars. Proteomic analyses have shown that Fra a 1 allergen and several enzymes of the flavonoid biosynthesis pathway were down-regulated. Later, suppression of three Fra a genes using RNAi approach alters phenolic compound levels in strawberry and led to decreased accumulation of main anthocyanins responsible for the red color of fruits [78]. Finally, Casanal et al. [79] suggested that Fra a proteins may play an important role in the control of flavonoid pathways by binding to metabolic intermediates. It can be assumed that a similar connection exists in raspberry, and that future studies in this direction are necessary. If this hypothesis is confirmed, yellow-colored raspberry cultivars will have less allergenic potential and are suitable for feeding children with increased allergenic sensitivity.

5. Conclusions

In this study, we developed a set of SSR markers representing structural and regulatory flavonoid biosynthesis genes of berry crops. These genic SSRs have been successful to identify allelic variations in *Fragaria* and *Rubus* species with contrasting color berry phenotypes. The results suggest that boreal *Rubus* species are more related to *Fragaria* compared to raspberry and blackberry. We demonstrated a clear relationship between transferability of SSR markers within and between the *Fragaria* and *Rubus* genera and location of primer binding sites and between the marker polymorphism and its location in coding and non-coding sequences. The SSR markers representing TF genes showed high allelic variability and may be good candidates for MAS in berry species. The genic SSRs developed in this study may be used for future genetic diversity and population genetics studies in *Fragaria* and *Rubus* species, as well as may be considered as candidate markers in breeding programs for improvement anthocyanin related traits.

Supplementary Materials: The following are available online at http://www.mdpi.com/2073-4425/11/1/11/s1, Figure S1: Alignment of the chitinase III amino acid sequences from strawberry and raspberry cultivars *F.* × *ananassa* cv. Chandler (AF134347), *R. idaeus* of unknown origin (Marzban et al. [33]), *R. idaeus* cvs. Zolotaya Osen (MK333194), Oranzhevoe Chudo (MK333195), and Babye Leto II (MK333196). Identical amino acid positions are highlighted by the same color, Table S1: Data on 25 SSR loci and their PCR primer pairs used to genotype *Fragaria* and *Rubus* specimens.

Author Contributions: Conceptualization, V.G.L. and K.A.S.; Methodology, V.G.L. and K.A.S.; Investigation, V.G.L., N.M.S., O.P.M. and K.A.S.; Formal Analysis, V.G.L., O.P.M. and T.N.L.; Visualization, T.N.L.; Data curation, V.G.L., K.V.K. and K.A.S.; Funding Acquisition, V.G.L., K.V.K. and K.A.S.; Project Administration, V.G.L. and K.A.S.; Resources, V.G.L. and K.A.S.; Supervision, V.G.L. and K.A.S.; Writing, V.G.L., K.V.K. and K.A.S. All authors have read and agreed to the published version of the manuscript.

Funding: The work was financially supported by the Ministry of Education and Science of the Russian Federation (grant No. 075-15-2019-1205 from 04.06.2019, unique project identifier RFMEFI57417X0149).

Acknowledgments: We thank I. A. Pozdniakov (OOO Microklon, Pushchino, Russia) for providing us with plant materials used in this study.

Conflicts of Interest: The authors declare no conflict of interest.

References

1. Potter, D.; Eriksson, T.; Evans, R.C.; Oh, S.; Smedmark, J.E.E.; Morgan, D.R.; Kerr, M.; Robertson, K.R.; Arsenault, M.; Dickinson, T.A.; et al. Phylogeny and classification of *Rosaceae*. *Plant System. Evol.* **2007**, *266*, 5–43. [CrossRef]

2. Foster, T.M.; Bassil, N.V.; Dossett, M.; Worthington, M.L.; Graham, J. Genetic and genomic resources for *Rubus* breeding: A roadmap for the future. *Hort. Res.* **2019**, *6*, 116. [CrossRef] [PubMed]

3. Mezzetti, B.; Giampieri, F.; Zhang, Y.-T.; Zhong, C.-F. Status of strawberry breeding programs and cultivation systems in Europe and the rest of the world. *J. Berry Res.* **2018**, *8*, 205–221. [CrossRef]

4. FAOSTAT. 2019. Available online: http://www.fao.org/faostat (accessed on 23 October 2019).

5. Graham, J.; Jennings, S.N. Raspberry breeding. In *Breeding Tree Crops*; Jain, S.M., Priyadarshan, M., Eds.; IBH & Science Publication: Oxford, UK, 2009; pp. 233–248.

6. Lerceteau-Kohler, E.; Moing, A.; Guerin, G.; Renaud, C.; Petit, A.; Rothan, C.; Denoyes, B. Genetic dissection of fruit quality traits in the octoploid cultivated strawberry highlights the role of homoeo-QTL in their control. *Theor. Appl. Genet.* **2012**, *124*, 1059–1077. [CrossRef]

7. Swanson, J.-D.; Weber, C.; Finn, C.E.; Fernández-Fernández, F.; Sargent, D.J.; Carlson, J.E.; Graham, J. Raspberries and blackberries. In *Genetics, Genomics and Breeding of Berries*; Folta, K.M., Kole, C., Eds.; CRC Press, Science Publishers: Manchester, NH, USA, 2011; pp. 64–113.

8. Liston, A.; Cronn, R.; Ashman, T.-L. *Fragaria*: A genus with deep historical roots and ripe for evolutionary and ecological insights. *Am. J. Bot.* **2014**, *101*, 1686–1699. [CrossRef]

9. Edger, P.P.; Poorten, T.J.; Vanburen, R.; Hardigan, M.A.; Colle, M.; Mckain, M.R.; Smith, R.D.; Teresi, S.J.; Nelson, A.D.L.; Wai, C.M.; et al. Origin and evolution of the octoploid strawberry genome. *Nat. Genet.* **2019**, *51*, 541–547. [CrossRef]

10. Hilmarsson, H.S.; Hytonen, T.; Isobe, S.; Goransson, M.; Toivainen, T.; Hallsson, J.H. Population genetic analysis of a global collection of *Fragaria vesca* using microsatellite markers. *PLoS ONE* **2017**, *12*, e0183384. [CrossRef]

11. Du, J.; Lv, Y.; Xiong, J.; Ge, C.; Iqbal, S.; Qiao, Y. Identifying genome-wide sequence variations and candidate genes implicated in self-incompatibility by resequencing *Fragaria viridis*. *Int. J. Mol. Sci.* **2019**, *20*, 1039. [CrossRef]

12. Davis, T.; Denoyes-Rothan, B.; Lerceteau-Köhler, E. Strawberry. In *Fruits and Nuts*; Kole, C., Ed.; Springer: Berlin/Heidelberg, Germany, 2007; pp. 189–205.

13. Bushakra, J.M.; Lewers, K.S.; Staton, M.E.; Zhebentyayeva, T.; Saski, C.A. Developing expressed sequence tag libraries and the discovery of simple sequence repeat markers for two species of raspberry (*Rubus L.*). *BMC Plant Biol.* **2015**, *15*, 258. [CrossRef]

14. Amsellem, L.; Dutech, C.; Billotte, N. Isolation and characterization of polymorphic microsatellite loci in *Rubus alceifolius* Poir (*Rosaceae*), an invasive weed in La Reunion island. *Mol. Ecol. Notes* **2001**, *1*, 33–35. [CrossRef]

15. James, C.M.; Wilson, F.; Hadonou, A.M.; Tobutt, K.R. Isolation and characterization of polymorphic microsatellites in diploid strawberry (*Fragaria vesca* L.) for mapping, diversity studies and clone identification. *Mol. Ecol. Notes* **2003**, *3*, 171–173. [CrossRef]

16. Mnejja, M.; Garcia-Mas, J.; Audergon, J.-M.; Arús, P. Prunus microsatellite marker transferability across rosaceous crops. *Tree Genet. Genomes* **2010**, *6*, 689–700. [CrossRef]

17. Bushakra, J.M.; Stephens, M.J.; Atmadjaja, A.N.; Lewers, K.S.; Symonds, V.V.; Udall, J.A.; Chagne, D.; Buck, E.J.; Gardiner, S.E. Construction of black (*Rubus occidentalis*) and red (*R.idaeus*) raspberry linkage maps and their comparison to the genomes of strawberry, apple, and peach. *Appl. Genet.* **2012**, *125*, 311–327. [CrossRef] [PubMed]

18. Graham, J.; Brennan, R. Introduction to the *Rubus* Genus. In *Raspberry: Breeding, Challenges and Advances*; Graham, J., Brennan, R., Eds.; Springer Nature Switzerland AG: Cham, Switzerland, 2018; pp. 1–16.

19. Woodhead, M.; Weir, A.; Smith, K.; McCallum, S.; Machenzie, K.; Graham, J. Functional markers for red raspberry. *J. Am. Soc. Hortic. Sci.* **2010**, *135*, 418–427. [CrossRef]

20. Lewers, K.S.; Weber, C.A. The trouble with genetic mapping of raspberry. *HortScience* **2005**, *40*, 1108. [CrossRef]

21. Zorrilla-Fontanesi, Y.; Cabeza, A.; Torres, A.M.; Botella, M.A.; Valpuesta, V.; Monfort, A.; Sanchez-Sevilla, J.F.; Amaya, I. Development and bin mapping of strawberry genic-SSRs in diploid *Fragaria* and their transferability across the Rosoideae subfamily. *Mol. Breed.* **2011**, *27*, 137–156. [CrossRef]

22. Varshney, R.K.; Graner, A.; Sorrells, M.E. Genic microsatellite markers in plants: Features and applications. *Trends Biotech.* **2005**, *23*, 48–55. [CrossRef]

23. Lebedev, V.G.; Subbotina, N.M.; Maluchenko, O.P.; Krutovsky, K.V.; Shestibratov, K.A. Assessment of genetic diversity in differently colored raspberry cultivars using SSR markers located in flavonoid biosynthesis genes. *Agronomy* **2019**, *9*, 518. [CrossRef]

24. Gonzalez, M.; Salazar, E.; Castillo, J.; Morales, P.; Mura-Jornet, I.; Maldonado, J.; Silva, H.; Carrasco, B. Genetic structure based on EST-SSR: A putative tool for fruit color selection in Japanese plum (*Prunus salicina* L.) breeding programs. *Mol. Breed.* **2016**, *36*, 1–15. [CrossRef]

25. Jaakola, L. New insights into the regulation of anthocyanin biosynthesis in fruits. *Trends Plant Sci.* **2013**, *18*, 477–483. [CrossRef]

26. Hall, H.K.; Kampler, C. *Raspberry Breeding. Fruit, Vegetable and Cereal Science and Biotechnology*; Isleworth: London, UK, 2011; Volume 5, pp. 44–62.

27. Martins, W.S.; Lucas, D.C.S.; Neves, K.F.S.; Bertioli, D.J. WebSat—A web software for microsatellite marker development. *Bioinformation* **2009**, *3*, 282–283. [CrossRef] [PubMed]

28. Castillo, N.R.F.; Reed, B.M.; Graham, J.; Fernandez-Fernandez, F.; Bassil, N.V. Microsatellite markers for raspberry and blackberry. *J. Am. Soc. Hortic. Sci.* **2010**, *135*, 271–278. [CrossRef]

29. Nunes, C.F.; Ferreira, J.L.; Nunes-Fernandes, M.C.; de Souza Breves, S.; Generoso, A.L.; Fontes-Soares, B.D.; Carvalho-Dias, M.S.; Pasqual, M.; Borem, A.; de Almeida Cancado, G.M. An improved method for genomic DNA extraction from strawberry leaves. *Ciência Rural* **2011**, *41*, 1383–1389. [CrossRef]

30. Liu, K.; Muse, S.V. PowerMarker: An integrated analysis environment for genetic marker analysis. *Bioinformatics* **2005**, *21*, 2128–2129. [CrossRef]

31. Meirmans, P.G.; Van Tienderen, P.H. GENOTYPE and GENODIVE: Two programs for the analysis of genetic diversity of asexual organisms. *Mol. Ecol. Notes* **2004**, *4*, 792–794. [CrossRef]

32. Grandin, U. PC-ORD version 5: A user-friendly toolbox for ecologists. *J. Veg. Sci.* **2006**, *17*, 843–844. [CrossRef]

33. Marzban, G.; Herndl, A.; Kolarich, D.; Maghuly, F.; Mansfeld, A.; Hemmer, W.; Katinger, H.; Laimer, M. Identification of four IgE-reactive proteins in raspberry (*Rubus ideaeus* L.). *Mol. Nutr. Food Res* **2008**, *52*, 1497–1506. [CrossRef]

34. Vidal, N.M.; Grazziotin, A.L.; Ramos, H.C.C.; Pereira, M.G.; Venancio, T.M. Development of a gene-centered SSR atlas as a resource for papaya (*Carica papaya*) marker-assisted selection and population genetic studies. *PLoS ONE* **2014**, *9*, e112654. [CrossRef]

35. Vieira, M.L.C.; Santini, L.; Diniz, A.L.; Munhoz, C.F. Microsatellite markers: What they mean and why they are so useful. *Genet. Mol. Biol.* **2016**, *39*, 312–328. [CrossRef]

36. Oyant, L.H.-S.; Crespel, L.; Rajapakse, S.; Zhang, L.; Foucher, F. Genetic linkage maps of rose constructed with new microsatellite markers and locating QTL controlling flowering traits. *Tree Genet. Genomes* **2008**, *4*, 11–23. [CrossRef]

37. Park, Y.H.; Ahna, S.G.; Choia, Y.M.; Oha, H.J.; Ahnb, D.C.; Kimb, J.G.; Kanga, J.S.; Choia, Y.W.; Jeong, B.R. Rose (*Rosa hybrida* L.) EST-derived microsatellite markers and their transferability to strawberry (*Fragaria* spp.). *Sci. Hortic.* **2010**, *125*, 733–739. [CrossRef]

38. Liu, W.; Jia, X.; Liu, Z.; Zhang, Z.; Wang, Y.; Liu, Z.; Xie, W. Development and characterization of transcription factor gene-derived microsatellite (TFGM) markers in *Medicago truncatula* and their transferability in leguminous and non-leguminous species. *Molecules* **2015**, *20*, 8759–8771. [CrossRef] [PubMed]

39. Sargent, D.J.; Rys, A.; Nier, S.; Simpson, D.W.; Tobutt, K.R. The development and mapping of functional markers in *Fragaria* and their transferability and potential for mapping in other genera. *Theor. Appl. Genet.* **2007**, *114*, 373–384. [CrossRef]

40. Cerezo, A.B.; Cuevas, E.; Winterhalter, P.; Garcia-Parrilla, M.C.; Troncoso, A.M. Isolation, identification, and antioxidant activity of anthocyanin compounds in *Camarosa strawberry*. *Food Chem.* **2010**, *123*, 574–582. [CrossRef]

41. Mazur, S.P.; Nes, A.; Wold, A.-B.; Remberg, S.F.; Aaby, K. Quality and chemical composition of ten red raspberry (*Rubus idaeus* L.) genotypes during three harvest seasons. *Food Chem.* **2014**, *160*, 233–240. [CrossRef]

42. Xu, W.; Dubos, C.; Lepiniec, L. Transcriptional control of flavonoid biosynthesis by MYB-bHLH-WDR complexes. *Trends Plant Sci.* **2015**, *20*, 176–185. [CrossRef]

43. Luo, P.; Ning, G.; Wang, Z.; Shen, Y.; Jin, H.; Li, P.; Huang, S.; Zhao, J.; Bao, M. Disequilibrium of flavonol synthase and dihydroflavonol-4-reductase expression associated tightly to white vs. red color flower formation in plants. *Front. Plant Sci.* **2016**, *6*, 1257. [CrossRef]

44. Zheng, D.; Schröder, G.; Schröder, J.; Hrazdina, G. Molecular and biochemical characterization of three aromatic polyketide synthase genes from *Rubus idaeus*. *Plant Mol. Biol.* **2001**, *46*, 1–15. [CrossRef]

45. Lin-Wang, K.; Bolitho, K.; Grafton, K.; Kortstee, A.; Karunairetnam, S.; McGhie, T.K.; Espley, R.V.; Hellens, R.P.; Allan, A.C. An R2R3 MYB transcription factor associated with regulation of the anthocyanin biosynthetic pathway in *Rosaceae*. *BMC Plant Biol.* **2010**, *10*, 50. [CrossRef]

46. Kassim, A.; Poette, J.; Paterson, A.; Zait, D.; McCallum, S.; Woodhead, M.; Smith, K.; Hackett, C.; Graham, J. Environmental and seasonal influences on red raspberry anthocyanin antioxidant contents and identification of quantitative traits loci (QTL). *Mol. Nutr. Food Res.* **2009**, *53*, 625–634. [CrossRef] [PubMed]

47. Tortora, M.L.; Díaz-Ricci, J.C.; Pedraza, R.O. Protection of strawberry plants (*Fragaria ananassa* Duch.) against anthracnose disease induced by *Azospirillum Bras*. *Plant Soil* **2012**, *356*, 279–290. [CrossRef]

48. Dobes, C.H.; Scheffknecht, S. Isolation and characterization of microsatellite loci for the *Potentilla* core group (Rosaceae) using 454 sequencing. *Mol. Ecol. Resour.* **2012**, *12*, 726–739. [CrossRef] [PubMed]

49. Lee, G.-A.; Song, J.Y.; Choi, H.-R.; Chung, J.-W.; Jeon, Y.-A.; Lee, J.-R.; Ma, K.-H.; Lee, M.-C. Novel microsatellite markers acquired from *Rubus coreanus* Miq. and cross-amplification in other *Rubus* species. *Molecules* **2015**, *20*, 6432–6442. [CrossRef] [PubMed]

50. Parida, S.K.; Verma, M.; Yadav, S.K.; Ambawat, S.; Das, S.; Garg, R.; Jain, M. Development of genome-wide informative simple sequence repeat markers for large-scale genotyping applications in chickpea and development of web resource. *Front. Plant Sci.* **2015**, *6*, 645. [CrossRef] [PubMed]

51. Wang, H.; Walla, J.A.; Zhong, S.; Huang, D.; Dai, W. Development and cross-species/genera transferability of microsatellite markers discovered using 454 genome sequencing in chokecherry (*Prunus virginiana* L.). *Plant Cell Rep* **2012**, *31*, 2047–2055. [CrossRef]

52. Du, Q.; Gong, C.; Pan, W.; Zhang, D. Development and application of microsatellites in candidate genes related to wood properties in the chinese white poplar (*Populus tomentosa* Carr.). *DNA Res.* **2013**, *20*, 31–44. [CrossRef]

53. Bosamia, T.C.; Mishra, G.P.; Thankappan, R.; Dobaria, J.R. Novel and stress relevant EST derived SSR markers developed and validated in peanut. *PLoS ONE* **2015**, *10*, e0129127. [CrossRef]

54. Aaby, K.; Mazur, S.; Nes, A.; Skrede, G. Phenolic compounds in strawberry (*Fragaria x ananassa* Duch.) fruits: Composition in 27 cultivars and changes during ripening. *Food Chem.* **2012**, *132*, 86–97. [CrossRef]

55. Aaby, K.; Skareta, J.; Røen, D.; Sønstebyc, A. Sensory and instrumental analysis of eight genotypes of red raspberry (*Rubus idaeus* L.) fruits. *J. Berry Res.* **2019**, *9*, 1–16. [CrossRef]

56. Leišová-Svobodová, L.; Phillips, J.; Martinussen, I.; Holubec, V. Genetic differentiation of *Rubus chamaemorus* populations in the Czech Republic and Norway after the last glacial period. *Ecol. Evol.* **2018**, *8*, 5701–5711. [CrossRef] [PubMed]

57. Kostamo, K.; Toljamo, A.; Antonius, K.; Kokko, H.; Karenlampi, S.L. Morphological and molecular identification to secure cultivar maintenance and management of self-sterile *Rubus* arcticus. *Ann. Bot.* **2013**, *111*, 713–721. [CrossRef] [PubMed]

58. Hirakawa, H.; Shirasawa, K.; Kosugi, S.; Tashiro, K.; Nakayama, S.; Yamada, M.; Kohara, M.; Watanabe, A.; Kishida, Y.; Fujishiro, T.; et al. Dissection of the octoploid strawberry genome by deep sequencing of the genomes of *Fragaria* species. *DNA Res.* **2014**, *21*, 169–181. [CrossRef]

59. Bassil, N.V.; Gunn, M.; Folta, K.M.; Lewers., K.S. Microsatellite markers for *Fragaria* from 'Strawberry Festival' expressed sequence tags. *Mol. Ecol. Notes* **2006**, *6*, 473–476. [CrossRef]

60. Graham, J.; Smith, K.; Woodhead, M.; Russell, J. Development and use of simple sequence repeat SSR markers in *Rubus* species. *Mol. Ecol. Notes* **2002**, *2*, 250–252. [CrossRef]

61. Fernandez-Fernandez, F.; Antanaviciute, L.; Govan, C.L.; Sargent, D.J. Development of a multiplexed microsatellite set for fingerprinting red raspberry (*Rubus idaeus*) germplasm and its transferability to other *Rubus* species. *J. Berry Res.* **2011**, *1*, 177–187. [CrossRef]

62. Stafne, E.T.; Clark, J.R.; Weber, C.A.; Graham, J.; Lewers, K.S. Simple sequence repeat (SSR) markers for genetic mapping of raspberry and blackberry. *J. Am. Soc. Hortic. Sci.* **2005**, *130*, 722–728. [CrossRef]

63. Qi, W.C.; Chen, X.; Fang, P.H.; Shi, S.C.; Li, J.J.; Liu, X.T.; Cao, X.Q.; Zhao, N.; Hao, H.Y.; Li, Y.J.; et al. Genomicand transcriptomic sequencing of *Rosa hybrida* provides microsatellite markers for breeding, flower trait improvement and taxonomy studies. *BMC Plant Biol.* **2018**, *18*, 119. [CrossRef]

64. Xiang, Y.; Huang, C.-H.; Hu, Y.; Wen, J.; Li, S.; Yi, T.; Chen, H.; Xiang, J.; Ma, H. Evolution of Rosaceae fruit types based on nuclear phylogeny in the context of geological times and genome duplication. *Mol. Biol. Evol.* **2017**, *34*, 262–281. [CrossRef]

65. Yoon, M.-Y.; Moe, K.T.; Kim, D.-Y.; Rho, I.-R.; Kim, S.; Kim, K.-T.; Won, M.-K.; Chung, J.-W.; Park, Y.-J. Genetic diversity and population structure analysis of strawberry (*Fragaria* × *ananassa* Duch.) using SSR markers. *Electron. J. Biotech.* **2012**, *15*, 1–16. [CrossRef]

66. Ashley, M.V.; Wilk, J.A.; Styan, S.M.N.; Craft, K.J.; Jones, K.L.; Feldheim, K.A.; Lewers, K.S.; Ashman, T.L. High variability and disomic segregation of microsatellites in the octoploid *Fragaria virginiana* Mill. (Rosaceae). *Theor. Appl. Genet.* **2003**, *107*, 1201–1207. [CrossRef] [PubMed]

67. Dossett, M.; Bassil, N.V.; Lewers, K.S.; Finn, C.E. Genetic diversity in wild and cultivated black raspberry (*Rubus occidentalis* L.) evaluated by simple sequence repeat markers. *Genet. Resour. Crop Evol.* **2012**, *59*, 1849–1865. [CrossRef]

68. Biswas, A.; Melmaiee, K.; Elavarthi, S.; Julian Jones, J.; Umesh Reddy, U. Characterization of strawberry (*Fragaria* spp.) accessions by genotyping with SSR markers and phenotyping by leaf antioxidant and trichome analysis. *Sci. Hortic.* **2019**, *256*, 108561. [CrossRef]

69. Vallarino, J.G.; de Abreu, E.; Lima, F.; Soria, C.; Tong, H.; Pott, D.M.; Willmitzer, L.; Fernie, A.R.; Nikoloski, Z.; Osorio, S. Genetic diversity of strawberry germplasm using metabolomic biomarkers. *Sci. Rep.* **2018**, *8*, 14386. [CrossRef]

70. Sanchez-Sevilla, J.F.; Horvath, A.; Botella, M.A.; Gaston, A.; Folta, K.; Kilian, A.; Denoyes, B.; Amaya, I. Diversity arrays technology (DArT) marker platforms for diversity analysis and linkage mapping in a complex crop, the octoploid cultivated strawberry (*Fragaria* × *ananassa*). *PLoS ONE* **2015**, *10*, e0144960. [CrossRef] [PubMed]

71. Simlat, M.; Ptak, A.; Kula, A.; Orzeł, A. Assessment of genetic variability among raspberry accessions using molecular markers. *Acta Sci. Pol. Hortorum Cultus* **2018**, *17*, 61–72. [CrossRef]

72. Biswas, M.K.; Nath, U.K.; Howlader, J.; Bagchi, M.; Natarajan, S.; Kayum, M.A.; Kim, H.-T.; Park, J.-I.; Kang, J.-G.; Nou, I.-S. Exploration and exploitation of novel SSR markers for candidate transcription factor genes in *Lilium* species. *Genes* **2018**, *9*, 97. [CrossRef]

73. Kujur, A.; Bajaj, D.; Saxena, M.S.; Tripathi, S.; Upadhyaya, H.D.; Gowda, C.L.; Singh, S.; Jain, M.; Tyagi, A.K.; Parida, S.K. Functionally relevant microsatellite markers from chickpea transcription factor genes for efficient genotyping applications and trait association mapping. *DNA Res.* **2013**, *20*, 355–374. [CrossRef]

74. Tuan, P.A.; Bai, S.; Yaegaki, H.; Tamura, T.; Hihara, S.; Moriguchi, T.; Oda, K. The crucial role of PpMYB10.1 in anthocyanin accumulation in peach and relationships between its allelic type and skin color phenotype. *BMC Plant Biol.* **2015**, *15*, 280. [CrossRef]

75. Leoni, C.; Volpicella, M.; Dileo, M.C.; Gattulli, B.A.; Ceci, L.R. Chitinases as food allergens. *Molecules* **2019**, *24*, 2087. [CrossRef]

76. Cao, S.; Wang, Y.; Li, Z.; Shi, W.; Gao, F.; Zhou, Y.; Zhang, G.; Feng, J. Genome-wide identification and expression analyses of the chitinases under cold and osmotic stress in *Ammopiptanthus nanus*. *Genes* **2019**, *10*, 472. [CrossRef] [PubMed]

77. Hjernø, K.; Alm, R.; Canbäck, B.; Matthiesen, R.; Trajkovski, K.; Björk, L.; Roespstorff, P.; Emanuelsson, C. Down-regulation of the strawberry Bet v 1-homologous allergen in concert with the flavonoid biosynthesis pathway in colorlessstrawberry mutant. *Proteomics* **2006**, *6*, 1574–1587. [CrossRef] [PubMed]

78. Muñoz, C.; Hoffmann, T.; Escobar, N.M.; Ludemann, F.; Botella, M.A.; Valpuesta, V.; Schwab, W. The strawberry fruit Fra a allergen functions in flavonoid biosynthesis. *Mol. Plant* **2010**, *3*, 113–124. [CrossRef] [PubMed]

79. Casanal, A.; Zander, U.; Munoz, C.; Dupeux, F.; Luque, I.; Botella, M.A.; Schwab, W.; Valpuesta, V.; Marquez, J.A. The strawberry pathogenesis-related 10 (PR-10) Fra a proteins control flavonoid biosynthesis by binding to metabolic intermediates. *J. Biol. Chem.* **2013**, *288*, 35322–35332. [CrossRef] [PubMed]

 *genes*

Article

The Molecular Determination of Hybridity and Homozygosity Estimates in Breeding Populations of Lettuce (*Lactuca sativa* L.)

Alice Patella, Fabio Palumbo *, Giulio Galla and Gianni Barcaccia

Department of Agronomy, Food, Natural Resources, Animals and Environment, University of Padova, 35020 Legnaro PD, Italy; alice.patella@phd.unipd.it (A.P.); giulio.galla@unipd.it (G.G.); gianni.barcaccia@unipd.it (G.B.)
* Correspondence: fabio.palumbo@unipd.it

Received: 23 September 2019; Accepted: 7 November 2019; Published: 9 November 2019

Abstract: The development of new varieties of horticultural crops benefits from the integration of conventional and molecular marker-assisted breeding schemes in order to combine phenotyping and genotyping information. In this study, a selected panel of 16 microsatellite markers were used in different steps of a breeding programme of lettuce (*Lactuca sativa* L., $2n = 18$). Molecular markers were first used to genotype 71 putative parental lines and to plan 89 controlled crosses designed to maximise recombination potentials. The resulting 871 progeny plants were then molecularly screened, and their marker allele profiles were compared with the profiles expected based on the parental lines. The average cross-pollination success rate was $68 \pm 33\%$, so 602 F1 hybrids were completely identified. Unexpected genotypes were detected in 5% of cases, consistent with this species' spontaneous out-pollination rate. Finally, in a later step of the breeding programme, 47 different F3 progenies, selected by phenotyping for a number of morphological descriptors, were characterised in terms of their observed homozygosity and within-population genetic uniformity and stability. Ten of these populations had a median homozygosity above 90% and a median genetic similarity above 95% and are, therefore, particularly suitable for pre-commercial trials. In conclusion, this study shows the synergistic effects and advantages of conventional and molecular methods of selection applied in different steps of a breeding programme aimed at developing new varieties of lettuce.

Keywords: pure lines; F1 hybrids; microsatellite markers; marker-assisted breeding; crop improvement; varieties

---

1. Introduction

Lettuce (*Lactuca sativa* L.) is a self-pollinating leafy vegetable species ($2n = 2x = 18$) of the Asteraceae family. It is cultivated on a large scale throughout the world for consumption as a fresh vegetable on its own or in combination with other ready-to-eat vegetable products [1]. Its growing economic importance has led seed companies to regularly develop new varieties with ever higher agronomic traits. However, breeding programmes are highly limited by the reproductive system of this species. The flower structure of lettuce determines a reproductive strategy known as cleistogamy, in which anther dehiscence and subsequent pollination take place before flower opening, resulting in a very high rate of self-pollination, very often equal to or close to 100% [2]. According to recent estimates, out-cross rates are limited to 1–6% [3]. These reproductive barriers mean that in natural conditions the species spontaneously constitutes pure lines, characterised by phenotypic uniformity and genotypic stability, due to their very high homozygosity. In conventional breeding programmes, developing segregating and recombinant F2 populations traditionally requires crosses to be hand pollinated while self-pollination is prevented by emasculating the flowers. The most popular emasculation and hand-pollination technique is that described by Olivier [4]. Known as the "wash method", it involves hand-spraying the inflorescence with water during pistil emergence to remove the pollen attached to the female part of the flower. The inflorescence is then left to dry for a short period, after which it is rubbed with a ripe flower of the pollinating variety [5]. A slightly different, but also widely used, technique is the "clip-and-wash method", which involves clipping the tip of the corolla before spraying with water. This guarantees more efficient pollen removal and cross rates close to 100% from the subsequent manual pollination [2]. However, these breeding techniques are time-consuming and technically highly demanding, and are only really effective if coupled with molecular analyses aimed at screening progeny plants and assessing their hybridity.

In recent years, many seed firms have begun using molecular markers to carry out assisted selection schemes and to speed up varietal development programmes [3]. Simple Sequence Repeat (SSR) markers are, so far, the most commonly used markers for these purposes [6–8] as they are codominant, have high reproducibility and multi-allelism, and can be detected at any stage of plant development, without being influenced by the environment [9]. There are a considerable number of SSR markers for lettuce in the literature [10]. Truco, et al. [11] produced an integrated genome map from 7 different linkage maps, which included 130 SSR loci organised in 9 linkage groups. Rauscher and Simko [10] augmented this genomic map with 54 genomic SSR and 52 EST-SSR (Expressed Sequence Tag) loci. Finally, with the publication of the *L. sativa* genome draft [12], tens of thousands of new SSR regions have become available for testing and use.

Given the availability of markers in lettuce, Marker-Assisted Selection (MAS) has started to be adopted in plant breeding programmes for various purposes, including identification of resistance genes [13,14] or Quantitative Trait Loci (QTLs) of phytopathogens [15,16], the study of QTLs controlling complex traits [17,18], and investigation of the genetic identity and purity of either experimental or commercial lines [19]. On the other hand, very few attempts have been made to prove the efficiency of molecular markers in Marker-Assisted Breeding (MAB) activities, where the genotypic background is molecularly investigated to complement traditional phenotypic selection [20].

In this work, SSR markers were used in three different steps of a conventional breeding scheme aimed at developing new varieties characterised by distinctiveness, uniformity, and stability (Figure 1).

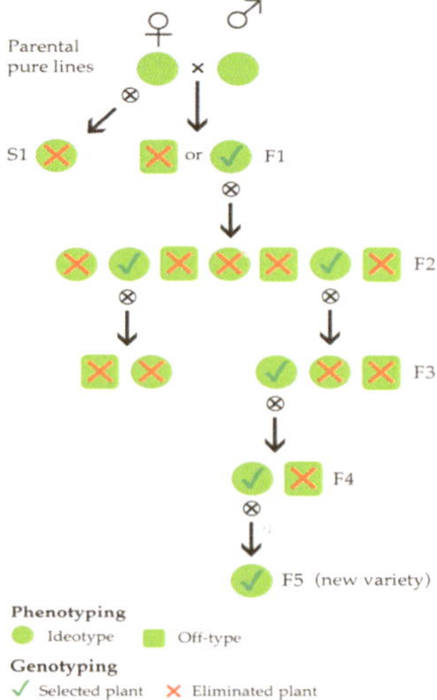

Figure 1. Simplified overview of a lettuce breeding scheme in which selection is based on both plant phenotyping and genotyping.

We first examined the genetic background of a number of superior pure lines in order to plan experimental matings to produce F1 hybrids and then derived F2 progenies manifesting morphological variability as a result of genetic segregation and recombination (Figure 1). Each offspring in the F1 generation was analysed to distinguish the individuals resulting from planned out-crosses from those resulting from accidental selfing (Figure 1). After genotyping, the S1 individuals were discarded, and the F1 individuals were self-pollinated. In the F3 generations (Figure 1); experimental populations, previously selected according to their morphological traits, were also characterised by molecular markers due to the need to assess their stability and uniformity in order to run pre-commercial trials.

2. Materials and Methods

2.1. Plant Materials and Breeding Techniques

Plant materials were developed and provided by Blumen Group SpA, Italy and belonged to five different lettuce cultivar types (Table S1). Seventy-one parental lines (germplasm composed of experimental, pre-commercial and commercial lines) were involved in 89 combinations of crosses, in which each progeny consisted of 6–12 individuals (871 progeny samples). Parental lines were grown in the spring of 2015, and the 89 programmed crosses were carried out in the summer using the clip-and-wash method [2]. This involved making an incision in the calyx and corolla and washing the anthers in the early morning before the pollen grains could settle naturally on the outermost stigmatic surface of pistils. The plants were then manually pollinated by rubbing anthers of the pollen donor on the stigma of the seed parent. For each planned cross, a bulk of 4/5 flowers from a pollinator parent was used to pollinate as many flowers of a seed plant. Seeds were collected from the seed plant and sown in early autumn for genotyping selection and agronomic evaluation (spring 2016).

Finally, to assess the uniformity of the 47 experimental lines, previously chosen for morpho-phenological traits and pathogen resistances, 940 samples belonging to the 47 F3 populations (labelled 1 to 47) were collected in the spring of 2018. Each experimental line comprised 20 individuals.

2.2. DNA Isolation

A total of 100 mg of fresh leaves was collected from each of the 1882 lettuce samples (71 parents, 871 progeny and 940 F3) and ground to a fine powder using Tissue Lyser II (Qiagen, Valencia, CA, USA). Genomic DNA (gDNA) was extracted with the Dneasy® 96 Plant Kit (Qiagen), according to the manufacturer's protocols. After extraction, the integrity of the gDNA was assessed by electrophoresis on 1% (*w/v*) agarose gel stained with SYBR Safe® 1 × DNA Gel Stain (Life Technologies, Carlsbad, CA, USA) in Tris-Acetate-EDTA (TAE) running buffer. Both the yield and purity of the extracted gDNA samples were evaluated using a NanoDrop 2000c UV-Vis Spectrophotometer (Thermo Scientific, Carlsbad, CA, USA). Following DNA quantification, all DNA samples were diluted to a final concentration of 20 ng/µL.

2.3. Primer Design and Testing of SSR Marker Amplification

Sixteen SSR marker loci were selected from those available in the scientific literature [10,21], according to (i) chromosomal location; (ii) polymorphism rate, expressed as PIC (Polymorphism Information Content); (iii) allele size range; (iv) annealing temperature of the locus-specific primers. Amplifications were performed according to the method previously described by Schuelke [22], with some minor modifications. Briefly, three primers were used to amplify each microsatellite locus: a pair of locus-specific primers, one with an oligonucleotide tail at the 5′ end (M13, PAN-1, PAN-2 or PAN-3, Table S2), and a third universal primer complementary to the tail and labelled with a fluorescent dye (6-FAM, VIC, NED, or PET). Primer pair efficiency was tested in silico using the PRaTo [23] web-tool and were organised in three multiplex reactions, as shown in Table 1.

Table 1. Microsatellite loci information. For each primer pair, the original simple sequence repeat (SSR) name, ID used in this work, linkage group [10,21], SSR motif, primer sequences (PAN1, PAN2, PAN3, or M13 tails at the 5′ end are indicated in square brackets; for further details see Table S2), dye and the multiplex to which the SSR marker locus belongs is shown.

Marker Name	ID	LG	Motif		Primer Sequence	Dye	Multiplex
LSSA27-2 [10]	Lsat1	1	$(AC)_7$	For	[M13]CACACTACCACCCAACACG	6-FAM	1
				Rev	ACCCTCTTCGCTTCTTCTT		
SML-045 [21]	Lsat2	2	$(AAG)_{9/12}$	For	ACAAAACCGTTTCACCCAAA	6-FAM	1
				Rev	[M13]AGCCCTGTCCTCTTCAGGAT		
LSSB54 [10]	Lsat3	8	$(GT)_{10}$	For	[PAN1]CTTGAGAGTGCTTGGAGAGGAT	VIC	1
				Rev	CACATACAACAAGACAAGTCCCA		
LSSA05 [10]	Lsat4	8	$(TC)_{18}$	For	AGAACAACGGTAGCTTGTTAAATTG	VIC	1
				Rev	[PAN1]ATCGTCGGTTAATCTTCGTCG		
LSSA04 [10]	Lsat5	4	$(TC)_{14}$	For	[PAN2]AAGGAAAGGAAGGGTTGACTTGT	NED	1
				Rev	TTGGTGAAGAAAAGAGAGAGTTT		
LSSA11 [10]	Lsat6	3	$(CT)_{20}$	For	[PAN2]ACTCCCACTATCCTCTTTGCAT	NED	1
				Rev	GCCCACATTCTTAATCTTGTCC		
LSSA14 [10]	Lsat7	9	$(AG)_{18}$	For	[PAN3]TGATGACTCCAGTCTTAGATACCA	PET	1
				Rev	AGTCCCCGACTATCAGTCTCA		
LSSB09 [10]	Lsat8	2	$(TG)_8$	For	AGAATGAGAAGGATGAAATGGCTG	6-FAM	2
				Rev	[M13]AAACACCTTTAGCATCAAAATACCC		
SML-029 [21]	Lsat9	9	$(GAG)_{7/8}$	For	[M13]AGCCCAGAAGAGCGTGATTA	6-FAM	2
				Rev	TGCAGGGCTCCTTGATCTAC		
LSSB17-1 [10]	Lsat10	7	$(GT)_{11}$	For	ACTAGGGCTCTAATACAACTTGT	VIC	2
				Rev	[PAN1]TTGGCTTACAGTTATGGATTAAATG		
LSSA17 [21]	Lsat11	3	$(AG)_{21}$	For	[PAN1]AATGTGCGTGAGAGTTTCCTTT	VIC	2
				Rev	CAAGAAGGCAGTGATGAAGTTG		
LSSA12 [10]	Lsat12	5	$(GT)_{11}$	For	[PAN2]ACAAGGCCCAATCCTTTTCT	NED	2
				Rev	TCGAAAATTTGGAGAGAGTTTCTT		
LSSA15 [10]	Lsat13	1	$(AC)_{11}$	For	GCCCAACCCAAGAAGAGGAG	PET	2
				Rev	[PAN3]TGGAGAGGAGTGGAGAGTGTT		
LSSA28-1 [10]	Lsat14	4	$(GA)_{28}$	For	TTCATCTCTCTCCTCCTTCAGC	6-FAM	3
				Rev	[M13]ATCCCCATTGTCCTCCC		
LSSA21-1 [10]	Lsat15	8	$(TC)_{19}$	For	[PAN2]TTGTACCCAGTTGTCCAAACAG	NED	3
				Rev	CAGATTGTTGCAGATTTCTTCG		
LSSB68 [10]	Lsat16	na	$(CT)_{20}$	For	GTCTGTGTGGTTTTGGT	PET	3
				Rev	[PAN3]TGTGGTGGAGTGTGATTT		

The 16 primer pairs were first tested individually (singleplex reactions) using three randomly chosen lettuce gDNA to evaluate primer efficiency and to check the correspondence between expected and actual size of the bands; they were then evaluated in multiplex PCRs to assess possible primer interactions.

All amplification reactions (both singleplex and multiplex) were performed in a 10 μL reaction volume containing 1× Platinum® Multiplex PCR Master Mix (Thermo Scientific), 10% GC Enhancer (Thermo Scientific), 0.25 μM of non-tailed primer, 0.75 μM of tailed primer, 0.50 μM of fluorophore-labelled primer (universal primer) and 20 ng of genomic DNA. Thermal cycling conditions were as follows for multiplex 1 and 2:94 °C for 5 minutes followed by 6 cycles of 94 °C for 30 seconds, 61 °C for 30 seconds, 72 °C for 45 seconds, with a 1 °C annealing temperature stepdown per cycle (from 61 °C to 56 °C). The annealing temperature for the following 35 cycles was set at 56 °C, with denaturation and extension phases as above and a final extension at 60 °C for 30 minutes. The multiplex 3 thermal cycling conditions were instead: 94 °C for 5 minutes followed by 6 cycles of 94 °C for 30 seconds, 56 °C for 30 seconds, 72 °C for 45 seconds, with a 1 °C annealing temperature stepdown per cycle (from 56 °C to 51 °C). The annealing temperature for the following 35 cycles was set at 51 °C with denaturation and extension phases as above and a final extension at 60 °C for 30 minutes. All amplifications were performed in a GeneAmp® PCR 9700 thermal cycler (Applied Biosystems, Carlsbad, CA, USA). PCR products were first checked on gel electrophoresis (2% Ultrapure™ Agarose in TAE 1×, SYBR Safe® 1×, Life Technologies) then run on capillary electrophoresis with ABI 3730 DNA Analyzer (Applied Biosystem), using LIZ500 as the molecular weight standard. The size of each peak was determined with the Peak Scanner 1.0 software (Applied Biosystems).

2.4. Genotyping and Data Analysis

The 71 potential parents were genotyped at 16 SSR loci and statistical analyses were performed using NTSYS (Numerical Taxonomy and Multivariate Analysis System) version 2.2 (Exeter Software) [24]. Rohlf's (or the simple matching) coefficient was used to calculate pairwise genetic similarity (GS) in all possible comparisons and to construct a genetic similarity matrix, according to the formula:

$$GS_{ij} = m/(m + n) \tag{1}$$

where "i" and "j" are two different individuals, while "m" and "n" represent the number of matching and non-matching attributes, respectively. An unweighted pair group method with an arithmetic mean (UPGMA) dendrogram and a Principal Coordinates Analysis (PCoA) of parental lines were carried out using the Jaccard coefficient in the PAST software v. 3.14 with 10,000 bootstrap repetitions [25]. The genetic structure of the lines was modelled using a Bayesian clustering algorithm implemented in STRUCTURE v. 2.2 [26]. Since no *a priori* knowledge of the origin of the populations under study was available, the admixture model and then the correlated allele frequencies model were used. Ten replicate simulations were conducted for each value of K, with the number of founding groups ranging from 1 to 8, using a burn-in of 200,000 and 1,000,000 iterations. The most likely K Estimates were determined using the method described by Evanno et al. [26]. Estimates of membership were plotted as a histogram in an Excel spreadsheet. Finally, observed homozygosity (Ho) was determined with the POPGENE software [27].

The 89 subsequent crosses were planned according to the following criteria: (i) high genetic dissimilarity values among parents within the same lettuce cultivar type and between them, (ii) availability of informative loci able to distinguish between the resulting offspring and individuals resulting from accidental self-pollitated. Only homozygous loci for different alleles were considered informative, whereas heterozygous loci were taken into account only if the origin of the parental alleles could be clearly discerned in the progenies. The resulting offspring (871 samples) were then screened, with the analysis restricted to those SSR loci which had previously proven to be informative for hybrid detection. This made it possible to determine whether individuals belonging to a given F1 population originated from cross-pollination or self-pollination. The successful crosses (S.C.) rate of 89 was calculated as follows:

$$S.C. = (No\ F1 \times 100)/(No\ Tot - No\ U.G.) \tag{2}$$

where "No F1" is the number of hybrid individuals, "No Tot" is the number of all individuals in a progeny population (No tot = No F1 + No U.G. + No SP) and "No U.G." is the number of unexpected genotypes deriving from unplanned crosses.

Finally, 940 samples from the 47 F3 populations were genotyped using the previously-described panel of SSR markers. The POPGENE software [27] was used to compute the mean values of observed homozygosity for each population (3), where n is the total number of samples). In addition, the median of genetic similarity between the 47 lines was calculated using Rohlf's coefficient, which was designed for codominant molecular markers [28,29]. Comparison of genetic similarity among ten selected populations was instead calculated using the Jaccard coefficient, in accordance with the literature [30]. Genetic similarity matrices were generated in the NTSYS software [24].

$$\overline{Ho} = \sum_{n} Ho/n \tag{3}$$

3. Results

3.1. Parental Lines

The 16 SSR markers, organised in three multi-locus PCRs, were used firstly to amplify and score the 71 parental lines. Fourteen of the 16 SSR markers proved to be polymorphic among plant accessions. The similarity matrix constructed using Rohlf's coefficient revealed genetic similarity values ranging from 53% to 100% (Figure S1). The resulting unweighted pair group method with an arithmetic mean (UPGMA) dendrogram showed the samples clustering into two main sub-groups. Eighteen parental lines were not fully distinguishable, while the remaining 53 had unique genotypic profiles. The first principal coordinate from the PCoA accounted for 22% of the total variation and divided the samples into two groups, analogous to the clustering in the tree. The second principal coordinate accounted for 12% of the total variation. These results were confirmed by investigation of the genetic structure of the 71 parental lettuce lines based on allele frequencies; the best estimate of population size was $K = 2$ (Figure S2), such that the samples were grouped into two genetically distinct clusters (Figure 2). The lettuce cultivar types were reported in Table S1, but they did not correspond to different clusters in the UPGMA tree.

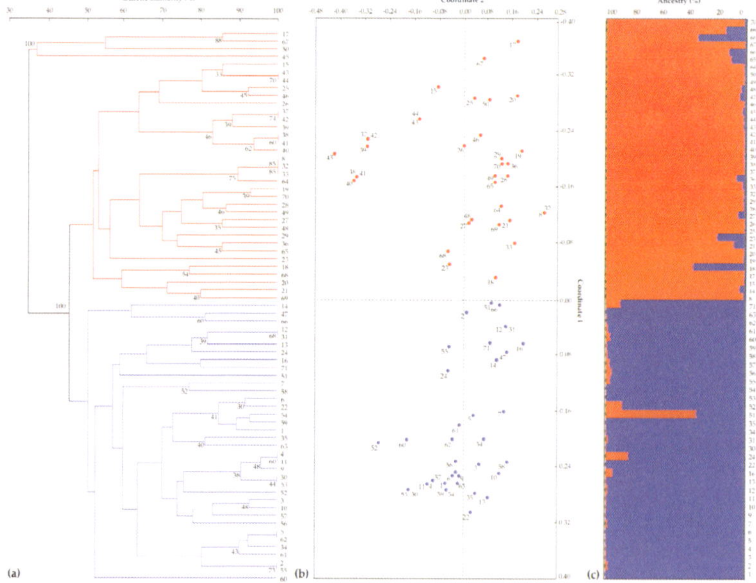

Figure 2. (a) Genetic similarity-based unweighted pair group method with an arithmetic mean (UPGMA) dendrogram of 71 parental lines calculated using the Jaccard coefficient. Bootstrap estimates ≥30% are reported next to the nodes (red and blue branches indicate the two clusters identified). (b) Principal coordinate analysis (PCoA). The 71 samples are shown in red or blue according to the clustering shown in the UPGMA tree. (c) The population genetic structure of the 71 lines as estimated by STRUCTURE. Each sample is represented by a vertical histogram partitioned into $K = 2$ coloured segments (red or blue, in accordance with (a,b)) representing the estimated membership. The proportion of subgroup membership (%) is reported on the ordinate axis, and the identification number of each accession is reported below each histogram.

The mean observed homozygosity was 82%, with a minimum value of 69% and a maximum of 100%. It is worth noting that 19 of the 71 parental lines (27%) had observed homozygosity values greater than 90%, while 30 of the 71 (42%) had a medium-high observed homozygosity (Ho) between

81% and 90%. Fourteen of the 71 parental lines (20%) had observed homozygosity ranging from 71% to 80%, and only 8 individuals had values lower than 70% (Figure 3a and Figure S1).

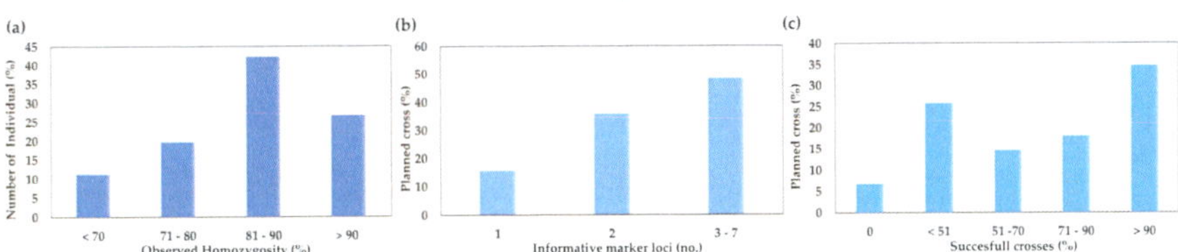

Figure 3. (**a**) Observed homozygosity of 71 lettuce parental individuals belonging to as many pure lines. (**b**) Histogram of discriminating loci in 89 cross combinations (in percentages). (**c**) Histogram of the percentages of pollination success in 89 programmed lettuce crosses.

3.2. Determination of Hybridity

Using a combination of genotypic and phenotypic data, 89 cross combinations were planned (Table S3). Before proceeding, we also checked the availability of informative loci able to distinguish between offspring resulting from out-cross and those obtained by accidental self-pollination. Screening identified 1 discriminant locus in 16% of cases, 2 informative loci in 36% of cases, and 3 to 7 informative molecular markers in 48% of the crosses (Figure 3b). The three most informative loci were Lsat3, Lsat7, and Lsat6, while Lsat4 and Lsat13 were monomorphic in almost all parental groups. It is worth noting that the Lsat8 marker was in a heterozygous state in all but four parental genotypes (7, 45, 58, and 60).

We were able to take advantage of these informative loci to calculate the success rate of each cross. In 30% of manual pollinations (27 out of 89), a success rate of 100% was achieved (i.e., all the offspring were hybrids), whereas in 18% of crosses (16 out of 89), the S.C. ranged from 71% to 90%. A hybridity rate fluctuating between 51% and 70% was reported in 15% of cases (13 out of 89), while 26% of the crosses produced fewer than 50% hybrids each. Finally, in only 7% of crosses (6 out of 89) were all the offspring the result of self-pollination (Figure 3c and Table S3). Overall, the mean hybridization rate (the average number of hybrids per crosses) was 68 ± 33%, and out of a total of 871 individuals, 602 (69%) were hybrids, and 556 were derived from programmed crosses. The remaining 46 individuals (5% of the total) had a unexpected genotypes (U.G.) compared with their putative parents (Table S3).

3.3. Lettuce Breeding Populations

The 47 F3 experimental lines were genotyped using the same set of 16 SSR loci as for the previous analyses. The homozygosity estimates of all samples (940) ranged from 67% to 93% (Figure 4a). Ten experimental populations had a median observed homozygosity ≥90%. Outliers—with homozygosity values consistently deviating from the median—were present in only three experimental populations (11, 14, and 32).

The median genetic similarity observed within each line was always greater than 90% (Figure 4b), and 37 experimental populations had a median genetic similarity ≥95%. Outliers were present in 21 of the 47 lines (Figure 4b).

After assembling the data, we found 10 breeding populations, belonging to butterhead type (Table S4), to have Ho values ≥ 90%, and a median genetic similarity ≥ 95%; the box-plots of these populations are labelled in red in Figure 4a,b. Finally, in the genetic similarity matrix calculated from all pairwise comparisons of these ten populations, the Jaccard's index ranged from 44% ± 3% (between populations 3 and 18) to 96 ± 5% (between populations 45 and 47, Figure S3). Moreover, the populations called 45 and 47 were constituted starting from the same parents (2 × 6, Table S4).

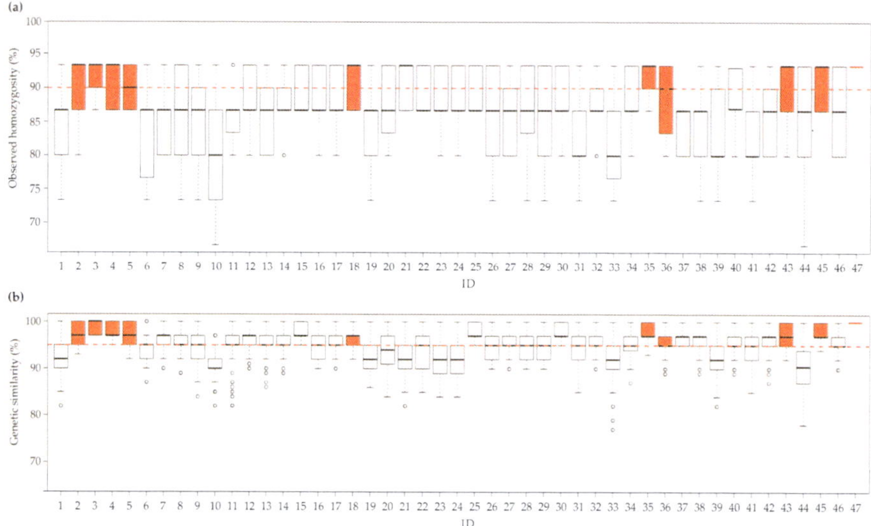

Figure 4. Statistics relating to the observed homozygosity and genetic similarity among lines. (**a**) Box-plot of the median observed homozygosity (in percentages) in each of the 47 populations. The red dotted line represents the homozygosity threshold set at 90%. (**b**) Box plot of the median genetic similarity in each experimental population (in percentages). The red dotted line represents the genetic similarity threshold set at 95%. The red box-plots represent the ten best experimental populations (observed homozygosity ≥90% and genetic similarity values ≥95%). The second and third quartiles are marked inside the square and are divided by a bold bar (median). Dots show outlier samples.

4. Discussion

The last decade has seen major advances in the acquisition of knowledge concerning the genetics of lettuce and, in particular, the development of molecular markers [1,11,21]. This has facilitated marker-assisted selection programmes, especially those aimed at countering the onset of new diseases. For example, several studies have dealt with identifying the QTLs associated with biotic and abiotic stress resistance [17,31,32]. Molecular markers have also been extensively used to assess genetic variation and relationships in lettuce germplasm [19] and to identify possible duplicate varieties [33]. However, although the benefits derived from exploitation of these molecular tools have also been discussed in marker-assisted breeding programmes [34] and demonstrated in several species [7,35], there are only a few studies on this type in lettuce [20]. The aim of our work, therefore, was to integrate conventional and biotechnological methods in three different steps of a breeding programme to show that this strategy is also effective in *L. sativa* (Figure 1). This is of pivotal importance if we consider the economic impact of lettuce (the world production of lettuce and chicory in 2017 was 26.8 million tons [36]) and the need to regularly develop new varieties.

Commercial lettuce varieties are usually characterised by pure lines due to the autogamous nature of this species. In order to introduce variability, manual pollination is usually carried out to cross genetically stable parent lines with agronomic traits of interest. Progeny selection is a crucial step, but despite the efficiency of some emasculation and hand pollination methods developed over the years [2], a major problem—distinguishing unequivocally and rapidly F1 individuals from self-pollinated progeny—still remains. The use of molecular assays to quickly and accurately screen progeny makes it possible to overcome most of the conventional breeding limits in this species.

In this context, our SSR-based analysis has (i) facilitated selection of the best parents to cross in order to maximise the variability of the progeny both within the same cultivar type and among them, (ii) allowed accurate evaluation of the resulting offspring, and (iii) sped up the screening of experimental F3 lines for their stability and uniformity.

The first part of our work focused on pre-screening 71 parental lines for crossing with the aim of maximising the gains obtained from each out-pollination within cultivar type and, in some cases, among them. As expected, the similarity matrix and the unweighted pair group method with an arithmetic mean (UPGMA) dendrogram showed varying levels of similarity among the different parental genotypes. Parental germplasm appeared to divide into two different groups, as revealed by the Principal Coordinates Analysis (PCoA) results and particularly by the genetic structure analysis. However, samples did not separate in UPGMA tree and PCoA according to the cultivar type, but we may assume that increasing the number of markers it could be possible to clarify this clustering. Although 53 parental lines were found to be fully distinguishable, with similarity values ranging from 53% to 98% and characterised by a unique genotypic profile, it was impossible to identify unequivocally the remaining 18. This is not surprising if we consider that some of the parental lines were closely related. We may speculate that increasing the number of SSRs would allow us to address these remaining issues. Given the aim of this study, these data were useful to avoid crosses between parents with 100% similarity. To introduce variability according to the phenotype and the lowest similarity values, we carried out 89 crossing combinations. Another aspect that needs to be considered when planning crosses is the stability of the parental line in terms of homozygosity. In our study, the median observed homozygosity of the parental lines was lower than expected (82%), especially in light of the strictly autogamous reproductive system of lettuce [37]. Overall, the fact that only one individual in four had homozygosity values greater than 90% showed that some of these lines were not entirely stable. However, it must be borne in mind that, although the observed homozygosity was not optimal, some of these lines, experimental lines, were chosen to produce F1 partly because they displayed resistance to multiple pathogens and had interesting phenotypic traits.

Before proceeding with hand pollination, in order to distinguish between F1 individuals resulting from cross-pollination and those resulting from self-pollination, we first examined the informative loci among the parental lines used in the crosses. Only homozygous loci for different alleles in parental lines were considered informative. Our analysis showed at least 2 informative loci in 84% of the programmed crosses. It is worth pointing out that restricting the analysis to the informative loci brought us considerable savings in terms of time and costs.

Overall, the molecular determination of hybridity was successful: F1 individuals represented at least 51% of the offspring in 67% of the manual crosses, and 100% of the offspring in 30% of the crosses (100% success rate), in agreement with the estimates originally reported by the developer of the pollination technique [2]. Unexpected genotypes (U.G.) were identified in 5% of the individual progeny. In these cases, the progeny genotypes appeared to diverge from what would be expected given the parents. This percentage is consistent with the spontaneous or undesired occurrences of cross-pollination (1–6%) reported in the literature for this species [3], mainly due to pollinator insects. However, we cannot exclude human error during manual pollination or seed collection.

Finally, at an advanced step of the breeding programme, we genetically assessed 47 different experimental F3 populations (940 samples), previously selected for their morpho-phenotypic traits and resistances of interest (Figure 1). Interestingly, the findings in terms of both homozygosity and intra-line similarity were very good. This would suggest that in strictly autogamous species, such as lettuce, three cycles of self-pollination may already be sufficient to reach desired outcomes in terms of genetic uniformity and homozygosity. This also confirms that the use of molecular markers could speed up the process by making it possible to select the best individuals on the basis of their genotype, thereby reducing the number of generations needed to develop new varieties. The ten experimental populations with the highest homozygosity estimates (≥90%) and the highest intra-genetic similarity values (≥95%) were considered suitable for pre-commercial trials (red box plot, Figure 4). However, a pairwise comparison of two of them (identified as 45 and 47) showed them to be genetically too similar (96% ± 5% genetic similarity, Figure S3), in agreement with phenotypic data and their common origin (Table S4), to be registered and marketed as distinct varieties. According to the most recent guidelines concerning the protection of new plant varieties, the similarity threshold to define two lettuce varieties as distinct is set at 96% [30]. The next step will be to integrate molecular data and morphological observations in order to the select the best genotypes (positive selection) for evaluation as pre-commercial varieties. In particular, the eligible genotypes will be self-pollinated to multiply the seed so that their agronomic performance can be compared in different locations and periods of the year, and with the best commercial varieties already on the market.

For the remaining experimental populations (white box blot, Figure 4), an attempt could be made to increase their genetic uniformity through negative selection to remove the most genetically divergent individuals (i.e., outlier samples). Moreover, if necessary, the remaining genotypes can undergo a further selfing cycle aimed at reaching optimum values of homozygosity.

5. Conclusions

The results of this study demonstrate the advantages of mutual integration of traditional and biotechnological methods and show the added value that molecular markers can give to breeding programmes. We used microsatellite markers in three different steps of a conventional lettuce breeding program (see Figure 1) and demonstrated, firstly, the efficiency of SSR markers not only in selecting the best parental plants for crossing based on their observed homozygosity and dissimilarity values, but also in screening the resulting F1 progeny to distinguish between the offspring resulting from cross-pollination and those resulting from self-pollination. Furthermore, using the same SSR panel, we were able to act downstream of the breeding scheme to assess the uniformity of some pre-commercial cultivars. Our molecular assay could therefore also be used by seed firms to assess newly developed varieties for distinctiveness, uniformity and stability (DUS test), three major requirements for registering plant materials [6]. Finally, molecular characterisation of a new variety could also be used to register it in national or international varietal catalogues. In fact, the genotype or molecular profile of a registered variety can be crucial in solving cases of fraudulent practices, and in curbing plagiarism and unfair free-riding on the original plant breeder's time and investment [30].

Supplementary Materials: The following are available online at http://www.mdpi.com/2073-4425/10/11/916/s1: Figure S1: Pairwise genetic similarity matrix of the 71 individuals analysed (in percentages) based on Rohlf's genetic similarity coefficient. High genetic similarity values are labelled in green, the low values in red, and intermediate values are coloured on a scale from green to red. The observed homozygosity values of the 71 putative parental lines are reported to the left of each ID name. Figure S2: Definition of the subgroup number of parental lines based on the SSR marker dataset. Mean ΔK is calculated as $|L''(K)|/(SD(L(K))$, following Evanno et al. [23]. The blue line represents the ΔK values; Figure S3: Pairwise genetic similarity matrix of ten selected populations (in percentages) based on the Jaccard coefficient. The high genetic similarity values are labelled in green, the low values in red, and intermediate values are coloured on a scale from green to red; Table S1: Lettuce parental lines information, including ID of accessions, cultivar type of materials and subpopulation classification based on STRUCTURE analysis (1 = blue and 2 = red).; Table S2: SSR primer tails and dyes. List of the primer tails used with their sequences and corresponding dyes; Table S3: Lettuce plant material information, including ID of accessions used in the crosses, total number of plants per programmed cross, number of informative marker loci, hybrid plants, selfed plants and unexpected genotypes, and the mean hybridisation values (in percentages) for all the programmed crosses; Table S4: Information about ten F3 selected populations, including ID of parental lines used in the crosses.

Author Contributions: Conceptualization, F.P. and G.B.; formal analysis, A.P. and G.G.; investigation, A.P. and G.G.; methodology, F.P. and G.B.; resources, A.P.; software, A.P. and G.G.; supervision, G.B.; Writing—Original Draft, A.P. and F.P.; Writing—Review and Editing, F.P. and G.B.

Funding: This project was financially supported by Blumen Group SpA (Bologna, Italy), within a Research Contract stipulated with DAFNAE, University of Padova (Italy): "Development of new varieties in horticultural species using molecular marker-assisted breeding methods" PI: Gianni Barcaccia.

Acknowledgments: This research was carried out in partial fulfillment of the Ph.D. Program of Alice Patella by taking advantage of a Doctoral Research Fellowship provided by Blumen Group SpA (Bologna, Italy).

Conflicts of Interest: The authors declare no conflict of interest.

References

1. Wang, S.; Wang, B.; Liu, J.; Ren, J.; Huang, X.; Zhou, G.; Wang, A. Novel polymorphic EST-based microsatellite markers characterized in lettuce (*Lactuca sativa*). *Biologia* **2017**, *72*, 1300–1305. [CrossRef]
2. Nagata, R. Clip-and-wash Method of Emasculation for Lettuce. *HortScience* **1992**, *27*, 907–908. [CrossRef]
3. Barcaccia, G.; Falcinelli, M. *Genetica e Genomica*; Liguori Editore, S.r.l.: Naples, Italy, 2006; Volume 3.
4. Oliver, G.W. New methods of plant breeding. *J. Hered.* **1910**, *1*, 21–30. [CrossRef]
5. Prohens-Tomás, J.; Nuez, F. *Vegetables I: Asteraceae, Brassicaceae, Chenopodicaceae, and Cucurbitaceae*; Springer Science & Business Media: Berlin, Germany, 2007; Volume 1.
6. Patella, A.; Scariolo, F.; Palumbo, F.; Barcaccia, G. Genetic Structure of Cultivated Varieties of Radicchio (*Cichorium intybus* L.): A Comparison between F1 Hybrids and Synthetics. *Plants* **2019**, *8*, 213. [CrossRef] [PubMed]
7. Palumbo, F.; Galla, G.; Vitulo, N.; Barcaccia, G. First draft genome sequencing of fennel (*Foeniculum vulgare* Mill.): Identification of simple sequence repeats and their application in marker-assisted breeding. *Mol. Breed.* **2018**, *38*, 122. [CrossRef]
8. Jha, N.; Jacob, S.; Nepolean, T.; Jain, S.; Kumar, M. SSR markers based DNA fingerprinting and it's utility in testing purity of eggplant hybrid seeds. *Qual. Assur. Saf. Crop. Foods* **2016**, *8*, 333–338. [CrossRef]
9. Singh, D.; Singh, C.K.; Tomar, R.S.S.; Taunk, J.; Singh, R.; Maurya, S.; Chaturvedi, A.K.; Pal, M.; Singh, R.; Dubey, S.K. Molecular Assortment of Lens Species with Different Adaptations to Drought Conditions Using SSR Markers. *PLoS ONE* **2016**, *11*, e0147213. [CrossRef] [PubMed]
10. Rauscher, G.; Simko, I. Development of genomic SSR markers for fingerprinting lettuce (*Lactuca sativa* L.) cultivars and mapping genes. *BMC Plant Biol.* **2013**, *13*, 11. [CrossRef]
11. Truco, M.J.; Antonise, R.; Lavelle, D.; Ochoa, O.; Kozik, A.; Witsenboer, H.; Fort, S.B.; Jeuken, M.J.W.; Kesseli, R.V.; Lindhout, P.; et al. A high-density, integrated genetic linkage map of lettuce (*Lactuca* spp.). *Theor. Appl. Genet.* **2007**, *115*, 735–746. [CrossRef]
12. Reyes-Chin-Wo, S.; Wang, Z.; Yang, X.; Kozik, A.; Arikit, S.; Song, C.; Xia, L.; Froenicke, L.; Lavelle, D.O.; Truco, M.-J.; et al. Genome assembly with in vitro proximity ligation data and whole-genome triplication in lettuce. *Nat. Commun.* **2017**, *8*, 14953. [CrossRef]
13. Bull, C.T.; Goldman, P.H.; Hayes, R.; Madden, L.V.; Koike, S.T.; Ryder, E. Genetic Diversity of Lettuce for Resistance to Bacterial Leaf Spot Caused by *Xanthomonas campestris* pv. *vitians*. *Plant Health Prog.* **2007**, *8*, 11. [CrossRef]
14. Giesbers, A.K.J.; Pelgrom, A.J.E.; Visser, R.G.F.; Niks, R.E.; Ackerveken, G.V.D.; Jeuken, M.J.W. Effector-mediated discovery of a novel resistance gene against *Bremia lactucae* in a nonhost lettuce species. *New Phytol.* **2017**, *216*, 915–926. [CrossRef] [PubMed]
15. Macias-González, M.; Truco, M.J.; Bertier, L.D.; Jenni, S.; Simko, I.; Hayes, R.J.; Michelmore, R.W. Genetic architecture of tipburn resistance in lettuce. *Theor. Appl. Genet.* **2019**, *132*, 2209–2222. [CrossRef] [PubMed]
16. Mamo, B.E.; Hayes, R.J.; Truco, M.J.; Puri, K.D.; Michelmore, R.W.; Subbarao, K.V.; Simko, I. The genetics of resistance to lettuce drop (*Sclerotinia* spp.) in lettuce in a recombinant inbred line population from Reine des Glaces × Eruption. *Theor. Appl. Genet.* **2019**, *132*, 2439–2460. [CrossRef] [PubMed]
17. Hartman, Y.; Hooftman, D.A.P.; Uwimana, B.; Schranz, M.E.; Van De Wiel, C.C.M.; Smulders, M.J.M.; Visser, R.G.F.; Michelmore, R.W.; Van Tienderen, P.H. Abiotic stress QTL in lettuce crop–wild hybrids: Comparing greenhouse and field experiments. *Ecol. Evol.* **2014**, *4*, 2395–2409. [CrossRef] [PubMed]
18. Su, W.; Tao, R.; Liu, W.; Yu, C.; Yue, Z.; He, S.; Lavelle, D.; Zhang, W.; Zhang, L.; An, G.; et al. Characterization of four polymorphic genes controlling red leaf colour in lettuce that have undergone disruptive selection since domestication. *Plant Biotechnol. J.* **2019**. [CrossRef]
19. El-Esawi, M.A. Molecular Genetic Markers for Assessing the Genetic Variation and Relationships in Lactuca Germplasm. *Annu. Res. Rev. Biol.* **2015**, *8*, 1–13. [CrossRef]
20. Hayes, R.; Simko, I. Breeding lettuce for improved fresh-cut processing. *Acta Hortic.* **2016**, *1141*, 65–76. [CrossRef]
21. Simko, I. Development of EST-SSR markers for the study of population structure in lettuce (*Lactuca sativa* L.). *J. Hered.* **2009**, *100*, 256–262. [CrossRef]

22. Schuelke, M. An economic method for the fluorescent labeling of PCR fragments. *Nat. Biotechnol.* **2000**, *18*, 233–234. [CrossRef]

23. Nonis, A.; Scortegagna, M.; Nonis, A.; Ruperti, B. PRaTo: A web-tool to select optimal primer pairs for qPCR. *Biochem. Biophys. Res. Commun.* **2011**, *415*, 707–708. [CrossRef] [PubMed]

24. Saitou, N.; Nei, M. The neighbor-joining method: A new method for reconstructing phylogenetic trees. *Mol. Biol. Evol.* **1987**, *4*, 406–425. [PubMed]

25. Hammer, Ø.; Harper, D.A.; Ryan, P.D. PAST: Paleontological statistics software package for education and data analysis. *Palaeontol. Electron.* **2001**, *4*, 1–9.

26. Evanno, G.; Regnaut, S.; Goudet, J. Detecting the number of clusters of individuals using the software structure: A simulation study. *Mol. Ecol.* **2005**, *14*, 2611–2620. [CrossRef]

27. Yeh, F.C.; Yang, R.-C.; Boyle, T. *POPGENE Version 1.31*; University of Alberta: Edmonton, AB, Canada, 1999.

28. Palumbo, F.; Galla, G.; Martinez-Bello, L.; Barcaccia, G. Venetian Local Corn (*Zea mays* L.) Germplasm: Disclosing the Genetic Anatomy of Old Landraces Suited for Typical Cornmeal Mush Production. *Diversity* **2017**, *9*, 32. [CrossRef]

29. Feng, J.Y.; Li, M.; Zhao, S.; Zhang, C.; Yang, S.T.; Qiao, S.; Tan, W.F.; Qu, H.J.; Wang, D.Y.; Pu, Z.G. Analysis of evolution and genetic diversity of sweetpotato and its related different polyploidy wild species I-trifida using RAD-seq. *BMC Plant Biol.* **2018**, *18*, 181. [CrossRef]

30. Lawson, C. *Plant Breeder's Rights and Essentially Derived Varieties: Still Searching for Workable Solutions*; European Intellectual Property Review 499: Griffith University Law School Research: Nathan, QLD, Australia, 2016; pp. 16–17.

31. Hand, P.; Kift, N.; McClement, S.; Lynn, J.; Grube, R.; Schut, J.; Van der Arend, A.; Pink, D. Progress towards mapping QTLs for pest and disease resistance in lettuce. In Proceedings of the Eucarpia Leafy Vegetables Conference, Centre for Genetic Resources, Wageningen, The Netherlands, 22–24 May 2003; pp. 31–35.

32. Šimko, I.; Pechenick, D.; McHale, L.; Truco, M.; Ochoa, O.; Michelmore, R.; Scheffler, B. Development of Molecular Markers for Marker-Assisted Selection of Dieback Disease Resistance in Lettuce (*Lactuca sativa*). *Acta Hortic.* **2010**, *859*, 401–408. [CrossRef]

33. Sochor, M.; Jemelková, M.; Doležalová, I. Phenotyping and SSR markers as a tool for identification of duplicates in lettuce germplasm. *Czech J. Genet. Plant Breed.* **2019**, *55*, 110–119. [CrossRef]

34. Bhat, J.A.; Shivaraj, S.M.; Ali, S.; Mir, Z.A.; Islam, A.; Deshmukh, R. Genomic Resources and Omics-Assisted Breeding Approaches for Pulse Crop Improvement. In *Pulse Improvement*; Springer: Berlin/Heidelberg, Germany, 2018; pp. 13–55.

35. Phing Lau, W.C.; Latif, M.A.; Rafii, M.Y.; Ismail, M.R.; Puteh, A. Advances to improve the eating and cooking qualities of rice by marker-assisted breeding. *Crit. Rev. Biotechnol.* **2016**, *36*, 87–98. [CrossRef]

36. FAO Site. Available online: http://www.fao.org/faostat/en/#data/QC (accessed on 11 September 2019).

37. Uwimana, B.; Smulders, M.J.M.; Hooftman, D.A.P.; Hartman, Y.; Van Tienderen, P.H.; Jansen, J.; McHale, L.K.; Michelmore, R.W.; Van De Wiel, C.C.M.; Visser, R.G.F. Hybridization between crops and wild relatives: The contribution of cultivated lettuce to the vigour of crop-wild hybrids under drought, salinity and nutrient deficiency conditions. *Theor. Appl. Genet.* **2012**, *125*, 1097–1111. [CrossRef]

 *genes*

Article

Molecular Analysis of the Official Algerian Olive Collection Highlighted a Hotspot of Biodiversity in the Central Mediterranean Basin

Benalia Haddad [1], Alessandro Silvestre Gristina [2,*], Francesco Mercati [2], Abd Elkader Saadi [3], Nassima Aiter [4,5], Adriana Martorana [2], Abdoallah Sharaf [2,6] and Francesco Carimi [2]

[1] Département de Productions Végétales, Laboratoire Amélioration Intégrative Des Productions Végétales (AIPV, C2711100), Ecole Nationale Supérieure Agronomique (ENSA), Hassan Badi, El Harrach, Algiers 16000, Algeria; b.haddad@ensa.dz

[2] Institute of Biosciences and BioResources, National Research Council (CNR), Research Division of Palermo, Corso Calatafimi 414, 90129 Palermo, Italy; francesco.mercati@ibbr.cnr.it (F.M.); adrianamartorana1982@gmail.com (A.M.); abdoallah.sharaf@agr.asu.edu.eg (A.S.); francesco.carimi@ibbr.cnr.it (F.C.)

[3] University Hassiba Benbouali, Faculty of Science of Nature and Life, Plant Biotechnology Laboratory, BP 151, Chlef 02000, Algeria; a.saadi@univ-chlef.dz

[4] Université Saad Dahleb-Blida 1, Faculté des Sciences de la Nature et de la Vie, Laboratoire de Biotechnologie des Productions Végétales, Département de Biotechnologies, Blida 09000, Algeria; haddad_nassima@hotmail.fr

[5] Laboratoire de culture in vitro, Département central, Institut Technique de l'Arboriculture Fruitière et de la Vigne, ITAFV, Algiers 16000, Algeria

[6] Institute of Molecular Biology of Plants, Biology Centre, CAS, Branišovská 31, 37005 České Budějovice, Czech Republic

* Correspondence: alessandro.gristina@ibbr.cnr.it; Tel.: +39-091-657-4578

Received: 30 January 2020; Accepted: 9 March 2020; Published: 13 March 2020

Abstract: Genetic diversity and population structure studies of local olive germplasm are important to safeguard biodiversity, for genetic resources management and to improve the knowledge on the distribution and evolution patterns of this species. In the present study Algerian olive germplasm was characterized using 16 nuclear (nuSSR) and six chloroplast (cpSSR) microsatellites. Algerian varieties, collected from the National Olive Germplasm Repository (ITAFV), 10 of which had never been genotyped before, were analyzed. Our results highlighted the presence of an exclusive genetic core represented by 13 cultivars located in a mountainous area in the North-East of Algeria, named Little Kabylie. Comparison with published datasets, representative of the Mediterranean genetic background, revealed that the most Algerian varieties showed affinity with Central and Eastern Mediterranean cultivars. Interestingly, cpSSR phylogenetic analysis supported results from nuSSRs, highlighting similarities between Algerian germplasm and wild olives from Greece, Italy, Spain and Morocco. This study sheds light on the genetic relationship of Algerian and Mediterranean olive germplasm suggesting possible events of secondary domestication and/or crossing and hybridization across the Mediterranean area. Our findings revealed a distinctive genetic background for cultivars from Little Kabylie and support the increasing awareness that North Africa represents a hotspot of diversity for crop varieties and crop wild relative species.

Keywords: *Olea europaea* L.; olive; cpSSR; nuSSR; genetic diversity; population structure; Mediterranean Region

1. Introduction

Olive (*Olea europaea* L.) is one of the most important fruit species of the Mediterranean region [1]. Ninety-eight percent of olive trees of the world are cultivated in this region [2], providing over 90% of World production [3]. Olive fruits and olive oil are central in the Mediterranean diet and symbols of the Mediterranean culture. It is commonly believed that olive domestication occurred in the Near East approximately 6000 years ago [4]. Phoenicians, Greeks and Romans later spread olive cultivation to the western Mediterranean region [5–8]. The hypothesis of a human-mediated diffusion of the olive tree from the eastern to western Mediterranean basin is supported by recent genetic studies [9], demonstrating that as many as 90% of current cultivars are characterized by the same chloroplast haplotype lineage [4,10]. Therefore, the spreading of the olive culture throughout the Mediterranean Basin by human migrations and commercial exchanges has played a key role in determining the pattern of olive germplasm [11,12]. The cultivated olive germplasm shows a high degree of diversity, with about 1250 recognized cultivars [13]. Olive cultivation in Algeria dates back to antiquity, and it has maintained great socio-economic importance until present days [14] and is mostly present along the Mediterranean coast. In this area, the mountainous region of Kabylie, geographically divided in two districts by the river Soummam, great Kabylie to the West, and little Kabylie to the East, can be considered an important reserve of local olive germplasm [14]. The olive sector is considered strategic for the Algerian economy and for this reason the Algerian Ministry of Agriculture and Rural Development recently set a strategy for the expansion of olive tree cultivation in different regions, aiming to reach one million hectares by 2019, using the local genetic resources. Therefore, the identification and characterization of local germplasm is a key step for future breeding programs, cultivar selection for new plantations and to preserve Algerian olive biodiversity from the risk of genetic erosion due to introduction of foreign cultivars. Despite that Hauville [15] reported 150 varieties in Algeria, according to the Algerian Ministry of Agriculture and Rural Development, 36 main varieties are officially recognized and cultivated in the experimental field of the Institut Technique de l'Arboriculture Fruitiere et de la Vigne (ITAFV, Takarietz, Bejaia). Previous studies on Algerian olive germplasm focused mainly on the genetic characterization of a subset of local cultivars [16,17], their population structure and their genetic relationship with wild olive trees [14]. In his study of the World Olive Germplasm Bank (WOGB) of Marrakech, Haouane et al. [18] analyzed some Algerian varieties with nuclear and chloroplast microsatellites (nuSSRs and cpSSRs), but only their cpSSRs profiles are publicly available.

The present study is the first genetic characterization of the official Algerian National Olive Germplasm Repository, using both nuSSRs and cpSSRs. Among the available molecular markers, nuSSRs were chosen for their highly reproducible and informative co-dominant and multi-allelic nature, which allowed to evaluate the genetic diversity in several plant crops, such as maize [19], rice [20], common bean [21], wheat [22], tomato [23,24], grape [25–28] and olive [29–33]. CpSSRs are maternally inherited in angiosperms, and they have been informative in unravelling the phylogenetic pattern in olive germplasm [4,34] but also in other crops such as grape [35,36].

In order to provide new insights in the origin and diffusion of olive cultivars around the Mediterranean basin, we analyzed the Algerian varieties with 16 nuSSR and six cpSSR markers and compared our results with the widest published datasets available, which includes olive varieties representative of Mediterranean Basin crop's biodiversity. The aims of this research were: (i) to evaluate the genetic diversity of the main Algerian olive varieties; (ii) to assess, for the first time, the genetic relationships between this germplasm and olive accessions from public datasets; (iii) to provide useful knowledge for future cultivation expansion and breeding programs.

2. Materials and Methods

2.1. Plant Material and Sampling

A total of 34 Algerian varieties from the ITAFV national olive germplasm collection (Table 1) were sampled for the genetic characterization. The ITAFV experimental field was created between 1947 and 1954, covering 0.95 ha. It is located 30 km off Bejaia Takarietz (latitude 36.24, longitude 6.57 and altitude 63.30), in the coastal area of the Sidi Aich district (Figure 1), an area with an arboriculture vocation, characterized by a Mediterranean climate [37]. Information on Algerian varieties including its Arabic name, meaning, synonyms, putative origin, diffusion and use are reported in Table S1, the catalogue illustrating the main features of each variety is presented in Table S2.

Table 1. List of Algerian varieties analyzed, collected at the Institut Technique de l'Arboriculture Fruitière et de la Vigne (ITAFV).

Accession Number in Figure 1	ID	Cultivar
1	OE-AL-001	Abani
2	OE-AL-002	Aberkane
3	OE-AL-003	Aeleh
4	OE-AL-004	Aghchren d'el Ousseur
5	OE-AL-005	Aghchren de Titest
6	OE-AL-006	Aghenfas
7	OE-AL-007	Agrarez
8	OE-AL-008	Aguenaou
9	OE-AL-009	Aimel
10	OE-AL-010	Akerma
11	OE-AL-011	Azeradj
12	OE-AL-012	Blanquette de Guelma
13	OE-AL-013	Bouchouk Guergour
14	OE-AL-014	Bouchouk Lafayette
15	OE-AL-015	Bouchouk Soummam
16	OE-AL-016	Boughenfous
17	OE-AL-017	Bouichret
18	OE-AL-018	Boukaila
19	OE-AL-019	Bouricha
20	OE-AL-020	Chemlal
21	OE-AL-021	Ferkani
22	OE-AL-022	Grosse du Hamma
23	OE-AL-023	Hamra
24	OE-AL-024	Limli
25	OE-AL-025	Longue de Miliana
26	OE-AL-026	Mekki
27	OE-AL-027	Neb Djemel
28	OE-AL-028	Ronde de Miliana
29	OE-AL-029	Rougette de Mitidja
30	OE-AL-030	Sigoise
31	OE-AL-031	Souidi
32	OE-AL-032	Tabelout
33	OE-AL-033	Takesrit
34	OE-AL-034	Tefah

Figure 1. Geographic origin of the Algerian olive cultivars sampled. The yellow point indicates the ITAFV (Takarietz, Bejaia). In brackets, the region of diffusion of the characterized cultivars is indicated. The numbers highlight the origin of cultivars: (1) Abani; (2) Aberkane; (3) Aeleh; (4) Aghchrend'el Ousseur; (5) Aghchren de Titest; (6) Aghenfas; (7) Agrarez; (8) Aguenaou; (9) Aimel; (10) Akerma; (11) Azeradj; (12) Blanquette de Guelma; (13) Bouchouk Guergour; (14) Bouchouk Lafayette; (15) Bouchouk Soummam; (16) Boughenfous; (17) Bouichret; (18) Boukaila; (19) Bouricha; (20) Chemlal; (21) Ferkani; (22) Grosse du Hamma; (23) Hamra; (24) Limli; (25) Longue de Miliana; (26) Mekki; (27) Neb Djemel; (28) Ronde de Miliana; (29) Rougette de Mitidja; (30) Sigoise; (31) Souidi; (32) Tabelout; (33) Takesrit; (34) Tefah.

2.2. Molecular Analyses (nuSSRs and cpSSRs)

Total genomic DNA was extracted from 0.1 g of dry leaves following the Doyle and Doyle [38] CTAB (cetyl-trimethylammonium bromide) method. The extract was treated with DNase-free RNase (Roche Diagnostics, Mannheim, Germany) and the quality and concentration were checked by a NanoDrop Spectrophotometer (Thermo Scientific—Waltham, MA, USA). .

Algerian varieties were analyzed by 16 nuSSRs available in current literature [28–31] (Table S3). The haplotype of each variety was evaluated by using six cpSSRs [34] (Table S3). Multiplexed amplification reactions were performed in 15 µL final volume reaction mixture as described by Garfi et al. [39]. The amplification products were solved on ABI PRISM 310 Genetic Analyzer (Applied Biosystems by Life Technologies, Foster City, CA, USA) and the alleles were sized by GENEMAPPER 4.0 (Applied Biosystems by Life Technologies).

Many articles analyzed large datasets of wild and cultivated olive nuSSR profiles, but unfortunately, only a few of them provide their genetic profiles. We compared our nuSSRs profiles with the largest published dataset available from the WOGB of Cordoba [40], using a common subset of seven SSRs (Table S3). Normalization among datasets was achieved using the common variety Chemlal de Kabylie, present in our dataset with the synonym Chemlal. For all the analyses, WOGB profiles were grouped geographically as follow: Spain and Portugal (Iberian Peninsula—IB); France (FRA), Italy (ITA), Morocco and Tunisia (Maghreb—MAG); Croatia, Albania and Greece (Balcanic Peninsula—BAL), Turkey and Cyprus (Turkey—TUR); Iran, Israel, Lebanon, Syria and Egypt (East Mediterranean—East-M). Due to the reduced number of SSRs, WOGB dataset was reduced to 351 accessions to consider only unique genetic profiles, and some Algerian varieties were grouped in single genetic profiles for the Structure analyses because with seven SSRs they were not able to differentiate, namely: Aguenau, including Agrarez and Hamra, and Aimel, including Aberkane. cpSSRs profiles were also compared with the available published dataset [4,34].

2.3. Data Analysis

For each microsatellite we estimated the principal genetic parameters, i.e., number of alleles (Na), expected (He) and observed (Ho) heterozygosity and Polymorphic Information Content (PIC) by using PowerMarker [41], Haplotype analysis software version 1.05 [42] and FreeNA [43] software.

To identify the number of genetic groups in the Algerian germplasm, cluster analysis was carried out for nuSSR and cpSSR separately according to the UPGMA (Unweighted Pair-Group Method with Arithmetical Averages) algorithm and two phylogenetic trees were generated using the R package Adegenet [44]. The levels of support for the nodes were estimated by bootstrap analysis (1000 replicates). The number of genetic groups within the Algerian collection alone and among Algerian and other Mediterranean and Near East cultivars, was inferred by means of Principal Coordinates Analysis (PCoA) in GenAlEx v.6.51b2 [45] and of Bayesian analysis in STRUCTURE [46]. The most likely number of genetic groups (K) in STRUCTURE was calculated following Evanno et al. [47]. Twenty independent runs (100,000 burn-in, 1,000,000 Marchov Chain Monte Carlo) for each K were carried out using the admixture model with correlated marker frequency and default parameters. The runs were averaged using CLUMPP (CLUster Matching and Permutation Program) [48] and the histograms were shown using DISTRUCT program [49]. Individuals with ancestry value < 0.65 were considered mosaics (Tables S4 and S5), while those with higher values were assigned to the corresponding cluster. Using nuSSR profiles, main genetic parameters, including pairwise Gst values [50], among Mediterranean populations were calculated in GenAlEx v.6.51b2 [45]. For Gst values, the significance of the differentiation between pairs of selected populations was tested by permutation procedures (9999 replicates).

For the comparison with the WOGB dataset [40], three STRUCTURE analyses were performed. The first analysis tested whether the number of Mediterranean olive genetic clusters (West, Central and Eastern Mediterranean) changed by the reduction of the SSRs marker panel. The second analysis identified the ancestry of the Algerian germplasm. Finally, a hierarchical analysis [51] following the procedure described above, was carried out only with samples belonging to Central and Eastern Mediterranean pool, showing an ancestry value higher than 0.80 (Table S5).

3. Results

3.1. Genetic Diversity Assessed by cp SSRs and nuSSRs

The six cpSSR markers showed a total of 12 alleles, with an average of two alleles per locus and major allele frequency values ranging from 0.647 to 1.000 (Table S6). Five out of six cpSSR used were polymorphic, while trnT-L-polyT locus was monomorphic in the analyzed collection (Table S6). The phylogenetic tree obtained through cpSSRs (Figure 2) highlighted three chlorotype groups, corresponding to the wild and cultivated lineages identified in the Mediterranean (Table 2) [4].

Table 2. Comparison of chlorotype lineages of Algerian varieties identified in this study and in published literature.

	Accession Name	Origin	Chlorotype Lineage		
			This Study	Besnard et al. [4]	Houane et al. [18]
1	Abani	OWGB-Marrakech - ITAFV- Takarietz	E2	E2-1	E3
2	Aberkane	OWGB-Marrakech - ITAFV- Takarietz	E1	E1-1	-
3	Aeleh	ITAFV- Takarietz	E1	-	-
4	Aghchren de Titest	OWGB-Marrakech - ITAFV- Takarietz	E1	E1-1	-
5	Aghchren d'Elousseur/Azeradj Tamorka	OWGB-Marrakech - ITAFV- Takarietz	E2	E1-1	-
6	Aghenfas	ITAFV- Takarietz	E1	-	-
7	Agrarez	OWGB-Marrakech - ITAFV- Takarietz	E1	E1-1	-
8	Aguenaou	OWGB-Marrakech - ITAFV- Takarietz	E1	E1-2	-
9	Aharoune	OWGB-Marrakech	-	E3-2	-
10	Ahia Ousbaa	OWGB-Marrakech	-	E2-1	E3
11	Aîmel	OWGB-Marrakech - ITAFV- Takarietz	E1	E1-1	-
12	Akenane	OWGB-Marrakech	-	E1-1	-
13	Akerma	OWGB-Marrakech - ITAFV- Takarietz	E3	E1-1	-
14	Azeboudj de Khirane	OWGB-Marrakech	-	E2-1	E3
15	Azeradj	OWGB-Marrakech - ITAFV- Takarietz	E1	E1-1	-
16	Blanquette de Castu	OWGB-Marrakech	-	E1-1	-
17	Blanquette de Guelma	OWGB-Marrakech - ITAFV- Takarietz	E1	E1-2	-
18	Bouchouk Lafayette	OWGB-Marrakech - ITAFV- Takarietz	E1	E1-1	-
19	Bouchouk Soummam	OWGB-Marrakech - ITAFV- Takarietz	E1	E1-1	-
20	Bouchouk_Guergour	ITAFV- Takarietz	E1	-	-
21	Boughenfous	ITAFV- Takarietz	E1	-	-
22	Bouichret	ITAFV- Takarietz	E1	-	-
23	Boukaïla	OWGB-Marrakech - ITAFV- Takarietz	E3	E1-1	-
24	Bouricha	OWGB-Marrakech - ITAFV- Takarietz	E1	E3-3	-
25	Chemlal de Kabylie	OWGB-Marrakech - ITAFV- Takarietz	E3	E3-2	-
26	Ferkani/Jemri bouchouka	OWGB-Marrakech - ITAFV- Takarietz	E1	E1-1	-
27	Grosse du Hamma	OWGB-Marrakech - ITAFV- Takarietz	E1	E1-2	-
28	Hamra	OWGB-Marrakech - ITAFV- Takarietz	E1	E3-3	-
29	Ifiri	OWGB-Marrakech	-	E1-1	-
30	Khadraïa	OWGB-Marrakech	-	E2-1	E3
31	Limli	OWGB-Marrakech - ITAFV- Takarietz	E2	E1-1	-
32	Longue de Meliana	OWGB-Marrakech - ITAFV- Takarietz	E1	E1-1	-
33	Mekki	OWGB-Marrakech - ITAFV- Takarietz	E1	E1-1	-
34	Neb jmel	OWGB-Marrakech - ITAFV- Takarietz	E2	E2-1	E3
35	Ronde de Meliana	OWGB-Marrakech - ITAFV- Takarietz	E2	E1-1	-
36	Rougette de Metidja	OWGB-Marrakech - ITAFV- Takarietz	E2	E1-1	E3
37	Sigoise	ITAFV- Takarietz	E1	-	-
38	Souidi	OWGB-Marrakech - ITAFV- Takarietz	E1	E2-1	E3
39	Tabelout	ITAFV- Takarietz	E2	-	-
40	Taksrit	OWGB-Marrakech - ITAFV- Takarietz	E1	E1-1	-
41	Tefah	OWGB-Marrakech - ITAFV- Takarietz	E1	E1-1	-
42	Zeboudj Boudoudan	OWGB-Marrakech	-	E1-1	-
43	Zeletni	OWGB-Marrakech	-	E2-1	E3

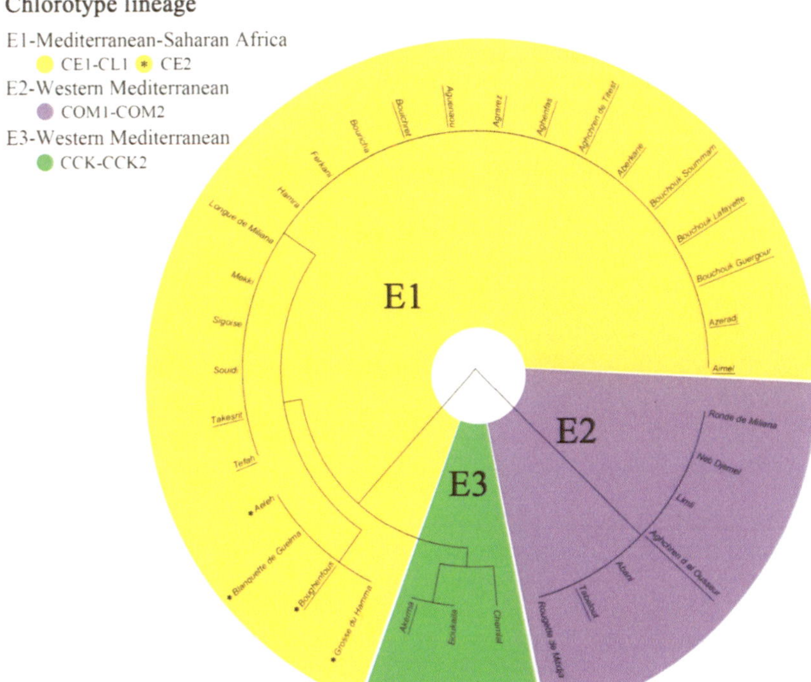

Figure 2. UPGMA (Unweighted Pair-Group Method with Arithmetical Averages) tree of Algerian germplasm based on chloroplast simple sequence repeats (cpSSRs). Original haplotypes (CE1, CE2, COM1-COM2, CCK-CCK2) obtained with cpSSR from Besnard et al. [34] are reported together with the corresponding haplotype lineage: E1, E2 and E3 following Besnard et al. [4].

NuSSRs markers amplified a total of 140 alleles, ranging from five to 11 for EMO90 and DCA07, respectively, with an average of 7.2 alleles per marker (Table S6), which is in agreement with previous studies [14,52]. The average PIC value for nuSSRs (0.659) indicates that the analyzed markers are highly informative and useful for variety screening. Among nuSSR loci, 11 (69%) showed high polymorphism with PIC values exceeding 0.6 (Table S6). In agreement with the PIC value, the average He value was 0.716, while the Ho value was greater than 0.500 for 10 loci (Table S6), underlining a remarkable rate of heterozygosity among the studied cultivars.

The UPGMA phylogenetic tree based on nuSSRs underlined a main genetic group (A) of 21 varieties, 13 of which belonged to Little Kabylie (LK), split in four subclusters. A second group (B) of 13 cultivars separated in two subclusters (Figure 3a). Within group A, subcluster A1 included only cultivars native to LK; A2 consisted of two varieties from LK plus Sigoise and Neb Djemel, considered native to the Mascara Plain (West from LK) and Cherchar (South-East from LK), respectively; A3 showed three varieties from LK, Hamra from the nearby coastal area of Jijel, and Mekki from the Aurès mountain region (close to the Sahara desert); A4 included Chemlal, an important variety that covers 30% of the Algerian olive orchard, and three cultivars with local distribution, i.e., Bouricha, Grosse du Hamma and Rougette de Mitidja. In the B group, the two subclusters accounted six varieties each, with three cultivars considered native to Kabylie, i.e., Akerma, Bouchouk Soummam and Tabelout. The last remaining cultivar, considered native to Kabylie, Aghchren d'el Osseur, clustered alone as an outgroup.

STRUCTURE analysis clearly assigned 30 varieties (ancestry value > 0.65) to one of the six identified genetic groups (Figure 3b, Table S4), while the remaining four cultivars (Aeleh, Blanquette de Guelma, Bouricha, Neb Djemel) showed a mosaic genetic pattern. The six genetic groups were in agreement with phylogenetic analysis, with group K1, K2, K4 and K5 included in cluster A, counting the 13 varieties from LK, while K3 and K6 belonged to group B. In particular, K4 and K5 exactly corresponded to group A1 and A3, respectively; K1 included the variety belonging to A2, plus Grosse du Hamma; in K2, two varieties of A4 (Chemlal and Rougette de Mitidja) plus the cultivar Aghchren d'el Osseur were included. Finally, K3 and K6 account respectively for three and seven varieties, but they did not correspond to the subclusters B.

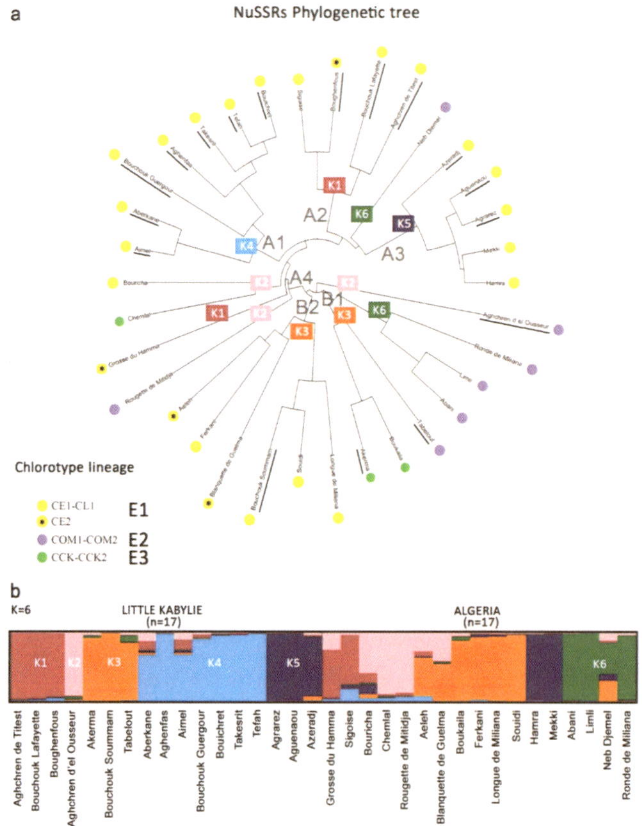

Figure 3. (**a**) UPGMA tree of 34 Algerian varieties based on nuclear simple sequence repeats (nuSSRs). Capital letters indicate the two main clusters (A and B) and relative subclusters; colored dots highlight the haplotype lineage of each variety, and colored rectangles indicate the genetic cluster identified by STRUCTURE analysis. Underlined varieties are native to Little Kabylie (LK). (**b**) STRUCTURE analysis of Algerian germplasm showing on the left side accessions from Little Kabylie and on the right cultivars from other Algerian regions; each color represents the identified genetic cluster and the length of the colored segment shows the estimated membership proportion of each sample to designed group.

We further analyzed the nuclear genetic profiles by PCoA (Figure S1). The resulting pattern reflected the genetic structure identified by UPGMA and STRUCTURE: the first axis separated most of the LK varieties with some other cultivars from group A. The second axis separated a group of five varieties, corresponding to A3.

3.2. Relationship among Algerian and Mediterranean and Near East Germplasm

In order to frame the genetic relationships of the Algerian cultivars within the three main Mediterranean lineages, we compared our profiles with a large dataset of cultivated accessions across the Mediterranean Basin and Near East [40] using hierarchical and Bayesian clustering by mean of a common set of seven nuSSRs. The genetic parameters for each population are shown in Table 3. We observed high values of genetic diversity for each population (ranging from 0.635—FRA—to 0.746—EAST-M, mean 0.698), and a mean of 0.772 for observed heterozygosity, with the Algerian cultivars that showed the lowest value (0.696). The inbreeding coefficient was negative for all populations, but can be considered in equilibrium.

Table 3. Genetic parameters of Algerian and Mediterranean germplasm obtained by nuSSR profiles.

Pop	N	Na	Ne	I	Ho	He	F
ALG	30.7	7.1	3.7	1.4	0.696	0.674	−0.024
	0.2	0.8	0.6	0.1	0.085	0.053	0.083
IB	205.4	11.0	4.1	1.5	0.796	0.684	−0.166
	0.3	2.4	0.8	0.2	0.077	0.066	0.023
FRA	8.7	4.7	3.2	1.2	0.732	0.635	−0.171
	0.2	0.7	0.5	0.2	0.075	0.067	0.071
ITA	25.9	7.4	4.5	1.6	0.857	0.734	−0.169
	0.1	1.3	0.8	0.2	0.057	0.047	0.031
MAG	9.0	5.3	3.5	1.3	0.762	0.660	−0.130
	0.0	0.4	0.5	0.1	0.095	0.068	0.080
BAL	31.0	8.0	4.6	1.6	0.793	0.725	−0.085
	0.0	1.4	0.7	0.2	0.087	0.068	0.037
TUR	17.0	8.3	4.7	1.7	0.756	0.728	−0.045
	0.0	1.2	0.7	0.2	0.076	0.068	0.052
EAST-M	52.6	10.3	5.2	1.8	0.785	0.746	−0.049
	0.3	1.8	1.0	0.2	0.070	0.062	0.030
Mean	47.5	7.8	4.2	1.5	0.772	0.698	−0.105
	8.2	0.5	0.3	0.1	0.027	0.021	0.020

Mean value over loci and standard errors for each population: N: Number of samples; Na: Number of different alleles; Ne: Number of effective alleles; I: Shannon's information index; He: Expected heterozygosity; Ho: Observed heterozygosity; F: Inbreeding coefficient. ALG: Algeria; IB: Iberian Peninsula—Spain and Portugal; FRA: France; ITA: Italy; MAG: Maghreb—Morocco and Tunisia; BAL: Balcanic Peninsula—Croatia, Albania and Greece; TUR: Turkey—Turkey and Cyprus; EAST-M: East Mediterranean—Iran, Israel, Lebanon, Syria and Egypt.

The pairwise Nei's genetic distances and relative Gst values (Table 4) indicated that the Algerian and Iberian germplasm were significantly ($p < 0.01$) the most unrelated, followed by Algeria vs. France and Algeria vs. Turkey. On the contrary, the most related cultivars were those from Turkey vs. East-Mediterranean, Italy vs. Balkan, Balkan vs. Turkey and Iberian vs. Maghreb. The PCoA analysis was able to separate a wide number of the Western varieties from the Central-Eastern group. The Algerian samples showed a bimodal distribution with a group of varieties located in the center of the graph with the Turkish and Near Eastern cultivars, and a second group clustering mainly with the central Mediterranean varieties from Italy and Balkan Peninsula (Figure 4a). The UPGMA phylogenetic tree differentiated Algerian germplasm in two subclusters (Figure S2), one with Western affinity consisting of Iberian Peninsula and Maghreb accessions and the other intermingled with mainly East Mediterranean cultivars and with a minor contribution of Iberian and Balkan varieties.

Table 4. Estimates of pairwise Gst values (below the diagonal) and Unbiased Nei's genetic distance (above the diagonal) among overall populations.

	IB	FR	ITA	MAG	ALG	BAL	TUR	EAST-M
IB		0.119	0.159	0.057	0.286	0.179	0.188	0.168
FR	0.024		0.134	0.139	0.272	0.125	0.073	0.136
ITA	0.030	0.024		0.153	0.242	0.051	0.087	0.119
MAG	0.017	0.027	0.023		0.226	0.111	0.137	0.163
ALG	0.051	0.045	0.040	0.044		0.193	0.261	0.245
BAL	0.033	0.022	0.008	0.013	0.032		0.052	0.094
TUR	0.033	0.015	0.013	0.015	0.044	0.009		0.025
EAST-M	0.031	0.024	0.018	0.022	0.040	0.014	0.004	

In bold significant values with $p \leq 0.01$ calculated over 999 permutations.

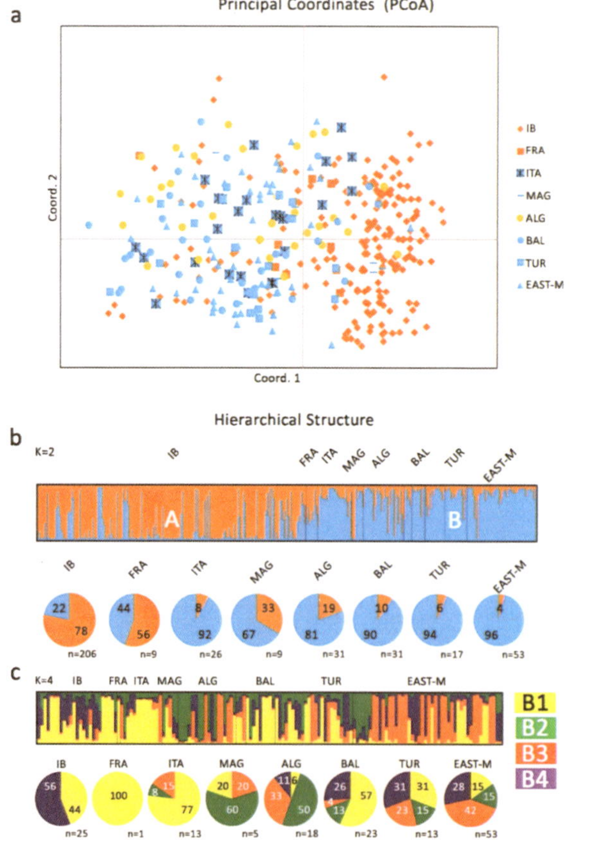

Figure 4. Genetic relationship among Algerian and Mediterranean cultivars. (**a**) Principal Coordinates Analysis (PCoA) and (**b**) STRUCTURE analysis showing unique profiles of 31 Algerian and 351 Mediterranean accessions from Trujillo et al. [40]. (**c**) Hierarchical STRUCTURE analysis of cluster B accessions with ancestry value > 0.80 following Emanuelli et al. [51]. Each color represents the identified genetic cluster (cluster A = orange; cluster B = blue) and the length of the colored segment shows the estimated membership proportion of each sample to the designed group.

STRUCTURE analysis of the whole dataset revealed that the most likely number of clusters of Mediterranean varieties ($k = 2$) was in agreement with previous results [40]. In particular, as reported in other studies [4,9,53], two main pools were identified (Figure 4b), the first (A—orange) accounting for the most part of Western Mediterranean varieties, and the second (B—light blue) including mainly Central and Eastern Mediterranean cultivars. Algerian varieties clustered mostly with cluster B (59%), 9% was grouped in cluster A and the remaining cultivars (32%) were mosaics between the two groups (Table S5). Interestingly, all the cultivars belonging to cluster B were from LK, i.e., Aghenfas, Tefah and Bouchouk Guergour, the latter already differentiated by UPGMA analysis.

Finally, to have a higher resolution of group B, the more heterogeneous pool, which includes samples from all population studied, we ran a second round of STRUCTURE analysis using only the samples closely associated to it (ancestry >0.8). In total, 18 Algerian cultivars and 126 accessions from the entire Mediterranean region were investigated. The analysis allowed to identify four subclusters, including 84 varieties with strong association (ancestry value >0.65), while the remainders 60 cultivars were mosaics (Figure 4c, Table S5). Subcluster B1 included an assorted group consisting of samples from the entire Mediterranean Basin dominated by Central-Western Mediterranean cultivars including 100% of French accessions, 77% Italian, 57% Balkan and 44% Iberian, with a minor contribution of Turkish (31%), Maghreb (20%), East-Mediterranean (15%) and Algerian (6%). Subcluster B2 was well represented by North African cultivars accounting for 50% of the Algerian germplasm and 60% of Maghreb accessions, with a minor contribution of Turkey (15%), East-Mediterranean (15%), Balkan (13%) and Italy (8%). Subcluster B3 included mainly Eastern and Central Mediterranean varieties, accounting for the 41% of East-Mediterranean accessions, 33% Algerian, 23% Turkish, 15% Italian, 20% Maghreb and 4% Balkan. Subcluster B4 included almost exclusively Western and Eastern cultivars, accounting for 56% of Iberian varieties, 31% Turkish, 28% East Mediterranean and 26% Balkan, with a smaller contribution of central Mediterranean accessions (11% of Algerian accessions).

4. Discussion

For thousands of years, olive cultivation has been central in the culture and economy of many Mediterranean and Middle Eastern regions. The ancient civilizations that thrived in this wide geographical area selected and diffused countless varieties across the different countries facing the Mediterranean Sea. Due to these complex historical events, an endless debate arose among scholars about olive domestication, and in particular whether there has been a single or multiple independent domestication events [54,55].

In the last decades, the development of molecular markers such as nuclear and chloroplast SSRs have made it possible to investigate the genetic fingerprint of cultivated and wild olives and disentangle the clues left by migrations and crossing among varieties across the entire olive distribution area. Despite many recent studies provided nuclear and chloroplast SSRs genetic profiles for hundreds of wild and cultivated olive accessions, the germplasm from Central and Southern Mediterranean regions, especially from the Maghreb area, is highly underrepresented. Here, for the first time, we characterized the official Algerian collection of olive varieties from ITAFV by mean of both nuSSRs and cpSSRs. Our results filled the gaps left by previous studies [14,16,17] as we provided the nuSSR genetic profiles for 34 out of 36 official Algerian varieties, 10 of which have never been described (Aimel, Bouchouk Guergour, Bouchouk Lafayette, Boukaila, Grosse du Hamma, Hamra, Longue de Miliana, Mekki, Neb Djemel, Ronde de Miliana) and, for the first time, cpSSR profiles for seven varieties (Aeleh, Aghenfas, Bouchouk Guergour, Boughenfous, Bouichret, Tabelout, Tefah). Overall, high genetic diversity was observed, in agreement with the range obtained in previous studies [5,40,56,57], indicating that the Algerian olive germplasm collection represents an important genetic reservoir for the species. Compared with previous studies on Algerian germplasm [14,17], we found a lower number of alleles but a remarkably higher observed and expected heterozygosity. These discrepancies are probably due to the different panel of varieties analyzed and the presence of wild germplasm in previous studies, which contained private alleles [14,17]. The 16 nuSSRs used here allowed us to discriminate all the Algerian cultivars and were powerful enough to resolve putative cases of synonymy. For example, we found that Agrarez and Azeradj had distinct profiles at eight loci and should not be considered as synonyms as previously suggested [14]. Most of the Algerian varieties (70%) belonged to the Mediterranean/Saharan Africa chlorotype olive lineage E1, widely represented in the cultivated and wild forms in the whole Mediterranean Basin. This cluster included two subclusters, CE1-CL1 and CE2 [34]. Subluster CE1-CL1 consisted of 13 varieties from LK region, Hamra from the nearby coastal area of Jijel, Longue de Miliana and Sigoise from the Central-West regions, Mekki, Ferkani and Souidi from Aurès. Subcluster CE2 accounted for one variety from LK (Boughenfous), and three varieties from the territory of Costantine (Grosse du Hamma), Guelma, (Blanquette de Guelma) and Aurès (Aeleh), respectively. Seven varieties, two from LK (Aghchren d'el Ousseur and Tabelout) and five from other regions (Abani, Limli, Neb Djemel, Ronde de Miliana, Rougette de Mitidja) grouped in the Central-Western Mediterranean lineage E2. Interestingly, this lineage is represented by wild olive mainly from Italy and Greece with a minor contribution from Spain and Morocco, and by few cultivars (*n* = 13) from different central Mediterranean regions (Corsica, France, Greece, Italy, Morocco, Sardinia, Spain and Tunisia) [58]. Chemlal, Akerma and Boukaila clustered in the other less common Western Mediterranean lineage E3, mainly found in wild olive from Spain and Morocco, and in cultivars from Maghreb except for three varieties from Corsica, France and Spain (i.e., Antonina, Olivière and Farga), respectively [58]. Our results mostly confirmed the chlorotypes identified in previous studies (*n* = 18), but highlighted some divergence (Table 2); in particular, we found that nine varieties chlorotypes were assigned differently compared to Besnard et al. [4], six when comparing Besnard et al. [4] and Haouane et al. [18], and two when comparing the three datasets. These results could be due to *i)* possible mislabeling errors in the WOGB collection of Marrakech; *ii)* errors in the published datasets, at least for the same six varieties coming from the above mentioned collection that showed different chlorotypes between the dataset of Besnard et al. [4] and Haouane et al. [18]; *iii)* different clones of the same varieties. The different clustering methods adopted in our study highlighted a clear genetic group mainly consisting of 13 LK cultivars, except for a few varieties (four) from other regions (emigrants). Conversely, a few cultivars (four) from LK clustered in other genetic groups (in-migrants). We can formulate some speculative hypotheses to explain these few exceptions to the general geographic and genetic division between LK region and other parts of Algeria. Emigrant varieties might share a wide genetic background with the LK group because they were selected in this region, but later, for some ecological/historical/agronomic reasons, their cultivation disappeared from the LK area and the knowledge of the original native region was lost. This could be the case of Mekki, Neb Djemel and Hamra, today cultivated only in the driest mountain area of the Aurès or Kenchela or in the coastal area, respectively, or in contrast diffuse throughout Algeria (Sigoise). It has been documented that the historical distribution of Mekki was connected to Phoenician, Greek and Roman dominations [37]. This variety share the chloroplast haplotype dominant in LK varieties, suggesting a common origin, thus, it is plausible that it was once distributed in this area and that it later disappeared in LK remaining confined in a restricted area near the old Roman town of Timgad. For in-migrant cultivars, we can speculate that the knowledge of the original native region was lost: these varieties could have been selected and genetically improved in LK by crossing with germplasm imported from other regions of Algeria/North Africa, thus explaining the genetic divergence from other LK varieties. In particular, nuSSR results were supported by cpSSR analysis indicating that *in-migrant* cultivars had different haplotypes as compared to other LK accessions. In a wider perspective, the combined use of nuSSR and cpSSR, that are differently affected by evolutionary processes (i.e., selection, mutation, recombination etc.), allowed us to investigate the genetic relationship of Algerian varieties with cultivars from other Mediterranean countries and shed light on the geographic origin of Algerian germplasm and relative patterns of crossing/migration across the Mediterranean Region. Our results are compatible with two different hypotheses: (i) local domestication from oleaster and Laperrine's olive; (ii) local selection by crossing of cultivars imported from other Mediterranean region with local germplasm. The gene flow between Algeria and the rest of the Mediterranean area has been probably limited, suggesting that development of new cultivars possibly proceeded through crossing of few imported varieties with local germplasm, namely oleaster and Laperrine's olive, as testified by cpSSRs pattern of genetic diversity. We found that the highest proportion of Algerian varieties shared the same haplotype with the majority of Mediterranean cultivars (E1) and the Laperrine's olive. Interestingly, 30% of Algerian varieties belonged to the other two lineages E2 and E3, unravelling the contribution of the Western Mediterranean olive lineage to the origin of North African cultivars. This result provides new evidence on the role of Algeria as possible and important secondary domestication/selection center, considering that only 4.9% and 4.4% of the Mediterranean cultivars belong to E2 and E3 lineages, respectively [4]. In particular, we can speculate that few varieties with chlorotype E2 probably represent locally domesticated cultivars or cultivars imported from a central Mediterranean region such as Italy, e.g., during Roman domination, as this haplotype is only found in few cultivated varieties from Central Mediterranean region and Maghreb but it is common in wild olive from Italy and Greece. In addition, the three cultivars belonging to lineage E3 represent the most likely candidates for secondary domestication events in this area, given that E3 haplotype is rare and found in wild oleaster from Spain and Morocco and cultivated varieties from Morocco ($n = 10$), France ($n = 2$) and in one variety from Italy and Spain each.

Finally, regardless of the true origin of the Algerian germplasm, LK varieties can be considered an exclusive genetic core, selected and developed during the different historical periods by the civilizations that thrived in this area from Phoenicians to Arabs until the present. Genetic differentiation parameters and results of the hierarchical STRUCTURE provided evidence that Algerian varieties are more genetically related to Central-Eastern Mediterranean cultivars than to the West. In particular, the second round of STRUCTURE highlighted four main subclusters among the group of Eastern varieties. The Algerian germplasm grouped mostly in two subclusters, both with Central and Eastern affinity. Subcluster B1 was particularly interesting because it was dominant in Algeria and Maghreb, suggesting gene flow between Near East and North Africa. We speculate that the two STRUCTURE clusters could correspond to bottleneck events due to the arrival of different civilization, i.e., Phoenician and Romans. According to the chlorotype lineages identified, 67% of Algerian varieties showed affinity to the Eastern Mediterranean germplasm and 33% to the West. Nuclear DNA confirmed this pattern, with 80% of varieties showing affinity to the Eastern cluster and 20% to the Western cluster. These results might depend on recurrent reticulation events during the diffusion of olive culture [54] and reflect the predominant role of Eastern germplasm in the development of olive cultivars around the whole Mediterranean Basin [54,59]. However, our study revealed an important contribution of Central-Western Mediterranean germplasm in the development of olive varieties, supporting the hypothesis of the existence of an independent domestication center in the Central Mediterranean area [9,55]. This hypothesis is supported by the fact that the wild Laperrine's olive tree share its haplotype (E1) with most of the world's olive varieties, including 67% of the Algerian varieties, whereas the remaining 33% share their haplotypes (E2 and E3) with wild olives from Spain, Morocco, Italy and Greece. In particular, given that North African germplasm is highly underrepresented in current literature, the role of this area in the history of olive domestication and cultivar development should be reconsidered to evaluate its real contribution.

Our present data do not allow discriminating between the two different hypotheses, namely the occurrence of a secondary domestication event or introgression of imported cultivars with local germplasm, including natural populations (oleaster and Laperrine's olive). We can hypothesize that before foreign civilizations arrived in Algeria, wild olive tree populations consisted of two taxa, oleaster (*Olea europaea* subsp. *sylvestris*) and Laperrine's olive (*O. europaea* subsp. *laperrinei*), that were already exploited by local human population, laying the foundations for the development of olive cultivation. Subsequently, phoenicians introduced Eastern Mediterranean olive trees to North Africa, triggering gene flow with the local germplasm (oleaster and Laperrine's olive) and with cultivated olive coming from other Mediterranean areas. Local people and settlers from abroad (e.g., Romans,

Arabs) eventually selected the cultivars better suited for the cultivation in the different environmental conditions of each Algerian region, thus originating the distribution pattern that we observe today.

5. Conclusions

Genetic studies of local olive varieties from different Mediterranean areas, in particular from North Africa, central Mediterranean area and Near East, can help to clarify the pathways of domestication and diffusion of this species along the history of civilizations. Due to its central position in the Mediterranean Basin, Algeria has played an important role for civilizations crossing the Mediterranean Basin, especially for Phoenicians, Romans and Arabs. To the best of our knowledge, this is the first study analyzing and characterizing the official Algerian olive germplasm collection by means of nuclear and chloroplast SSRs, comparing the results with available published datasets. An exclusive genetic group of 13 varieties from little Kabylie has been identified among the main Algerian olive cultivars and it can be considered a valuable genetic resource for future cultivation and breeding programs. Nuclear and chloroplast genetic profiles provided here will be useful for future program of plant material certification in Algeria. Bayesian and hierarchical cluster analyses allowed to develop inferences on the different patterns of genetic diversity observed. A detailed evolutionary view of Algerian germplasm has been defined, highlighting its genetic relationship with reference cultivars from the whole Mediterranean Basin and Near East. Our findings are compatible with the hypothesis of the existence of an independent olive domestication area in the center of the Mediterranean Basin, but further analysis with more extended datasets are needed to verify this hypothesis. Unfortunately, in contrast to other species, such as grapevine (Vitis International Variety Catalogue, European Vitis Database), there is no international database of olive varieties to use as a reference, and the genetic profiles are not always available. The creation of a public database for olive germplasm would greatly foster the elucidation of the history of domestication for this important crop species.

Supplementary Materials: The following are available online http://www.mdpi.com/2073-4425/11/3/303/s1, Figure S1: PCoA analysis of Algerian germplasm based on nuSSRs, Figure S2: UPGMA tree of the whole dataset covering the Mediterranean genetic background, Table S1: Name, synonym, Arabic name, meaning, geographic diffusion and origin, use of fruits and oil percentage of the analyzed Algerian varieties, Table S2: Description sheets of the analyzed Algerian varieties, Table S3: nuSSR and cpSSR markers used in the present study, genetic profiles for Algerian varieties, and the public Mediterranean dataset at seven SSRs (Trujillo et al. [40]), Table S4: Posterior membership coefficients following a STRUCTURE analysis and K = 6 for the studied germplasm, Table S5: Posterior membership coefficients following a STRUCTURE analysis and K = 2 for the studied germplasm, Table S6: Genetic parameters at 16 nuSSR and 6 cpSSR loci used to genotype the ITAFV Algerian germplasm collection.

Author Contributions: Conceptualization, B.H., A.S.G., F.M. and F.C.; Formal analysis, B.H., A.S.G., F.M., A.M. and A.S.; Investigation, B.H., A.S.G., F.M., A.M. and A.S.; Methodology, B.H., A.S.G., F.M. and F.C.; Project administration, B.H., A.S.G., F.M. and F.C.; Resources, B.H., N.A., A.E.S. and F.C.; Supervision, F.C.; Visualization, A.S.G. and F.M.; Writing—original draft, B.H., A.S.G., F.M. and F.C.; Writing—review and editing, B.H., A.S.G., F.M. and F.C. All authors have read and agreed to the published version of the manuscript.

Funding: This research was partially supported by the Italian Ministry of Foreign Affairs and International Cooperation (MAECI), project n B56J13000320006 'Scienze per la DIPLOMAzia'.

Acknowledgments: We are grateful to Roberto De Michele and Angela Carra for reviewing the first draft of the manuscript and providing useful suggestion and to Francesca La Bella for helping in the laboratory work.

Conflicts of Interest: The authors declare that they have no conflict of interest.

References

1. Loumou, A.; Giourga, C. Olive groves: "The life and identity of the Mediterranean". *Agric. Hum. Values* **2003**, *20*, 87–95. [CrossRef]
2. Wiesman, Z. *Desert Olive Oil Cultivation: Advanced Bio Technologies*; Academic Press: Cambridge, MA, USA, 2009.
3. International Olive Oil Council. Available online: http://www.internationaloliveoil.org/estaticos/view/131-world-olive-oil-figuresS\backslashS (accessed on 22 January 2019).
4. Besnard, G.; Khadari, B.; Navascués, M.; Fernández-Mazuecos, M.; El Bakkali, A.; Arrigo, N.; Baali-Cherif, D.; Brunini-Bronzini de Caraffa, V.; Santoni, S.; Vargas, P.; et al. The complex history of the olive tree: From

Late Quaternary diversification of Mediterranean lineages to primary domestication in the northern Levant. *Proc. R. Soc. B Biol. Sci.* **2013**, *280*, 20122833. [CrossRef]

5. Breton, C.; Tersac, M.; Berville, A. Genetic diversity and gene flow between the wild olive (oleaster, Olea europaea L.) and the olive: Several Plio-Pleistocene refuge zones in the Mediterranean basin suggested by simple sequence repeats analysis. *J. Biogeogr.* **2006**, *33*, 1916–1928. [CrossRef]

6. Kaniewski, D.; Van Campo, E.; Boiy, T.; Terral, J.-F.; Khadari, B.; Besnard, G. Primary domestication and early uses of the emblematic olive tree: Palaeobotanical, historical and molecular evidence from the Middle East. *Biol. Rev.* **2012**, *87*, 885–899. [CrossRef]

7. Terral, J.-F.; Arnold-Simard, G. Beginnings of olive cultivation in eastern Spain in relation to Holocene bioclimatic changes. *Quat. Res.* **1996**, *46*, 176–185. [CrossRef]

8. Zohary, D.; Hopf, M.; Weiss, E. *Domestication of Plants in the Old World: The Origin and Spread of Domesticated Plants in Southwest Asia, Europe, and the Mediterranean Basin*; Oxford University Press on Demand: Durham, NC, USA, 2012.

9. Diez, C.M.; Trujillo, I.; Martinez-Urdiroz, N.; Barranco, D.; Rallo, L.; Marfil, P.; Gaut, B.S. Olive domestication and diversification in the Mediterranean Basin. *New Phytol.* **2015**, *206*, 436–447. [CrossRef]

10. Besnard, G.; Berville, A. On chloroplast DNA variations in the olive (*Olea europaea* L.) complex: Comparison of RFLP and PCR polymorphisms. *Theor. Appl. Genet.* **2002**, *104*, 1157–1163. [CrossRef] [PubMed]

11. Baldoni, L.; Tosti, N.; Ricciolini, C.; Belaj, A.; Arcioni, S.; Pannelli, G.; Germana, M.A.; Mulas, M.; Porceddu, A. Genetic structure of wild and cultivated olives in the central Mediterranean basin. *Ann. Bot.* **2006**, *98*, 935–942. [CrossRef] [PubMed]

12. Besnard, G.; Terral, J.-F.; Cornille, A. On the origins and domestication of the olive: A review and perspectives. *Ann. Bot.* **2018**, *121*, 385–403. [CrossRef] [PubMed]

13. Bartolini, G.; Prevost, G.; Messeri, C.; Carignani, G. *Olive Germplasm: Cultivars and World-Wide Collections*; FAO: Rome, Italy, 1998.

14. Boucheffa, S.; Miazzi, M.; di Rienzo, V.; Mangini, G.; Fanelli, V.; Tamendjari, A.; Pignone, D.; Montemurro, C. The coexistence of oleaster and traditional varieties affects genetic diversity and population structure in Algerian olive (*Olea europaea*) germplasm. *Genet. Resour. Crop Evol.* **2017**, *64*, 379–390. [CrossRef]

15. Hauville, A. La repartition des varietes d'olivier en Algerie et ses consequences pratiques. *Bull. Soc. Agric. D'algerie* **1953**, *580*, 1–8. (In French)

16. Dominguez-Garcia, M.; Laib, M.; De la Rosa, R.; Belaj, A. Characterisation and identification of olive cultivars from North-eastern Algeria using molecular markers. *J. Hortic. Sci. Biotechnol.* **2012**, *87*, 95–100. [CrossRef]

17. Abdessemed, S.; Muzzalupo, I.; Benbouza, H. Assessment of genetic diversity among Algerian olive (Olea europaea L.) cultivars using SSR marker. *Sci. Hortic.* **2015**, *192*, 10–20. [CrossRef]

18. Haouane, H.; El Bakkali, A.; Moukhli, A.; Tollon, C.; Santoni, S.; Oukabli, A.; El Modafar, C.; Khadari, B. Genetic structure and core collection of the World Olive Germplasm Bank of Marrakech: Towards the optimised management and use of Mediterranean olive genetic resources. *Genetica* **2011**, *139*, 1083–1094. [CrossRef] [PubMed]

19. Reif, J.; Warburton, M.; Xia, X.; Hoisington, D.; Crossa, J.; Taba, S.; Muminović, J.; Bohn, M.; Frisch, M.; Melchinger, A. Grouping of accessions of Mexican races of maize revisited with SSR markers. *Theor. Appl. Genet.* **2006**, *113*, 177–185. [CrossRef] [PubMed]

20. Thomson, M.J.; Septiningsih, E.M.; Suwardjo, F.; Santoso, T.J.; Silitonga, T.S.; McCouch, S.R. Genetic diversity analysis of traditional and improved Indonesian rice (*Oryza sativa* L.) germplasm using microsatellite markers. *Theor. Appl. Genet.* **2007**, *114*, 559–568. [CrossRef]

21. Blair, M.W.; Díaz, L.M.; Buendía, H.F.; Duque, M.C. Genetic diversity, seed size associations and population structure of a core collection of common beans (*Phaseolus vulgaris* L.). *Theor. Appl. Genet.* **2009**, *119*, 955–972. [CrossRef]

22. Laidò, G.; Mangini, G.; Taranto, F.; Gadaleta, A.; Blanco, A.; Cattivelli, L.; Marone, D.; Mastrangelo, A.M.; Papa, R.; De Vita, P. Genetic diversity and population structure of tetraploid wheats (Triticum turgidum L.) estimated by SSR, DArT and pedigree data. *PLoS ONE* **2013**, *8*. [CrossRef]

23. Mazzucato, A.; Papa, R.; Bitocchi, E.; Mosconi, P.; Nanni, L.; Negri, V.; Picarella, M.E.; Siligato, F.; Soressi, G.P.; Tiranti, B.; et al. Genetic diversity, structure and marker-trait associations in a collection of Italian tomato (*Solanum lycopersicum* L.) landraces. *Theor. Appl. Genet.* **2008**, *116*, 657–669. [CrossRef]

24. Mercati, F.; Longo, C.; Poma, D.; Araniti, F.; Lupini, A.; Mammano, M.M.; Fiore, M.C.; Abenavoli, M.R.; Sunseri, F. Genetic variation of an Italian long shelf-life tomato (*Solanum lycopersicon* L.) collection by using SSR and morphological fruit traits. *Genet. Resour. Crop Evol.* **2015**, *62*, 721–732. [CrossRef]

25. Bacilieri, R.; Lacombe, T.; Le Cunff, L.; Di Vecchi-Staraz, M.; Laucou, V.; Genna, B.; Péros, J.-P.; This, P.; Boursiquot, J.-M. Genetic structure in cultivated grapevines is linked to geography and human selection. *BMC Plant Biol.* **2013**, *13*, 25. [CrossRef] [PubMed]

26. Riaz, S.; De Lorenzis, G.; Velasco, D.; Koehmstedt, A.; Maghradze, D.; Bobokashvili, Z.; Musayev, M.; Zdunic, G.; Laucou, V.; Walker, M.A.; et al. Genetic diversity analysis of cultivated and wild grapevine (*Vitis vinifera* L.) accessions around the Mediterranean basin and Central Asia. *BMC Plant Biol.* **2018**, *18*, 137. [CrossRef] [PubMed]

27. De Michele, R.; La Bella, F.; Gristina, A.S.; Fontana, I.; Pacifico, D.; Garfi, G.; Motisi, A.; Crucitti, D.; Abbate, L.; Carimi, F. Phylogenetic relationships among wild and cultivated grapevine in Sicily: A hotspot in the middle of the Mediterranean Basin. *Front. Plant Sci.* **2019**, *10*, 1506. [CrossRef] [PubMed]

28. Cipriani, G.; Spadotto, A.; Jurman, I.; Di Gaspero, G.; Crespan, M.; Meneghetti, S.; Frare, E.; Vignani, R.; Cresti, M.; Morgante, M.; et al. The SSR-based molecular profile of 1005 grapevine (*Vitis vinifera* L.) accessions uncovers new synonymy and parentages, and reveals a large admixture amongst varieties of different geographic origin. *Theor. Appl. Genet.* **2010**, *121*, 1569–1585. [CrossRef] [PubMed]

29. Sefc, K.M.; Lopes, M.S.; Mendonça, D.; Rodrigues Dos Santos, M.; Da Câmara Machado, M.; Da Câmara Machado, A. Identification of microsatellites loci in olive (*Olea europaea* L.) and their characterisation in Italian and Iberian olive trees. *Mol. Ecol.* **2000**, *9*, 1171–1193. [CrossRef] [PubMed]

30. Carriero, F.; Fontanazza, G.; Cellini, F.; Giorio, G. Identification of simple sequence repeats (SSRs) in olive (*Olea europaea* L.). *Theor. Appl. Genet.* **2002**, *104*, 301–307. [CrossRef]

31. De la Rosa, R.; James, C.; Tobutt, K. Isolation and characterization of polymorphic microsatellites in olive (Olea europaea L.) and their transferability to other genera in the Oleaceae. *Mol. Ecol. Notes* **2002**, *2*, 265–267.

32. Baldoni, L.; Cultrera, N.G.; Mariotti, R.; Ricciolini, C.; Arcioni, S.; Vendramin, G.G.; Buonamici, A.; Porceddu, A.; Sarri, V.; Ojeda, M.A.; et al. A consensus list of microsatellite markers for olive genotyping. *Mol. Breed.* **2009**, *24*, 213–231. [CrossRef]

33. Díez, C.M.; Trujillo, I.; Barrio, E.; Belaj, A.; Barranco, D.; Rallo, L. Centennial olive trees as a reservoir of genetic diversity. *Ann. Bot.* **2011**, *108*, 797–807. [CrossRef]

34. Besnard, G.; Rubio de Casas, R.; Vargas, P. Plastid and nuclear DNA polymorphism reveals historical processes of isolation and reticulation in the olive tree complex (*Olea europaea*). *J. Biogeogr.* **2007**, *34*, 736–752. [CrossRef]

35. Arroyo-García, R.; Lefort, F.; Andrés, M.T.d.; Ibáñez, J.; Borrego, J.; Jouve, N.; Cabello, F.; Martínez-Zapater, J.M. Chloroplast microsatellite polymorphisms in Vitis species. *Genome* **2002**, *45*, 1142–1149. [CrossRef] [PubMed]

36. Arroyo-García, R.; Ruiz-Garcia, L.; Bolling, L.; Ocete, R.; Lopez, M.; Arnold, C.; Ergul, A.; Söylemezo Lu, G.; Uzun, H.; Cabello, F.; et al. Multiple origins of cultivated grapevine (*Vitis vinifera* L. ssp. *sativa*) based on chloroplast DNA polymorphisms. *Mol. Ecol.* **2006**, *15*, 3707–3714.

37. Mendil, M.; Sebai, A. *Catalogue des variétés algériennes de l'olivier*; Ministère de l'agriculture et du développement rural, ITAF Alger: Tessala El Merdja, Algeria, 2006; p. 98. (In French)

38. Doyle, J.; Doyle, J. Isolation of DNA from fresh plant tissue. *Focus* **1987**, *12*, 13–15.

39. Garfi, G.; Mercati, F.; Fontana, I.; Collesano, G.; Pasta, S.; Vendramin, G.G.; De Michele, R.; Carimi, F. Habitat features and genetic integrity of wild grapevine *Vitis vinifera* L. subsp. *sylvestris* (CC Gmel.) Hegi populations: A case study from Sicily. *Flora-Morphol. Distrib. Funct. Ecol. Plants* **2013**, *208*, 538–548.

40. Trujillo, I.; Ojeda, M.A.; Urdiroz, N.M.; Potter, D.; Barranco, D.; Rallo, L.; Diez, C.M. Identification of the Worldwide Olive Germplasm Bank of Córdoba (Spain) using SSR and morphological markers. *Tree Genet. Genomes* **2014**, *10*, 141–155. [CrossRef]

41. Liu, K.; Muse, S.V. PowerMarker: An integrated analysis environment for genetic marker analysis. *Bioinformatics* **2005**, *21*, 2128–2129. [CrossRef]

42. Eliades, N.; Eliades, D. *HAPLOTYPE ANALYSIS: Software for analysis of haplotype data*; Distributed by the authors; Forest Genetics and Forest Tree Breeding, Georg-August University Goettingen: Goettingen, Germany; Available online: http://www.uni-goettingen.de/en/134935 (accessed on 30 January 2020).

43. Chapuis, M.-P.; Estoup, A. Microsatellite null alleles and estimation of population differentiation. *Mol. Biol. Evol.* **2007**, *24*, 621–631. [CrossRef]

44. Jombart, T. Adegenet: A R package for the multivariate analysis of genetic markers. *Bioinformatics* **2008**, *24*, 1403–1405. [CrossRef]

45. Peakall, R.; Smouse, P.E. GENALEX 6: Genetic analysis in Excel. Population genetic software for teaching and research. *Mol. Ecol. Notes* **2006**, *6*, 288–295. [CrossRef]

46. Pritchard, J.; Stephens, M.; Donnelly, P. Inference of population structure using multilocus genotype data. Genetics. *Genetics Soc Am.* **2000**, *155*, 945–959.

47. Evanno, G.; Regnaut, S.; Goudet, J. Detecting the number of clusters of individuals using the software STRUCTURE: A simulation study. *Mol. Ecol.* **2005**, *14*, 2611–2620. [CrossRef] [PubMed]

48. Jakobsson, M.; Rosenberg, N.A. CLUMPP: A cluster matching and permutation program for dealing with label switching and multimodality in analysis of population structure. *Bioinformatics* **2007**, *23*, 1801–1806. [CrossRef] [PubMed]

49. Rosenberg, N.A. DISTRUCT: A program for the graphical display of population structure. *Mol. Ecol. Notes* **2004**, *4*, 137–138. [CrossRef]

50. Weir, B.; Cockerham, C. Estimating f-statistics for the analysis of population structure. *Evolution* **1984**, *38*, 1358–1370. [PubMed]

51. Emanuelli, F.; Lorenzi, S.; Grzeskowiak, L.; Catalano, V.; Stefanini, M.; Troggio, M.; Myles, S.; Martinez-Zapater, J.M.; Zyprian, E.; Moreira, F.M.; et al. Genetic diversity and population structure assessed by SSR and SNP markers in a large germplasm collection of grape. *BMC Plant Biol.* **2013**, *13*, 39. [CrossRef] [PubMed]

52. Muzzalupo, I.; Vendramin, G.G.; Chiappetta, A. Genetic biodiversity of Italian olives (*Olea europaea*) germplasm analyzed by SSR markers. *Sci. World J.* **2014**, *2014*. [CrossRef] [PubMed]

53. Mousavi, S.; Mariotti, R.; Bagnoli, F.; Costantini, L.; Cultrera, N.G.; Arzani, K.; Pandolfi, S.; Vendramin, G.G.; Torkzaban, B.; Hosseini-Mazinani, M.; et al. The eastern part of the Fertile Crescent concealed an unexpected route of olive (Olea europaea L.) differentiation. *Ann. Bot.* **2017**, *119*, 1305–1318. [CrossRef]

54. Besnard, G.; Rubio de Casas, R. Single vs multiple independent olive domestications: The jury is (still) out. *New Phytol.* **2016**, *209*, 466–470. [CrossRef]

55. Díez, C.M.; Gaut, B.S. The jury may be out, but it is important that it deliberates: A response to Besnard and Rubio de Casas about olive domestication. *New Phytol.* **2016**, *209*, 471–473. [CrossRef]

56. Khadari, B.; Charafi, J.; Moukhli, A.; Ater, M. Substantial genetic diversity in cultivated Moroccan olive despite a single major cultivar: A paradoxical situation evidenced by the use of SSR loci. *Tree Genet. Genomes* **2008**, *4*, 213–221. [CrossRef]

57. Cicatelli, A.; Fortunati, T.; De Feis, I.; Castiglione, S. Oil composition and genetic biodiversity of ancient and new olive (Olea europea L.) varieties and accessions of southern Italy. *Plant Sci.* **2013**, *210*, 82–92. [CrossRef] [PubMed]

58. Besnard, G.; Hernández, P.; Khadari, B.; Dorado, G.; Savolainen, V. Genomic profiling of plastid DNA variation in the Mediterranean olive tree. *Bmc Plant Biol.* **2011**, *11*, 80. [CrossRef] [PubMed]

59. Besnard, G.; El Bakkali, A.; Haouane, H.; Baali-Cherif, D.; Moukhli, A.; Khadari, B. Population genetics of Mediterranean and Saharan olives: Geographic patterns of differentiation and evidence for early generations of admixture. *Ann. Bot.* **2013**, *112*, 1293–1302. [CrossRef] [PubMed]

 *genes*

Article

Diversity Analysis of Sweet Potato Genetic Resources Using Morphological and Qualitative Traits and Molecular Markers

Fabio Palumbo, Aline Carolina Galvao, Carlo Nicoletto *, Paolo Sambo and Gianni Barcaccia

Department of Agronomy, Food, Natural resources, Animals and Environment (DAFNAE) University of Padova, Agripolis Campus, Viale dell'Università, 16-35020 Legnaro, Italy; fabio.palumbo@unipd.it (F.P.); aline.galvao@unipd.it (A.C.G.); paolo.sambo@unipd.it (P.S.); gianni.barcaccia@unipd.it (G.B.)
* Correspondence: carlo.nicoletto@unipd.it; Tel.: +39-049-827-2826

Received: 24 September 2019; Accepted: 22 October 2019; Published: 24 October 2019

Abstract: The European Union (EU) market for sweet potatoes has increased by 100% over the last five years, and sweet potato cultivation in southern European countries is a new opportunity for the EU to exploit and introduce new genotypes. In view of this demand, the origins of the principal Italian sweet potato clones, compared with a core collection of genotypes from Central and Southern America, were investigated for the first time. This was accomplished by combining a genetic analysis, exploiting 14 hypervariable microsatellite markers, with morphological and chemical measurements based on 16 parameters. From the molecular analyses, Italian accessions were determined to be genetically very similar to the South American germplasm, but they were sub-clustered into two groups. This finding was subsequently confirmed by the morphological and chemical measurements. Moreover, the analysis of the genetic structure of the population suggested that one of the two groups of Italian genotypes may have descended from one of the South American accessions, as predicted on the basis of the shared morphological characteristics and molecular fingerprints. Overall, the combination of two different characterization methods, genetic markers and agronomic traits, was effective in differentiating or clustering the sweet potato genotypes, in agreement with their geographical origin or phenotypic descriptors. This information could be exploited by both breeders and farmers to detect and protect commercial varieties, and hence for traceability purposes.

Keywords: *Ipomoea batatas*; genetic diversity; SSR markers; qualitative traits

1. Introduction

Sweet potato (*Ipomoea batatas* Lam.) is a root crop of the Convolvulaceae family, originating in Central and South America, which spread through the world with great ease due to its prominent productive efficiency. This crop plays a vital role in food production because it is one of the most important root and tuber crops in the world. Its ability to produce energy is very efficient and it can provide a significant quantity of protein and sugars per hectare in a short time [1].

In the European context, this crop had an enormous rise in consumption and, according to the Center for the Promotion of Imports from Developing Countries (CBI) [2], its importation has doubled in recent years. The European Union (EU) market for sweet potatoes has increased by 100% over the last five years (CBI, 2015), and sweet potato cultivation in southern European countries presents a new opportunity for the EU to exploit and introduce new genotypes. In Italy, it is considered a niche and ethnic product, and the recent immigration flow has created a market with increasing domestic demand [3] and many future opportunities for growth and profitability.

Despite the historical and commercial importance of sweet potatoes, to date, no study has investigated the origin, the conservation, or the genetic background of this species in Italy. On a larger scale, several works have been published [4–8] on the genetic characterization of sweet potato accessions, mainly to investigate the dispersal of New World sweet potato landraces from the center of origin (Tropical America, [9]). One of the main obstacles to the understanding of the dispersal dynamics of sweet potato throughout the world is probably the genetics of this hexaploid species ($2n = 6x = 90$) [10], which severely complicates any genomic approach. In particular, sweet potato is an allohexaploid species (AABBBB), most likely derived from the interspecific hybridization between a diploid and tetraploid species followed by chromosomal doubling [11,12]. As a consequence, its inheritance model is admixed, including both disomic (AA) and tetrasomic (BBBB) pairings. On the other hand, it must be recognized that this polyploidy could represent an important source of genetic diversity [13]. According to Silva Ritschel and Huamán [14], the vast genetic diversity that characterizes the sweet potato germplasm is also due to sexual reproduction (i.e., genetic segregation and recombination) and asexual propagation (i.e., fixation of specific genetic combinations), as well as to the exchange and introduction of plants from all over the world. This diversity provides a valuable source for potentially useful traits and allows plant breeders and farmers to adapt the crop to heterogeneous and changing environments [15]. The evolving climate conditions and the staggering expansion of the world population together represent pressing challenges for agriculture.

As already seen in other crops, morphological, agronomic, and molecular marker approaches are often used in combination to complement the information provided singularly in order to investigate the heterogeneity described in a species [16]. Molecular markers such as microsatellites or SSRs play a central role in the assessment and conservation of genetic diversity due to their efficiency, reliability, and reproducibility. Several studies based on the application of SSRs have recently attempted to monitor and prevent genetic erosion of local crop varieties in Italian scenarios [15,17–19]. Estimating the allelic dosage at each locus represents a critical question in polyploid species, even when using co-dominant markers such as microsatellites [20]. For this reason, as has already been done in previous studies [21–23], SSRs were scored as dominant markers and organized in binary matrixes, similarly to the process with AFLP or RAPD markers. According to Silva Ritschel and Huamán [14], morphological and chemical characterization is an indirect measure of population genetic diversity. Morphological markers for sweet potatoes are accessible and easy to use, making the technique one of the most used for this kind of analysis [24–26].

In this study, the geographical and genetic origins of the principal Italian sweet potato accessions were compared with a core collection of accessions from Central and South America for the first time. As has already been achieved in previous works on sweet potato [27,28], this was achieved by combining the high polymorphism and reproducibility of SSR markers with the high information value of strategic morphological and qualitative traits.

2. Materials and Methods

2.1. Plant Material

The cultivation was carried out at the experimental farm "L. Toniolo" of Padova University (45°21′ N; 11°58′ E; 8 m a.s.l.) in the 2016 spring/fall growing cycle. The propagation material used in the experiment was obtained from the germplasm bank of the Padova University (Table 1). The pedigree information of the plant materials derived from Central and Sothern America is unknown; as to the origin of Italian genotypes, we only know that they were introduced into Tuscany in 1630, cultivated until the end of the 1800s exclusively in botanical gardens, and only spread to the Northern Italy cultivation areas from 1880. In January 2016, sweet potato cuttings were produced in a glass greenhouse set with a temperature of 25 °C and 18 °C during the day and night, respectively. Thirty storage roots for each genotype, 40 mm to 80 mm in diameter, were placed in PVC pots (three roots per pot) filled with a peaty substrate (Klasmann Potgrond H) integrated with 20% perlite. In May, the cuttings were suitable for transplanting (0.30–0.35 m tall). Before transplanting, the soil was plowed and fertilized with 80, 70, and 210 kg ha^{-1} of N, P_2O_5, and K_2O, respectively [29]. Cuttings were planted 0.10 m deep on the built-up rows, spacing the plants 0.35 m apart in the row and 0.80 m between rows. After transplanting, approximately 100 mL of water was provided for each cutting. The crop was irrigated three times during the growing cycle, at a rate of 30 mm m^{-2} for each irrigation. Sweet potatoes were harvested at the end of September 2016.

Table 1. List of sweet potato genotypes used.

Genetic Material	Plant Type	Country of Origin	Flesh Color	Skin Color	Root Shape
BR_1	Extremely spreading	Brazil	Purple	Dark purple	Elliptical
BR_11	Spreading	Brazil	Cream	Pink	Round elliptical
BR_13	Extremely spreading	Brazil	White	Cream	Elliptical
BR_25	Semi-erect	Brazil	Purple	Cream	Long oblong
BR_30	Spreading	Brazil	White	Pink	Long irregular
BR_32	Semi-erect	Brazil	Pale orange	Pink	Oblong
BR_33	Semi-erect	Brazil	Purple	Cream	Oblong
BR_51	Extremely spreading	Brazil	White	Cream	Long elliptical
BR_53	n.a.	Brazil	Purple	White	Oblong
BR_54	Extremely spreading	Brazil	Intermediate orange	Yellow	Elliptical
BR_66	n.a.	Brazil	White	White	Irregular
BR_78	Semi-erect	Brazil	Cream	Pink	Long irregular
BR_79	Spreading	Brazil	Purple	Pink	Obovate
BR_80	n.a.	Brazil	Purple	Dark purple	Obovate
IT_41	Semi-erect	Italy	Cream	Cream	Long irregular
IT_43	Spreading	Italy	Pale yellow	Pink	Obovate
IT_44	Semi-erect	Italy	White	Cream	Elliptical
IT_49	Semi-erect	Italy	Pale yellow	Pink	Round elliptical
alIT_81	Erect	Italy	Cream	Cream	Obovate
US_45	Semi-erect	USA	Purple	Dark purple	Long oblong
US_85	n.a.	USA	Pale yellow	Cream	Round elliptical
HO_86	n.a.	Honduras	Deep Orange	Purple red	Round elliptical

n.a.: not available.

2.2. Molecular Analysis

2.2.1. Genomic DNA Isolation

Leaves were collected from 1 month old transplants, snap-frozen in liquid nitrogen upon harvesting, and stored at −20 °C until further processing. Approximately 100 mg of leaf tissue was employed for the isolation of genomic DNA using the DNeasy plant kit (Qiagen, Valencia, CA, USA), according to the manufacturer's instructions. Extracted DNA samples were run on 0.8% agarose/1× TAE gel containing 1× SYBR Safe DNA stain (Life Technologies, Carlsbad, CA, USA) to evaluate their integrity. Both the purity and quantity were assessed with a NanoDrop 2000c UV-Vis spectrophotometer (Thermo Scientific, Pitsburgh, PA, USA).

2.2.2. SSR Genotyping

For the SSR analysis, microsatellite markers belonging to 14 distinct genomic loci from both coding regions (EST-SSR) and non-coding regions (nSSR) were obtained from different sources [4,5,10,30] (Table 2). These markers were chosen due to the high polymorphism they showed in the reference studies. In order to evaluate the efficiency and the polymorphism degree of this SSR set, a preliminary test was performed using three different clones randomly chosen from the sweet potato collection, and each was analyzed in two biological replicates (i.e., two distinct plants for each clone). Moreover, each biological replicate was, in turn, analyzed in two technical replicates, to evaluate the reproducibility of the SSR.

Table 2. List of SSR marker loci used in this study along with their basic information. For each SSR region, locus name, melting temperature (Tm), primer sequences, microsatellite motif, number of alleles according to the reference study, amplicon length in base pairs (bp), microsatellite type (including nSSR = SSR from non-coding regions and EST−SSR = SSR from coding regions), multiplex organization (for the simultaneous run of PCR products from different SSR marker loci in capillary electrophoresis) and the sources from which the SSR data were obtained are reported. Colored bases at the 5' end of each forward primer represent four universal sequences (designated as M13, in blue, PAN1, in green, PAN2, in yellow and PAN3, in red) complementary to as many fluorophore-labeled oligonucleotides (fluorophores adopted were 6-FAM, VIC, NED and PET).

Locus Name	Tm (°C) *	Primer Sequence (5'–3')	SSR Motif	Alleles	Length (bp)	Type	Multiple ×	Source
IBSSR04	62 62	GAGGTAGTTATTGTGGAGGACCTCCTTTGCCTCCTTTCATGC CCTTGCTCCCCATTTTCTTCTTG	(GA)11	7	216	nSSR	1	[30]
J263	62 61	GGAATTAACCGCTCACTAAAGCTCTGCTTCTCCTGCTGCTT GTGCGGCACTTGTCTTTGATA	(AAC)6	7	156–171	nSSR	1	[5]
J544b	61 59	TTGTAAAACGACGGCCAGTAGCAGTTGAGGAAAGCAAGG CAGGATTTACAGCCCCAGAA	(TCT)6	8	174–194	nSSR	1	[5]
Ib318	60 60	GAGGTAGTTATTGTGGAGGACAGAACGCATGGGCATTGA CCCACCGTGTAAGGAAATCA	n.a.	5	125-135	nSSR	1	[4]
Ib-255F1	61 59	GGAATTAACCGCTCACTAAAGCGTCCATGCTAAAGGTGTCAA ATAGGGGATTGTGCGTAATTTG	n.a.	8	210-245	nSSR	2	[4]
Ib297	59 60	GAGGTAGTTATTGTGGAGGACGCAATTTCACACACAAACACG CCCTTCTTCCACCACTTTCA	(CT)13	24	129–167	nSSR	2	[5]
Ib286	62 57	TTGTAAAACGACGGCCAGTAGCCACTCCAACAGCACATA GGTTTCCCAATCAGCAATTC	n.a.	10	90–122	nSSR	2	[4]
IbS11	58 61	GGAATTAACCGCTCACTAAAGCCCTGCGAAATCGAAATCT GGACTTCCTCTGCCTTGTTG	(TTC)10	13	218–248	nSSR	3	[5]
J116a	57 60	GAGGTAGTTATTGTGGAGGACTCTTTTGCATCAAAGAAATCCA CCTCAGCTTCTGGGAAACAG	(CCT)7	15	187–227	nSSR	3	[5]
J206A	59 57	TTGTAAAACGACGGCCAGTATCAGGGAGAGAGGACAGTAA TAGGCAAACCATAAACAGAGA	(GAT)6	9	103–121	nSSR	3	[5]
GDAAS0615	56 57	TGTAGAAAGACGAAGGGAAGGCCACATACAGACTACAACTTAC GGAGGAGCGTATTATGAACA	(GA)10	7	230	EST-SSR	4	[10]
GDAAS0757	56 56	TTGTAAAACGACGGCCAGTGAGATGATGACGATAGTGTTG GGAAGATTCATTGGCAGAAG	(GAA)11	9	293	EST-SSR	4	[10]
IBSSR27	56 55	GGAATTAACCGCTCACTAAAGGTGTTTATCACATCGTTTTCTG GGCTCGTACAATTTTCAAAG	(TA)6(CA)16	9	149	nSSR	4	[30]
GDAAS0156	54 55	GAGGTAGTTATTGTGGAGGACTCCAAATACCATACCCAAC CGCTTTCAAATAGAATCGTC	(TC)10	8	118	EST-SSR	4	[10]

* Tm of the forward primers does not take into account the tail sequence.

The PCRs were carried out via the three-primer strategy reported by Schuelke [31], with a major modification first described by Palumbo et al. [32]; Instead of using only M13, three additional universal sequences (designated as PAN1, PAN2, and PAN3) were used to tag the 5′ end of the forward primer of each couple (colored sequences in Table 2) and adopted in combination with M13, PAN1, PAN2, and PAN3 fluorophore-labeled oligonucleotides. Fluorophores adopted were 6-FAM, VIC, NED, and PET, respectively. Due to the genetic complexity of the species and thus the possibility of obtaining up to six alleles per SSR locus, PCRs were performed in single reactions. Each reaction contained approximately 40 ng of genomic DNA template, 1× Platinum® Multiplex PCR Master Mix (Applied Biosystems, Carlsbad, CA, USA), GC enhancer 10% (Applied Biosystems), 0.05 µM tailed forward primer (Invitrogen Corporation, Carlsbad, CA, USA), 0.1 µM reverse primer (Invitrogen Corporation), 0.23 µM universal primer (Invitrogen Corporation), and sterile water to volume. Amplifications were performed in a 96 well plate using a 9600 thermal cycler (Applied Biosystems), adopting the following conditions: after initial denaturation for 2 min at 95 °C, a touch-down PCR was undertaken with six cycles consisting of 30 s denaturation at 95 °C, 1 min annealing at 60 °C decreasing by 1.0 °C with each cycle and 30 s elongation at 72 °C; then 35 cycles at 95 °C for 30 s, 55 °C for 60 s, and 72 °C for 30 s. A final extension at 60 °C for 30 min terminated the reaction and filled in any protruding ends of the newly synthesized strands. The amplicons were visualized and quantified by agarose gel electrophoresis (2% agarose/1× TAE gel containing 1× Sybr Safe DNA stain (Life Technologies)), and the gel pictures were acquired with an UVITEC UV Transilluminator (Cambridge, UK) equipped with a digital camera. Subsequently, 10 ng of each PCR product was pooled and organized according to the four multiplexes reported in Table 2 and subjected to capillary electrophoresis on an ABI PRISM 3130xl Genetic Analyzer (Thermo Fisher) using LIZ500 (Applied Biosystems) as molecular weight standard and G5 (Applied Biosystems) as filter. Peak Scanner software v. 2.0 (Applied Biosystems) was used to determine the size of each peak, and each SSR was handled as a dominant marker. From the total of 14 SSR primer pairs initially selected for sweet potato genome analysis and tested for polymorphisms, Ib318 [4], J263 [5], and GDAAS0156 [10] showed weak resolution or screening errors and were excluded from our study.

2.2.3. Marker Data Analysis

Data were coded as (0,1) vectors, where 1 indicated the presence and 0 the absence of a peak/allele at a specific position in the electropherogram. The polymorphic information content (PIC) of each SSR locus over its n marker alleles was computed as [33],

$$\text{PIC} = 1 - \sum p_i^2 \tag{1}$$

where p_i is the frequency of the marker allele i.

Genetic similarity between the clones was estimated by applying Dice's coefficient [34] in all possible pairwise comparisons, and a triangular similarity data matrix was generated. The first two principal components of the matrix were thus computed through a principal coordinate analysis (PCoA). All calculations were conducted using NTSYS-pc v. 2.21q software [35]. Taking advantage of the genetic similarity data, PAST software v. 3.14 [36] was used to construct a dendrogram through the unweighted pair group arithmetic average (UPGMA) method and by applying the Dice's coefficient [34]. To measure the stability of the computed branches, a statistical bootstrap analysis was conducted with 1000 resampling replicates. GenAlEX software v. 6.5 [37] estimated the number of observed alleles and the presence of private allele throughput in all the samples, purposely grouped as "Italian clones ($N = 5$)" and "foreign clones ($N = 17$)" according to their putative origin. Marker alleles were scored as "private" when shared by at least 60% of the individuals of one group and simultaneously absent from the other group.

A Bayesian clustering algorithm implemented in STRUCTURE v. 2.2 software [38] was used to model the genetic structure of the considered *I. batatas* accessions' haploid genotypes. The "admixture model" and "correlated allele frequencies model" were selected, because no prior knowledge about their origin was available (first model) and to guarantee the identification of a previously undetected correlation without affecting the results if no correlation existed [39] (second model). The number of founding groups ranged from 2 to 10, and 10 replicate simulations were performed for each K value, setting a burn-in of 2×10^5 and a final run of 10^6 Markov chain Monte Carlo (MCMC) steps [40]. Finally, the most likely estimation of K was decided by evaluating the rate of change in the log probability of data between successive K values (ΔK method), according to Evanno et al. [41]. In particular, one 2D Excel vertical histogram for each accession, conveniently divided into K colored segments, was used to represent the estimated membership in each hypothesized ancestral genotype. Each color correlated to a putative ancestor.

2.3. Morphological and Chemical Analyses

The morphological characterization was performed in August 2016. Root shape, root skin color, root flesh color, and the general outline of the leaf were scored according to the morphological descriptors available from International Board for Plant Genetic Resources (IBPGR) [42]. All the morphological traits considered for each genotype were transformed into numbers using the CIP scale [42] in order to process them with statistics. Three biological replicates were performed for the chemical analyses in order to recover representative data about the sweet potato samples.

2.3.1. Extraction of Phenols for Analysis

Freeze-dried samples (1 g) were extracted in methanol (20 mL) with an Ultra Turrax T25 (IKA-Labortechnik, Staufen, Germany) at 1018 rpm until a uniform consistency was achieved. Samples were filtered (589 filter paper; Whatman, Germany) and appropriate aliquots of extracts were assayed by the Folin–Ciocalteu (FC) method [43] for total phenolic (TP) content, and by the ferric reducing antioxidant power (FRAP method) for antioxidant activity [44]. For HPLC analyses, extracts were further filtered with cellulose acetate syringe filters (0.45 μm porosity).

2.3.2. Determination of TP Content by the FC Assay

The TP content was determined according to the FC assay, using gallic acid as a calibration standard and a UV-1800 spectrophotometer (Shimadzu, Columbia, MD, USA). The FC assay was carried out by putting 200 μL of sweet potato extract into a 10 mL test tube, followed by the addition of FC reagent (1 mL). The mixture was vortexed for 20–30 s and 800 μL of filtered 20% sodium carbonate solution was added 1–8 min after the FC reagent addition. The mixture was then vortexed for 20–30 s (time 0). The absorbance of the colored reaction product was measured at 765 nm after two hours at room temperature. The TP content in the extracts was calculated from a standard calibration curve obtained with different concentrations of gallic acid, ranging from 0 to 600 μg mL^{-1} (coefficient of determination: $r^2 = 0.9992$). The results have been expressed as mg gallic acid equivalent (GAE) kg^{-1} dry weight.

2.3.3. Determination of Total Antioxidant Activity by FRAP

Freshly prepared FRAP reagent contained 1 mmol L^{-1} 2,4,6-tripyridyl-2-triazine and 2 mmol L^{-1} ferric chloride in 0.25 mol L^{-1} sodium acetate (pH 3.6). A methanol extract aliquot (100 μL) was added to the FRAP reagent (1900 μL) and accurately mixed. Absorbance was determined at 593 nm after leaving the mixture at 20 °C for 4 min. The calibration was performed with a standard curve (0–1200 μg mL^{-1} ferrous ion) (coefficient of determination: $r^2 = 0.9985$) obtained by the addition of freshly prepared ammonium ferrous sulfate. FRAP values were calculated as μg mL^{-1} ferrous ion (ferric reducing power) from three determinations and have been reported as mg kg^{-1} of Fe^{2+} (ferrous ion equivalent) of dry matter.

2.3.4. Quantitative Determination of Ions by IC and Organic Nitrogen

For the estimation of anions and cations, a freeze-dried sample (200 mg) was extracted in water (50 mL) and shaken at 150 rpm for 20 min. Samples were filtered in sequence through filter paper (589 Schleicher), and the extracts were further filtered through cellulose acetate syringe filters (0.20 mm) before analysis by ion chromatography (IC). The IC was performed using an ICS-900 ion chromatography system (Dionex Corporation) equipped with a dual piston pump, a model AS-DV autosampler, an isocratic column at room temperature, a DS5 conductivity detector, and an AMMS 300 suppressor (4 mm) for anions and CMMS 300 suppressor (4 mm) for cations. A Dionex Ion-Pac AS23 analytical column (4 × 250 mm) and a guard column (4 × 50 mm) were used for anion separations, whereas a Dionex IonPac CS12A analytical column (4 × 250 mm) and a guard column (4 × 50 mm) were used for cation separations. The eluent consisted of 4.5 mM sodium carbonate and 0.8 mM sodium bicarbonate at a flow rate of 1 mL min^{-1} for anions, and 20 mM metansolfonic acid for cations at the same flow rate. Chromeleon 6.5 chromatography management software was used for system control and data processing. Anions and cations were quantified following a calibration method. Dionex solutions containing seven anions and five cations at different concentrations were taken as standards, and the calibration curves for anions and cations were generated with concentrations ranging from 0.4 mg L^{-1} to 20 mg L^{-1} and from 0.5 mg L^{-1} to 50 mg L^{-1} of standards, respectively. The Kjeldahl method (ISO 1656) was used for organic nitrogen.

2.3.5. Brix Content

Approximately 0.5 mL of defrosting liquid of the product was used for the determination of the Brix content, carried out using a Hanna Instruments HI 96801 portable digital refractometer.

2.3.6. Starch

Starch analysis was performed by chromatographic analysis according to AOAC Official Method 996.11 (University of Florida, IFAS, Bulletin 339-2000 "Starch Gelatinization & Hydrolysis Method" Boehringer Mannheim, Starch determination, cat. N° 207748).

2.3.7. Quantitative Determination of Sugars by HPLC

Freeze-dried sweet potato root samples (0.2 g) were homogenized in demineralized water (20 mL) with an Ultra Turrax T25 until a uniform consistency was achieved at 1018 *g*. Samples were filtered in sequence through filter paper (589; Schleicher), and the extracts were further filtered through cellulose acetate syringe filters (0.45 mm) and analyzed by HPLC. The liquid chromatography apparatus utilized in this analysis was a Jasco X.LC system consisting of a model PU-2080 pump, a model RI-2031 refractive index detector, a model AS-2055 autosampler, and a model CO-2060 column. ChromNAV Chromatography Data System software was used for analysis of the results. The separation of sugars was achieved on a Hyper-Rez XP Carbohydrate Pb^{++} analytical column (7.7 × 300 mm; Thermo Scientific, Waltham, MA, USA), operating at 80 °C. Isocratic elution was effected using water at a flow rate of 0.6 mL min^{-1}. D-(+)-glucose, D-(−)-fructose, and maltose were quantified by a calibration method. All standards utilized in the experiments were accurately weighed and dissolved in water; the calibration curves were generated with concentrations ranging from 100 to 1000 mg L^{-1} of standards.

2.4. Statistical Analysis

Chemical and morphological data were finally used to construct a constrained UPGMA dendrogram using PAST software v. 3.14 [36], applying the Euclidean similarity index and keeping the position of the samples fixed throughout the tree according to the clustering resulting from the SSR-based dendrogram. To measure the stability of the branches, a statistical bootstrap analysis was conducted with 1000 resampling replicates.

The complete set of data for each variety was used for random combinations using the bootstrap method. For each variety, a set of 1000 combinations was produced, and the data were analyzed by the PCoA procedure using the software Statgraphics Centurion 18.1.06 (Statgraphics Technologies, Inc.). All qualitative trait data were processed by ANOVA and, in case of significant differences, average values were separated by Tukey HSD test (CoStat 6.400–CoHort Software, CA, USA).

3. Results and Discussion

Overall, 117 marker alleles were detected in 11 SSR loci analyzed throughout the accession pool, ranging from a minimum of 6 (J206A) to a maximum of 16 (GDAAS0757), with an average number equal to 10.5 per locus (Table 3 and supplementary Table S1). According to Botstein et al. [33], all examined marker loci were found to be highly informative and variable across the accessions, with a mean PIC value equal to 0.79, spanning from 0.61 (J206A) to 0.93 (GDAAS0615), as reported in Table 3. Since private polymorphisms are recognized as an efficient molecular tool for food traceability, the presence of marker alleles able to discriminate the accessions according to their putative origin was investigated. As many as 58 out of the total 117 marker alleles scored (Table 3, blue boxes) were exclusively identified only within the foreign accessions pool, but only two of them, i.e., loci IBSSR04 and J116a (Table 3, underlined percentages) were scored in at least 60% of the accessions. In contrast, six marker alleles were exclusively associated with the Italian pool (Table 3, red boxes) and only one, i.e., locus IBSSR27, was found in at least 60% of the Italian sweet potatoes (Table 3, underlined percentage).

According to the genetic similarity matrix calculated in all possible pairwise comparisons among the 22 accessions, Dice's coefficient ranged from 0.28 (between BR_78 and BR_30) to 0.97 (between IT_41 and IT_44, Table S2). US_45 resulted to be the most divergent genotype: the average genetic similarity value calculated against the rest of the accessions was as low as 0.41, highlighting a clear-cut differentiation from the rest of the pool. The mean genetic similarity value calculated among the five Italian accessions was 0.70. The same value calculated in all pairwise comparisons within the Brazilian accessions was considerably lower (0.54), consistent with the great morphological variability observed within the South American core collection. The accession BR_66 was the most closely related to the Italian clones, scoring an average genetic similarity value of 0.70 and a maximum of 0.74 when compared with IT_44. From the principal coordinate analysis (PCoA), the first coordinate separated a group composed of eight Brazilian entries (namely, BR_25, BR_53, BR_33, BR_32, BR_11, BR_78, BR_79, and BR_80) from a group including all of the Italian clones (Figure 1).

The first principal coordinate also underlined the separation of the two US accessions (US_45 and US_85), according to their contrasting phenotype and in agreement with their low genetic similarity value (0.38). The main result of the variation explained by the second principal coordinate was the split of three Italian accessions showing a cream skin color (namely, IT_44, IT_81, and IT_41) from two Italian accessions distinguishable for their pink skin color (IT_49 and IT_43). This finding was supported not only by a contrasting phenotype, but also by a relatively low estimate of mean genetic similarity (0.60) calculated between these two groups, suggesting a different origin of the Italian accessions. The UPGMA analysis confirmed the sharp detachment, already seen in the genetic similarity matrix (Table S2), of US_45 from the rest of the genotypes (Figure 2A), supported by a bootstrap value of 100.

Table 3. Marker allele size, overall allele frequencies (F.o.), allele frequencies in the foreign accessions (F.f.), allele frequencies in the Italian accessions (F.i.), and polymorphism information content (PIC) values are reported for the 11 SSR loci. The blue boxes highlight private marker alleles found only in the foreign accessions, while red boxes denote marker alleles detected only in the Italian accessions. Percentages underlined designate private marker alleles shared by at least 60% of the foreign samples, as well as private marker alleles shared by at least 60% of the Italian clones.

IBSSR04

Size	F.o.	F.f.	F.i.	PIC
216	9%	12%	0%	
218	18%	24%	0%	
220	59%	59%	60%	
222	45%	65%	0%	0.75
224	18%	12%	40%	
226	68%	71%	60%	
228	50%	47%	60%	
230	82%	76%	100%	

J544b

Size	F.o.	F.f.	F.i.	PIC
184	91%	88%	100%	
178	18%	24%	0%	
189	18%	24%	0%	
192	18%	24%	0%	0.63
195	59%	47%	100%	
198	82%	76%	100%	
209	82%	76%	100%	

Ib-255F1

Size	F.o.	F.f.	F.i.	PIC
233	18%	24%	0%	
237	23%	24%	20%	
239	5%	6%	0%	
243	5%	6%	0%	
251	50%	53%	40%	
253	45%	53%	20%	
255	73%	65%	100%	0.87
257	14%	18%	0%	
259	64%	53%	100%	
261	9%	12%	0%	
263	32%	41%	0%	
265	27%	35%	0%	
267	9%	0%	40%	

Ib297

Size	F.o.	F.f.	F.i.	PIC
137	9%	6%	20%	
147	5%	6%	0%	
149	59%	53%	80%	
151	14%	18%	0%	
155	64%	71%	40%	
159	14%	18%	0%	
161	55%	41%	100%	0.87
163	64%	65%	60%	
167	27%	35%	0%	
169	5%	0%	20%	
171	9%	0%	40%	
175	5%	6%	0%	
177	27%	35%	0%	
187	32%	41%	0%	

Ib286

Size	F.o.	F.f.	F.i.	PIC
97	5%	6%	0%	
99	5%	6%	0%	
101	18%	24%	0%	
105	32%	41%	0%	
107	14%	18%	0%	
109	100%	100%	100%	0.77
113	77%	71%	100%	
115	45%	41%	60%	
121	59%	71%	20%	
123	9%	12%	0%	

ibS11

Size	F.o.	F.f.	F.i.	PIC
224	9%	12%	0%	
227	55%	41%	100%	
230	18%	24%	0%	
233	9%	6%	20%	
236	55%	53%	60%	
239	32%	29%	40%	
242	27%	18%	60%	0.87
245	64%	71%	40%	
248	41%	53%	0%	
251	18%	24%	0%	
254	59%	59%	60%	
257	9%	12%	0%	
260	14%	18%	0%	
263	9%	12%	0%	

J116a

Size	F.o.	F.f.	F.i.	PIC
194	23%	29%	0%	
200	77%	71%	100%	
203	14%	12%	20%	
206	91%	88%	100%	
209	32%	41%	0%	0.69
212	55%	47%	80%	
215	82%	76%	100%	
218	9%	12%	0%	
221	50%	65%	0%	

J206A

Size	F.o.	F.f.	F.i.	PIC
118	82%	76%	100%	
124	55%	59%	40%	
127	100%	100%	100%	0.61
130	5%	6%	0%	
133	59%	53%	80%	
136	9%	12%	0%	

GDAAS0615

Size	F.o.	F.f.	F.i.	PIC
214	36%	29%	60%	
220	14%	18%	0%	
226	9%	12%	0%	
230	68%	76%	60%	
232	5%	6%	0%	
234	9%	12%	0%	0.93
236	23%	24%	20%	
246	9%	12%	0%	
248	14%	18%	0%	
250	18%	18%	20%	
256	5%	0%	20%	

GDAAS0757

Size	F.o.	F.f.	F.i.	PIC
272	18%	24%	0%	
281	23%	12%	60%	
284	18%	24%	0%	
290	9%	12%	0%	
293	14%	6%	40%	
296	41%	53%	0%	
299	5%	6%	0%	
302	50%	41%	80%	0.89
305	27%	35%	0%	
308	5%	6%	0%	
311	14%	12%	20%	
314	36%	29%	60%	
317	95%	100%	80%	
320	9%	12%	0%	
323	27%	24%	40%	
326	18%	6%	60%	

IBSSR27

Size	F.o.	F.f.	F.i.	PIC
139	24%	29%	0%	
155	5%	6%	0%	
161	14%	0%	60%	
163	10%	12%	0%	
165	95%	88%	100%	0.88
167	10%	12%	0%	
169	5%	6%	0%	
171	5%	0%	20%	
177	29%	29%	20%	

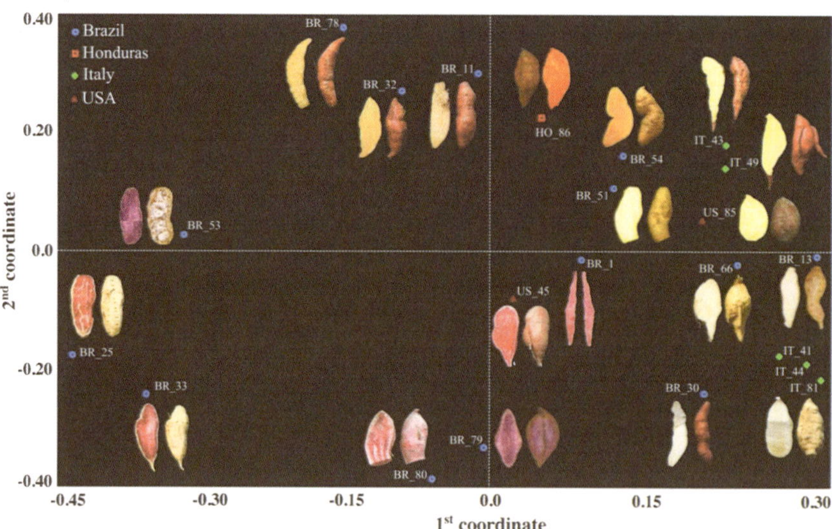

Figure 1. Principal coordinate analysis (PCoA) of the sweet potato core collection based on molecular markers: two-dimensional centroids derived from the genetic similarity estimates computed among accessions in all possible pairwise comparisons using the whole SSR marker data set. The first two coordinates were able to explain 54% of the total variation, accounting for 31% and 23% of the total, respectively. Four different colors have been used to distinguish the accessions based on their geographical origin: blue = Brazil, red = Honduras, green = Italy, and brown = USA.

Four main clustering patterns already highlighted with the PCoA (Figure 1) were also observed throughout the dendrogram (Figure 2 panel A), with bootstrap values always higher than 90%. In detail, BR_79 and BR_80 grouped together and scored the 91% of similarity; BR_33, BR_25, and BR_53 shared a mean genetic similarity value of 82%; IT_44, IT_41, and IT_81 showed, on average, 88% the genetic similarity and, finally, BR_11 grouped with BR_78 according to a genetic similarity of 78%. In particular, it is worth highlighting that although the five Italian accessions were all part of the same macrobranch of the tree, they subclustered into two groups, according to what was observed in the PCoA analysis. In fact, in the PCoA, the second coordinate alone explained 23% of the total variation, clearly separating IT_44, IT_41, and IT_81 from a second group including IT_43 and IT_49 (Figure 1). A noteworthy consideration involving BR_66 is the close association between this Brazilian accession and the first group of Italian accessions in both the PCoA and the UPGMA tree, with a similarity value of 0.74.

The ΔK criterion suggested by Evanno et al. [41] gave the highest value for the SSR analysis at three groups (for $K = 3$, ΔK resulted 58.53, Figure S1). According to this estimate, the whole group putatively originated from three ancestral genotypes (indicated in grey, orange, and blue in Figure 2B). One vertical histogram for each accession, conveniently divided into $K = 3$ colored segments, has been used to represent the estimated membership in each hypothesized ancestral genotype, and 90% was the threshold set for the admixed ancestry (Figure 2B). This finding seems to be in agreement with those reported by Roullier et al. ([45] Figure 2 in Appendix S1) and Wadl et al. [8], both supporting the $K = 3$ in the sweet potato clones from tropical America. Overall, one of the three ancestors (the one in grey) was predominant: 13 samples from USA, Honduras, Italy, and Brazil showed a membership to this ancestral genotype higher than 95%, suggesting a probable common origin. Five accessions (BR_66, IT_81, BR_53, BR_79, and BR_80) were admixed and one of the three ancestors (the one in grey) was always recurrent; interestingly, no example of hybridization between the other two ancestors (blue and orange, Figure 2) was observed.

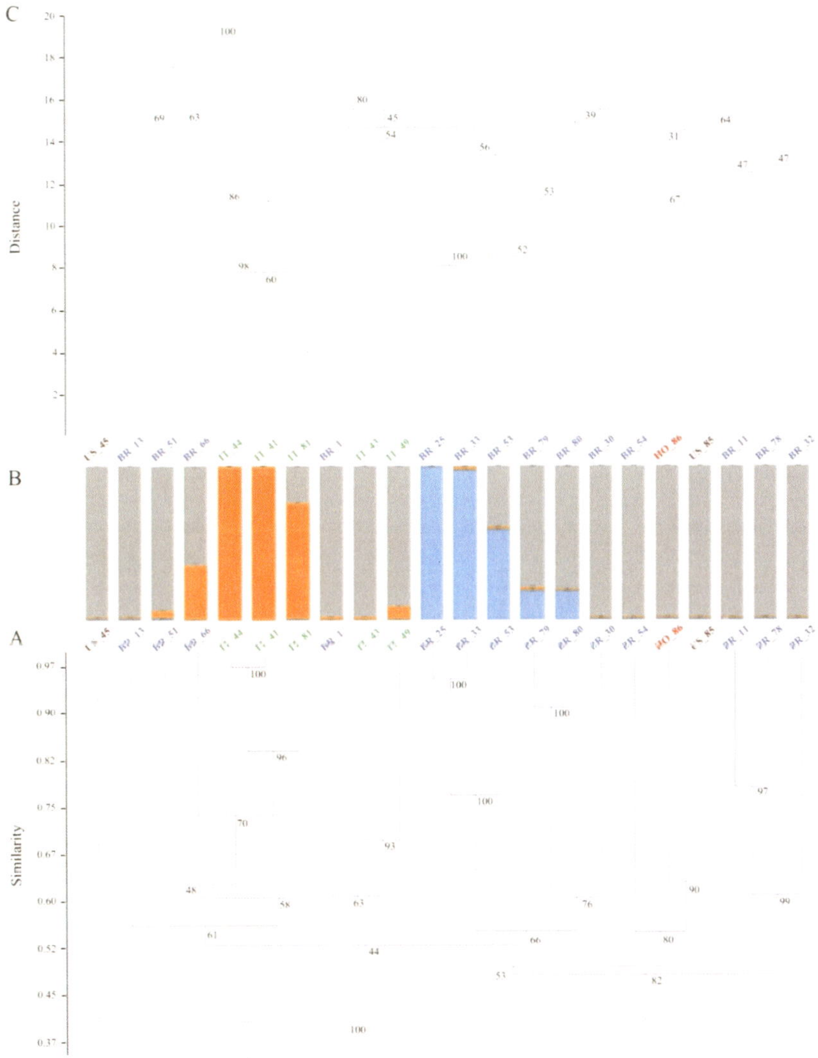

Figure 2. Genetic structure analysis of the sweet potato core collection: (**A**) The unweighted pair group method with arithmetic average means (UPGMA) tree of the genetic similarity estimates computed among pairwise comparisons of sweet potato accessions using the whole simple sequence repeat (SSR) marker data set, with nodes of the main subgroups supported by bootstrap values. The color scheme for this figure is the same as that used in Figure 1 (Blue = Brazil, red = Honduras, green = Italy, and brown = USA). (**B**) Population genetic structure of a core collection of $N = 22$ sweet potato accessions estimated using 11 microsatellite markers. Each sample is represented by a vertical bar partitioned into $K = 3$ colored segments representing the estimated membership. The proportion of ancestry (%) is reported on the ordinate axis, and the identification number of each accession is indicated below each histogram. (**C**) The unweighted pair group method with arithmetic average means (UPGMA)-constrained tree was built by applying the Euclidean similarity index and using the morpho-qualitative measurements of a subset of the *I. batatas* core collection. The positions of the samples throughout the dendrogram were kept fixed according to those ones resulting from the dendrogram in Figure 2A, and the bootstrap values supporting each node of the main subgroups were calculated.

In detail, two of three Italian accessions characterized by a cream skin color (IT_44 and IT_41) and strictly associated with the same branch of the UPGMA tree (similarity = 0.97) also shared the same marker allele cluster, both with accession scores of individual membership higher than 99%. Although these two Italian accessions were collected in two different areas, these results suggest a case of synonymy and we may suppose that the same genotype is cultivated in different regions with a different name. IT_81 and BR_66, although counted as admixed, shared the same ancestral genotype as IT_44 and IT_41, with percentages of 75% and 35%, respectively. Considering the PCoA and the UPGMA tree, it is probable that all four shared the same ancestral genotype (the one indicated in orange), and it is not to be ruled out that the three Italian accessions derived from BR_66. However, it is possible that BR_66 has undergone continuous events of hybridization with individuals descending from the "grey" progenitor, which would explain the admixed pattern of this Brazilian accession. In contrast, the three Italian accessions, whose memberships ranged from 75% (IT_81) to 99% (IT_44 and IT_41) may have preserved their "ancestral purity" thanks to repeated crosses with accessions of the same lineage or to asexual propagation breeding schemes. A second cluster included BR_25 and BR_33 (membership > 97%) and, to a lesser extent, BR_53 (membership = 59%), according to their sharp detachment from the rest of the pool highlighted by the PCoA analysis. The remaining accessions, including two Italian accessions (IT_49 and IT_43), were all part of a third cluster, except for BR_79 and BR_80, which were admixed (Figure 2B). The Euclidean-index-based UPGMA dendrogram emphasized the clear-cut detachment of US_45 from the rest of the pool (Figure 2C), as previously established by the SSR-based UPGMA (Figure 2A). In this case, the bootstrap support was 100%. The other samples were all clustered in three main branches, with a bootstrap value of 80%. In the first branch, BR_79, BR_80, BR_33, BR_25, and BR_53 clustered together with IT_43, IT_49, and BR_1. This finding was different from that observed in the SSR-based UPGMA tree (Figure 2A), where these two subgroups clustered separately, and it was not consistent with the PCoA analysis and the morphological data (Figure 1). Nevertheless, it must be noted that in the morphological and chemical-marker-based dendrogram (Figure 2C), the bootstrap support of this specific node was quite low (45%). Moreover, despite the different clusterization highlighted in the two UGPMA trees (Figure 2A,C), the two subgroups maintained the same composition. In particular, all of the samples in the first subgroups (BR_79, BR_80, BR_33, BR_25, and BR_53) were characterized by a typical purple color of flesh that was not detected in other samples. This observation, along with a full or partial membership to the same cluster (Figure 2 panel B), reinforces the hypothesis of a common ancestor for this group of samples, in agreement with their geography. Regarding the second subgroup (IT_43, IT_49, and BR_1), it is very hard to make hypotheses about its origin, although genetic data would presuppose a certain degree of kinship with the other three Italian accessions. Considering the second main branch of the Euclidean-distance-based UPGMA, two considerations are needed. First, IT_44, IT_41, IT_81, and BR_66 grouped together according to all the previous analyses (bootstrap support 86%). The fact that the genetic and the morpho-agronomic markers led to identical results may suggest that this first group of Italian genotypes was descended from BR_66, as initially predicted on the basis of their highly similar morphology (Figure 1). All the accessions were also analyzed by means of morpho-agronomic qualitative markers (Table 4).

Table 4. Qualitative measurements performed on the sweet potato accessions.

Genotype	K (mg/kg dw)	Mg (mg/kg dw)	Ca (mg/kg dw)	Total Soluble Solids (°Brix)	TP (mg GAE kg⁻¹ fw)	TAA (mg Fe²⁺E kg⁻¹ fw)	β-Carotene (mg/kg fw)	Vit C (mg/kg dw)	Sucrose (mg/kg dw)	Glucose (mg/k dw)	Fructose (mg/kg dw)	Starch (%)	Dry Matter (%)
US_45	7381 klm	2426 bcde	1810 abc	9 bc	6442 a	6706 a	n.d.	658 c	36430 l	35280 cd	33561 b	81 a	36 ab
BR_79	12151 b	1841 def	2481 ab	12 a	2875 bc	2563 cd	n.d.	2967 ab	109542 defg	12903 jklm	11357 fghi	71 ab	37 ab
BR_80	5907 n	2181 cde	2405 ab	11 ab	2369 bc	2599 cd	n.d.	3966 ab	93156 fghi	9487 klmn	8501 fghi	74 ab	43 a
BR_33	6828 lm	1460 ef	1582 c	8 c	1753 c	2095 cd	n.d.	3887 ab	37464 l	26310 de	29451 bcd	83 a	36 ab
BR_25	7074 lm	1580 ef	2104 ab	9 bc	3240 bc	3578 bc	n.d.	2988 ab	55692 kl	31784 cd	31623 bc	72 ab	36 ab
BR_53	8965 g	1732 def	2308 ab	11 ab	3358 bc	2869 cd	n.d.	1933 bc	116776 cde	21199 efg	22460 de	70 b	34 b
BR_13	7328 klm	2640 abc	1981 abc	9 bc	1839 c	2210 cd	n.d.	2950 b	106360 def	8561 lmn	6028 hij	65 bc	32 b
BR_1	5316 no	1484 ef	1908 abc	11 ab	4425 ab	4651 b	n.d.	4682 a	81680 hij	21667 efg	15816 ef	69 b	36 ab
BR_66	10647 de	2270 bcd	1795 bc	13 a	2078 bc	2156 cd	23.8	2305 b	149539 a	6364 mn	6803 ij	64 bc	42 a
IT_44	8045 ij	2658 abc	1922 abc	9 bc	875 c	834 e	n.d.	1690 bc	109205 def	4396 lmn	3763 j	79 ab	37 ab
IT_41	10626 de	2156 cde	1709 bc	10 b	1595 c	1335 de	n.d.	2112 b	99035 fg	8673 lmn	7189 hij	83 a	37 ab
IT_81	12176 b	2697 abc	1925 abc	10 b	1067 c	862 e	n.d.	2544 b	124873 bcd	5007 n	3671 j	72 ab	36 ab
BR_51	9403 g	2056 bcd	1816 abc	9 bc	1186 c	2019 cd	34.2	3856 ab	113505 def	46730 a	46568 a	70 b	30 b
IT_43	15129 a	1729 cde	1954 abc	11 ab	2085 bc	1906 de	n.d.	1867 bc	144240 ab	12253 ijklm	10429 fgh	75 ab	34 b
IT_49	7855 ijk	2302 bcd	1563 c	10 b	1453 c	1607 de	208	1389 bc	92224 ghi	36301 bc	24108 cd	80 a	30 b
BR_30	6766 lm	2118 cde	1848 bc	8 c	1657 c	1743 de	90.8	1427 bc	68226 jk	45548 a	33478 b	70 b	31 b
BR_54	6666 m	2619 abc	2624 a	9 bc	1317 c	1577 de	571	2285 b	138292 bc	13282 hijk	10709 fgh	78 ab	37 ab
HO_86	11914 bc	2192 bcd	2282 ab	9 bc	1189 c	1179 de	512	2477 b	95520 fgh	42100 ab	45446 a	52 c	36 ab
US_85	8970 g	2023 cde	2264 ab	8 c	720 c	799 e	n.d.	1595 c	69641 jk	25645 def	30837 bc	60 c	36 ab
BR_32	8096 hi	1645 def	2298 ab	11 ab	1551 c	1516 de	811	4709 a	138295 bc	17282 ghij	14933 efg	72 ab	32 b
BR_11	10431 f	1452 f	1763 bc	9 bc	2477 bc	1073 e	n.d.	3087 ab	88841 ghi	19294 efgh	14664 efg	72 ab	35 b
BR_78	6850 lm	2352 bcd	1808 bc	10 b	2119 bc	1809 de	n.d.	1256 c	100268 efg	17742 fghi	13919 fg	70 b	42 a

Within each parameter, values without common letters significantly differed at $F < 0.05$ according to Tukey's HSD test (n.d., not determined).

In addition to genetic characteristics, quality traits are generally preferred for nutritive value and market demand [46–48]. With regard to the qualitative aspects, the dataset obtained for the different parameters is summarized in Figure 3 by multivariate analysis.

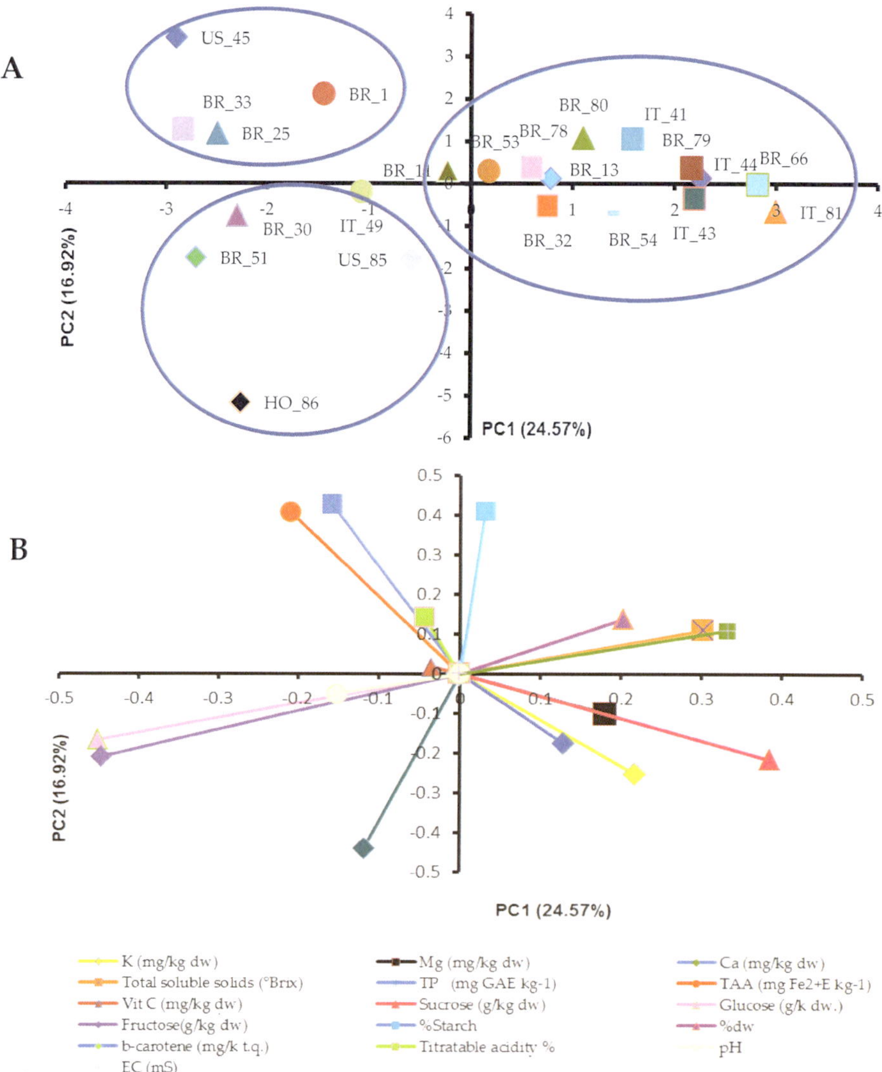

Figure 3. Principal components analysis (PCA) of the sweet potato core collection based on qualitative traits: (**A**) Score plot of the first two principal components (PC1 and PC2) for the 22 sweet potatoes. (**B**) Eigenvectors of the variables measured for the first two principal components. Loadings (eigenvalues) for the first and second principal components were equal to 25% and 17%, respectively. TP: total phenols; TAA: total antioxidant activity; dw: dry weight.

The first identified the different genotypes positioned on the basis of the measured qualitative characteristics, whereas the second vectorially highlighted the main qualitative traits that determined the genotype positioning. The results obtained from this elaboration allowed us to identify interesting cues regarding grouping of the different genotypes according to their qualitative peculiarities. In particular, it was possible to group the 22 genotypes into three macrogroups (A, B, and C). Group A comprised those genotypes characterized by a high anthocyanin content in both the skin and the flesh (US_45, BR_1, BR_33, and BR_25). The genotypes belonging to this group were characterized by high total antioxidant capacity and high total polyphenol content (Figure 3). The anthocyanins belong to the antioxidant family, in particular to the polyphenols, and have positive effects on human health [49]. Anthocyanin composition was determined in purple-fleshed sweet potatoes [50,51], highlighting that cyanidin and peonidin glycosides acylated with phenolic acids were the primary anthocyanin components. The second group, B, comprised genotypes phenotypically characterized by high beta-carotene content, thus featuring orange pulp, and by high simple sugar concentrations (BR_51, BR_30, IT_49, US_85, and HO_86). Carotenoids are secondary plant compounds that form lipid-soluble yellow, orange, and red pigments. Carotene-rich vegetables are associated with decreased risk of chronic diseases related to vision, skin, infection, and reproduction, in addition to being active oxygen species scavengers [52]. The most abundant carotenoid in sweet potato roots is usually β-carotene, which comprises more than 77% of total carotenoid content and can reach more than 99% in sweet potatoes characterized by orange flesh [53]. The colored genotypes considered in this experiment were characterized by a β-carotene content ranging from 23.8 to 811 μg g^{-1}. This range was wider than those measured by Simonne et al. [54] (1–190 μg g^{-1}) and by Grace et al. [50] (1–253.3 μg g^{-1}) in several sweet potato varieties. Finally, group C contained genotypes characterized by different flesh colors, but they shared a high content of sucrose and minerals (BR_32, BR_54, IT_43, and IT_81) or a high percentage of dry matter, soluble solids, and starch (BR_53, BR_78, BR_80, IT_41, BR_79, and BR_66). The high presence of starch and sucrose makes these varieties particularly sweet after slow cooking processes (i.e., boiling, oven, steam), following the production of maltose [3,55]. Genotypes belonging to group A and B may be more suitable for faster cooking methods (i.e., frying) because they are characterized by less starch, and they are more appealing for the consumer from a color point of view.

As shown from the comparison between grouping results of the core collection of sweet potato accessions (see Figures 1 and 3), some inconsistencies arose between the clustering based on molecular markers and that derived from qualitative parameters. This was understandable since both the molecular markers and qualitative parameters evaluated in this study not only represented a subset of the evaluable genotypic and phenotypic traits, but also because they did not show any known linkage. Consequently, a full overlap with what was detected by the principal components based on qualitative traits is not possible, as demonstrated by the centroids derived from the principal coordinates of molecular markers. An example of this behavior can be found between IT_43 and IT_49: these genotypes were grouped together based on molecular markers, whereas they differed according to the qualitative profiles (e.g., the former scored a higher concentration of potassium, whereas the latter a higher content of simple sugars).

4. Conclusions

We investigated, for the first time, the genetic structure and qualitative composition of the principal Italian sweet potato clones, along with their relatedness to a core collection of accessions from Central and Southern America. It is worth mentioning that the sweet potato accessions analyzed in this study represent the most cultivated clones in Italy and, to the best of our knowledge, no other locally adapted varieties are commercially available to Italian farmers. In fact, these materials, grown mainly in the Veneto region, are known to possess a high adaptation to the natural and anthropological environment in which they have been introduced and are still cultivated.

From the molecular analyses, Italian accessions were sub-clustered into two groups and were found to be genetically very similar to the South American germplasm. This finding was also supported by the morphological and chemical measurements affecting their principal qualitative traits.

Summarizing our results and considering both the morphological and qualitative and the genetic–molecular data, it is evident how the combination of these two different approaches was very effective in differentiating or clustering the different clonal genotypes, as expected by their geographical origin or their phenotypic characteristics. Moreover, because the molecular and the chemical results were often comparable, it was possible to make robust speculations on the common origins of sweet potato accessions. The experiment demonstrated not only a good relationship between genetic and morphological and qualitative aspects, but also allowed us to indirectly highlight the good level of adaptation of South American genotypes to European conditions. This last information allows us to suggest that breeders use South American germplasm, characterized above all by colored pulp, for the constitution of new genotypes in Europe, useful for the renewal and innovation of the European market as well as for providing new opportunities for farmers. On the whole, this information could be exploited by both breeders and farmers to detect and protect commercial varieties, and hence to certify the genetic identity of their propagation materials and overall quality of their food derivatives.

Supplementary Materials: The following are available online at http://www.mdpi.com/2073-4425/10/11/840/s1, Figure S1: Definition of the number of ancestral sweet potato genotypes. Mean ΔK is calculated as L"(K)/(SD(L(K))) following Evanno et al. (Evanno et al. 2005), where mean LnP(D) $\pm$ SD over 10 runs is a function of K, being L'(K) = ΔLnP(D), Table S1: SSR data recorded for the different sweet potato genotypes, Table S2: Genetic similarity matrix for the sweet potato accessions.

Author Contributions: Conceptualization, C.N., G.B. and P.S.; methodology, A.C.G., F.P.; validation, C.N., G.B. and P.S.; formal analysis, A.C.G., F.P. and C.N.; investigation, A.C.G., F.P. and C.N.; resources, G.B. and P.S.; data curation, A.C.G., F.P., C.N., G.B. and P.S.; writing—original draft preparation, A.C.G., F.P. and C.N.; writing—review and editing, F.P. and C.N.; supervision, G.B. and P.S.

Funding: This research received no external funding.

Conflicts of Interest: The authors declare no conflict of interest.

References

1. Woolfe, J.A. *Sweet Potato: An Untapped Food Resource*; Cambridge University Press: Cambridge, UK, 1992; Volume 366–372.

2. *CBI Market Intelligence CBI Product Factsheet: Fresh Sweet Potatoes in Europe*; CBI Market Intelligence: The Hague, The Netherlands, 2015.

3. Nicoletto, C.; Vianello, F.; Sambo, P. Effect of different home-cooking methods on textural and nutritional properties of sweet potato genotypes grown in temperate climate conditions. *J. Sci. Food Agric.* **2018**, *98*, 574–581. [CrossRef] [PubMed]

4. Veasey, E.A.; Borges, A.; Rosa, M.S.; Queiroz-Silva, J.R.; Bressan, E.D.A.; Peroni, N. Genetic diversity in Brazilian sweet potato (*Ipomoea batatas* (L.) Lam., Solanales, Convolvulaceae) landraces assessed with microsatellite markers. *Genet. Mol. Biol.* **2008**, *31*, 725–733. [CrossRef]

5. Roullier, C.; Rossel, G.; Tay, D.; McKey, D.; Lebot, V. Combining chloroplast and nuclear microsatellites to investigate origin and dispersal of New World sweet potato landraces. *Mol. Ecol.* **2011**, *20*, 3963–3977. [CrossRef] [PubMed]

6. Roullier, C.; Duputié, A.; Wennekes, P.; Benoit, L.; Fernández Bringas, V.M.; Rossel, G.; Tay, D.; McKey, D.; Lebot, V. Disentangling the Origins of Cultivated Sweet Potato (*Ipomoea batatas* (L.) Lam.). *PLoS ONE* **2013**, *8*, e62707. [CrossRef]

7. Yang, X.S.; Su, W.J.; Wang, L.J.; Lei, J.; Chai, S.S.; Liu, Q. chang Molecular diversity and genetic structure of 380 sweetpotato accessions as revealed by SSR markers. *J. Integr. Agric.* **2015**, *14*, 633–641. [CrossRef]

8. Wadl, P.A.; Olukolu, B.A.; Branham, S.E.; Jarret, R.L.; Yencho, G.C.; Jackson, D.M. Genetic Diversity and Population Structure of the USDA Sweetpotato (*Ipomoea batatas*) Germplasm Collections Using GBSpoly. *Front. Plant Sci.* **2018**, *9*, 1–13. [CrossRef]

9. Austin, D. The taxonomy, evolution and genetic diversity of sweet potatoes and related wild species. In *Exploration, Maintenance and Utilization of Sweet Potato Genetic Resources: Report of the First Sweet Potato Planning Conference*; International Potato Center (CIP): Lima, Perù, 1987.

10. Wang, Z.; Li, J.; Luo, Z.; Huang, L.; Chen, X.; Fang, B.; Li, Y.; Chen, J.; Zhang, X. Characterization and development of EST-derived SSR markers in cultivated sweetpotato (*Ipomoea batatas*). *BMC Plant Biol.* **2011**, *11*. [CrossRef]

11. Gao, M.; Ashu, G.M.; Stewart, L.; Akwe, W.A.; Njiti, V.; Barnes, S. Wx intron variations support an allohexaploid origin of the sweetpotato [*Ipomoea batatas* (L.) Lam]. *Euphytica* **2011**, *177*, 111–133. [CrossRef]

12. Magoon, M.; Krishnan, R.; Vijaya Bai, K. Cytological evidence on the origin of sweet potato. *Theor. Appl. Genet.* **1970**, *40*, 360–366. [CrossRef]

13. Soltis, D.E.; Soltis, P.S. Polyploidy: Recurrent formation and genome evolution. *Trends Ecol. Evol.* **1999**, *14*, 348–352. [CrossRef]

14. Silva Ritschel, P.; Huamán, Z. Variabilidade morfologica da coleçao de germoplasma de batata doce da Embrapa Centro Nacional de Pesquisas de Hortaliças. *Pesqui. Agropecu. Bras.* **2002**, *37*, 485–492. [CrossRef]

15. Palumbo, F.; Galla, G.; Martinez-Bello, L.; Barcaccia, G. Venetian Local Corn (Zea mays L.) Germplasm: Disclosing the Genetic Anatomy of Old Landraces Suited for Typical Cornmeal Mush Production. *Diversity* **2017**, *9*, 32. [CrossRef]

16. Cruz, C.D.; Carneiro, P.C.S.; Regazzi, A.J. *Modelos Biométricos Aplicados ao Melhoramento Genético—Volume 1*, 3rd ed.; Editora UFV: Viçosa, Brazil, 2012; ISBN 9788572694339.

17. Palumbo, F.; Galla, G.; Barcaccia, G. Developing a molecular identification assay of old landraces for the genetic authentication of typical agro-food products: The case study of the barley "agordino". *Food Technol. Biotechnol.* **2017**, *55*, 29–39. [CrossRef] [PubMed]

18. Crinò, P.; Pagnotta, M.A. Phenotyping, genotyping, and selections within italian local landraces of romanesco globe artichoke. *Diversity* **2017**, *9*, 14. [CrossRef]

19. Mercati, F.; Longo, C.; Poma, D.; Araniti, F.; Lupini, A.; Mammano, M.M.; Fiore, M.C.; Abenavoli, M.R.; Sunseri, F. Genetic variation of an Italian long shelf-life tomato (Solanum lycopersicon L.) collection by using SSR and morphological fruit traits. *Genet. Resour. Crop Evol.* **2014**, *62*, 721–732. [CrossRef]

20. Andreakis, N.; Kooistra, W.H.C.F.; Procaccini, G. High genetic diversity and connectivity in the polyploid invasive seaweed Asparagopsis taxiformis (Bonnemaisoniales) in the Mediterranean, explored with microsatellite alleles and multilocus genotypes. *Mol. Ecol.* **2009**, *18*, 212–226. [CrossRef]

21. Cordeiro, G.M.; Pan, Y.B.; Henry, R.J. Sugarcane microsatellites for the assessment of genetic diversity in sugarcane germplasm. *Plant Sci.* **2003**, *165*, 181–189. [CrossRef]

22. Pinto, L.R.; Oliveira, K.M.; Marconi, T.; Garcia, A.A.F.; Ulian, E.C.; De Souza, A.P. Characterization of novel sugarcane expressed sequence tag microsatellites and their comparison with genomic SSRs. *Plant Breed.* **2006**, *125*, 378–384. [CrossRef]

23. Babaei, A.; Tabaei-Aghdaei, S.R.; Khosh-Khui, M.; Omidbaigi, R.; Naghavi, M.R.; Esselink, G.D.; Smulders, M.J.M. Microsatellite analysis of Damask rose (Rosa damascena Mill.) accessions from various regions in Iran reveals multiple genotypes. *BMC Plant Biol.* **2007**, *7*, 1–6. [CrossRef]

24. Veasey, E.A.; Silva, J.R.D.Q.; Rosa, M.S.; Borges, A.; Bressan, E.D.A.; Peroni, N. Phenology and morphological diversity of sweet potato (*Ipomoea batatas*) landraces of the Vale do Ribeira. *Sci. Agric.* **2007**, *64*, 416–427. [CrossRef]

25. Yada, B.; Tukamuhabwa, P.; Alajo, A.; Mwanga, R.O.M. Morphological characterization of Ugandan sweetpotato germplasm. *Crop Sci.* **2010**, *50*, 2364–2371. [CrossRef]

26. Elameen, A.; Larsen, A.; Klemsdal, S.S.; Fjellheim, S.; Sundheim, L.; Msolla, S.; Masumba, E.; Rognli, O.A. Phenotypic diversity of plant morphological and root descriptor traits within a sweet potato, Ipomoea batatas (L.) Lam., germplasm collection from Tanzania. *Genet. Resour. Crop Evol.* **2011**, *58*, 397–407. [CrossRef]

27. Karuri, H.; Ateka, E.; Amata, R.; Nyende, A.; Muigai, A.; Mwasame, E.; Gichuki, S. Evaluating diversity among Kenyan sweet potato genotypes using morphological and SSR markers. *Int. J. Agric. Biol.* **2010**, *12*, 33–38.

28. Som, K.; Vernon, G.; Isaac, A.; Eric, Y.D.; Jeremy, T.O.; Tignegre, J.B.; Belem, J.; Tarpaga, M.V. Diversity analysis of sweet potato (*Ipomoea batatas* [L.] Lam) germplasm from Burkina Faso using morphological and simple sequence repeats markers. *Afr. J. Biotechnol.* **2014**, *13*, 729–742. [CrossRef]

29. Perelli, M.; Graziano, P.L.; Calzavara, R. *Nutrire le Piante*; ARVAN: Venice, Italy, 2009.

30. Hu, J.; Nakatani, M.; Mizuno, K.; Fujimura, T. Development and Characterization of Microsatellite Markers in Sweetpotato. *Breed. Sci.* **2004**, *54*, 177–188. [CrossRef]

31. Schuelke, M. An economic method for the fluorescent labeling of PCR fragments. *Nat. Biotechnol.* **2000**, *18*, 233. [CrossRef]

32. Palumbo, F.; Galla, G.; Vitulo, N.; Barcaccia, G. First draft genome sequencing of fennel (Foeniculum Vulgare Mill.): Identification of simple sequence repeats and their application in marker-assisted breeding. *Mol. Breed.* **2018**, *38*, 1–17. [CrossRef]

33. Botstein, D.; White, R.L.; Skolnick, M.; Davis, R.W. Construction of a genetic linkage map in man using restriction fragment length polymorphisms. *Am. J. Hum. Genet.* **1980**, *32*, 314–331.

34. Dice, L.R. Measures of the Amount of Ecologic Association Between Species. *Ecology* **1945**, *26*, 297–302. [CrossRef]

35. Rohlf, F.J. *NTSYS-pc: Numerical Taxonomy and Multivariate Analysis System*; Applied Biostatistics Inc.: Setauket, NY, USA, 2008.

36. Hammer, Ø.; Harper, D.A.; Ryan, P.D. PAST: Paleontological Statistics Software Package for Education and Data Analysis. *Palaeontol. Electron.* **2001**, *4*, 9.

37. Peakall, R.; Smouse, P.E. GenALEx 6.5: Genetic analysis in Excel. Population genetic software for teaching and research-an update. *Bioinformatics* **2012**, *28*, 2537–2539. [CrossRef] [PubMed]

38. Falush, D.; Stephens, M.; Pritchard, J.K. Inference of population structure using multilocus genotype data: Linked loci and correlated allele frequencies. *Genetics* **2003**, *164*, 1567–1587. [PubMed]

39. Porras-Hurtado, L.; Ruiz, Y.; Santos, C.; Phillips, C.; Carracedo, Á.; Lareu, M.V. An overview of STRUCTURE: Applications, parameter settings, and supporting software. *Front. Genet.* **2013**, *4*, 1–13. [CrossRef]

40. Pritchard, J.K.; Stephens, M.; Donnelly, P. Inference of population structure using multilocus genotype data. *Genetics* **2000**, *155*, 945–959. [PubMed]

41. Evanno, G.; Regnaut, S.; Goudet, J. Detecting the number of clusters of individuals using the software STRUCTURE: A simulation study. *Mol. Ecol.* **2005**, *14*, 2611–2620. [CrossRef] [PubMed]

42. *IBPGR Descriptors for Sweet Potato*; Huamán, Z. (Ed.) International Board for Plant Genetic Resources: Rome, Italy, 1991.

43. Singleton, V.L.; Orthofer, R.; Lamuela-Raventós, R.M. Analysis of total phenols and other oxidation substrates and antioxidants by means of folin-ciocalteu reagent. *Methods Enzymol.* **1999**, *299*, 152–178.

44. Benzie, I.F.F.; Strain, J.J. The ferric reducing ability of plasma (FRAP) as a measure of "antioxidant power": The FRAP assay. *Anal. Biochem.* **1996**, *239*, 70–76. [CrossRef]

45. Roullier, C.; Benoit, L.; McKey, D.B.; Lebot, V. Historical collections reveal patterns of diffusion of sweet potato in Oceania obscured by modern plant movements and recombination. *Proc. Natl. Acad. Sci. USA* **2013**, *110*, 2205–2210. [CrossRef]

46. Nicolle, C.; Simon, G.; Rock, E.; Amouroux, P.; Rémésy, C. Genetic Variability Influences Carotenoid, Vitamin, Phenolic, and Mineral Content in White, Yellow, Purple, Orange, and Dark-orange Carrot Cultivars. *J. Am. Soc. Hortic. Sci.* **2004**, *129*, 523–529. [CrossRef]

47. Tsegaye, E.; Dechassa, N.; Sastry, D.E.V. Genetic Variability for Yield and other agronomic traits in sweet potato. *J. Agron.* **2007**, *6*, 94–99.

48. Solankey, S.S.; Singh, P.K.; Singh, R.K. Genetic Diversity and Interrelationship of Qualitative and Quantitative Traits in Sweet Potato. *Int. J. Veg. Sci.* **2015**, *21*, 236–248. [CrossRef]

49. Lim, S.; Xu, J.; Kim, J.; Chen, T.-Y.; Su, X.; Standard, J.; Carey, E.; Griffin, J.; Herndon, B.; Katz, B.; et al. Role of Anthocyanin-enriched Purple-fleshed Sweet Potato P40 in Colorectal Cancer Prevention. *Mol. Nutr. Food Res.* **2013**, *57*, 1908–1917. [CrossRef] [PubMed]

50. Grace, M.H.; Yousef, G.G.; Gustafson, S.J.; Truong, V.-D.; Yencho, G.C.; Lila, M.A. Phytochemical changes in phenolics, anthocyanins, ascorbic acid, and carotenoids associated with sweetpotato storage and impacts on bioactive properties. *Food Chem.* **2014**, *145*, 717–724. [CrossRef] [PubMed]

51. Lee, M.J.; Park, J.S.; Choi, D.S.; Jung, M.Y. Characterization and quantitation of anthocyanins in purple-fleshed sweet potatoes cultivated in Korea by HPLC-DAD and HPLC-ESI-QTOF-MS/MS. *J. Agric. Food Chem.* **2013**, *61*, 3148–3158. [CrossRef] [PubMed]

52. Gorusupudi, A.; Bernstein, P.S. Macular Carotenoids: Human Health Aspects. In *Carotenoids: Nutrition, Analysis, and Technology*; Kaczor, A., Baranska, M., Eds.; Wiley Blackwell: Pondicherry, India, 2016; pp. 59–74.

53. Park, S.Y.; Lee, S.Y.; Yang, J.W.; Lee, J.S.; Oh, S.D.; Oh, S.; Lee, S.M.; Lim, M.H.; Park, S.K.; Jang, J.S.; et al. Comparative analysis of phytochemicals and polar metabolites from colored sweet potato (*Ipomoea batatas* L.) tubers. *Food Sci. Biotechnol.* **2016**, *25*, 283–291. [CrossRef]

54. Simonne, A.H.; Kays, S.J.; Koehler, P.E.; Eitenmiller, R.R. Assessment of β-Carotene content in sweetpotato breeding lines in relation to dietary requirements. *J. Food Compos. Anal.* **1993**, *6*, 336–345. [CrossRef]
55. Wei, S.; Lu, G.; Cao, H. Effects of cooking methods on starch and sugar composition of sweetpotato storage roots. *PLoS ONE* **2017**, *12*, e0182604. [CrossRef]

MDPI

Article

Serendipitous In Situ Conservation of Faba Bean Landraces in Tunisia: A Case Study

Elyes Babay [1,2,†], Khalil Khamassi [3,†], Wilma Sabetta [4], Monica Marilena Miazzi [5], Cinzia Montemurro [5], Domenico Pignone [4], Donatella Danzi [4], Mariella Matilde Finetti-Sialer [4,*] and Giacomo Mangini [5,*]

1   National Gene Bank of Tunisia (BNG), Street Yesser Arafet, Charguia 1, Tunis 1080, Tunisia; e.babay@yahoo.fr
2   Institut National de la Recherche Agronomique de Tunisie (INRAT), Agricultural Applied Biotechnology Laboratory (LR16INRAT06), University of Carthage, Rue Hédi Karray, Menzah 1, Tunis PC 1004, Tunisia
3   Institut National de la Recherche Agronomique de Tunisie (INRAT), Field Crop Laboratory (LR16INRAT02), University of Carthage, Rue Hédi Karray, Menzah 1, Tunis PC 1004, Tunisia; khalilkhamassi@iresa.agrinet.tn
4   Institute of Biosciences and Bioresources of the National Research Council (IBBR-CNR), Via Amendola 165/A, 70126 Bari, Italy; wilma.sabetta@ibbr.cnr.it (W.S.); domenico.pignone@ibbr.cnr.it (D.P.); donatella.danzi@ibbr.cnr.it (D.D.)
5   Department of Soil, Plant and Food Sciences (DiSSPA), Sect. Genetics and Plant Breeding, University of Bari Aldo Moro, Via Amendola 165/A, 70126 Bari, Italy; monicamarilena.miazzi@uniba.it (M.M.M.); cinzia.montemurro@uniba.it (C.M.)
*   Correspondence: mariella.finetti@ibbr.cnr.it (M.M.F.-S.); giacomo.mangini@uniba.it (G.M.); Tel.: +39-080-5442-992 (G.M.)
†   These authors contributed equally.

Received: 21 January 2020; Accepted: 16 February 2020; Published: 24 February 2020

Abstract: Cultivation of faba bean (*Vicia faba* L.) in Tunisia is largely based on improved varieties of the crop. However, a few farmers continue to produce local cultivars or landraces. The National Gene Bank of Tunisia (NGBT) recently launched a collection project for faba bean landraces, with special focus on the regions of the North West, traditionally devoted to cultivating grain legumes, and where around 80% of the total national faba bean cultivation area is located. The seed phenotypic features of the collected samples were studied, and the genetic diversity and population structure analyzed using simple sequence repeat markers. The genetic constitution of the present samples was compared to that of faba bean samples collected by teams of the International Center for Agricultural Research in the Dry Areas (ICARDA) in the 1970s in the same region, and stored at the ICARDA gene bank. The results of the diversity analysis demonstrate that the recently collected samples and those stored at ICARDA largely overlap, thus demonstrating that over the past 50 years, little genetic change has occurred to the local faba bean populations examined. These findings suggest that farmers serendipitously applied international best practices for in situ conservation of agricultural crops.

Keywords: *Vicia faba* L., genetic diversity; SSR markers; in situ conservation

---

1. Introduction

Faba bean (*Vicia faba* L., 2n = 2x = 12) is a facultative cross-pollinating species with outcrossing rates varying between 1 and 55% depending on its environment; it belongs to the *Fabaceae* family, *Faboideae* subfamily, tribe of *Fabeae*, and is not interfertile with any other *Vicia* species [1]. The wild progenitor of *V. faba* is unknown, but recent archaeological excavations have allowed, in the Mont-Carmel (Mediterranean Levantine), the discovery of fossilized seeds that are compatible with a wild progenitor of this crop, dating as back as 14,000 ybp [2]. Considering other archaeological evidences, such as those relating to findings in Tel el Kerkh [3], northwest Syria, it is possible to hypothesize that this species has been domesticated since the Neolithic era, and that the wild progenitor, possibly distributed in small habitats, was entirely domesticated and then became extinct [2–5]. According to Cole [6] and Cubero [7], the spread of faba bean from its center of origin to other countries could have involved five routes. In the Mediterranean, in particular, faba bean mainly spread through two routes: the first across Anatolia to Greece, the Illyric coast (possibly the Danubian regions), and then to Italy; the second, beginning at the Nile Delta, moving towards the West, along the North African Mediterranean coast, to the Maghreb and then to the Iberian Peninsula. It is worth mentioning that, in this regard, North Africa and Tunisia in particular constitute a center of primary and secondary diversification of several agricultural and wild species [8].

In Tunisia, faba bean covers more than 70% (59,583 ha) of the total area annually devoted to grain legume crops [9]. Approximately half of the area is cultivated with grain legume types meant for fresh pod consumption; the rest as forage (plant and seeds) is mainly based on small seeded types. The average productivity in Tunisia is 1.03 t/ha, 40% below the world average [10]. This is mainly due to the parasitic weed broomrape (*Orobanche crenata* Forssk. and *O. foetida* Poir.) and drought stress occurring in faba bean-growing areas [11]. Until the last century, most crops consisted of landraces, often named after the farmer who selected them or after areas where they were grown [12]. Some landraces ('Batata', 'Malti', 'Chemchali' or 'Masri local') are still grown by farmers and seeds can be bought from local informal markets.

In recent years, thanks to significant achievements in crop breeding, modern high-yielding varieties are widely used in cultivation and have almost completely replaced local populations and landraces [11]. The increase in yield was obtained mainly with breeding programs targeted at tolerance to abiotic (heat and drought) and biotic (foliar diseases and parasitic weeds) stresses. Moreover, recent breeding efforts are directed towards the development of new cultivars with low anti-nutritional compounds (vicine and convicine), to improve the quality and utilization efficiency of faba in human diet and for livestock feed [13]. Different molecular tools were used to investigate genetic diversity in grain legume species [14–17]. Some attempts to evaluate genetic variability have also been made for Tunisian faba bean germplasm. The analysis of isozyme [18] and sequence specific amplified polymorphism (SSAP) [19] markers analyzed in nine Tunisian *Vicia faba* collections has indicated a certain degree of genetic cohesiveness. Using simple sequence repeat (SSR) markers, 16 faba bean accessions, selected from 42 populations collected across eight southern oases arid agro-ecosystems, were analyzed, evidencing genetic cohesiveness among the studied samples [19–21], together with a low level of variability among accessions. In both reports, the authors stated that intense seed exchange among farmers had led to a leveled degree of genetic diversity among those populations.

In the 1960s, major concerns focused on the genetic erosion of biodiversity, eventually leading to fostering of ex situ conservation efforts and the creation of gene banks [22]. As a result, in the latest decades of the 20th century, several different research centers organized collection missions for crop diversity in order to secure the local germplasm before it was completely lost [23]. Within the frame of "emergency" collections, from the seventies until the early nineties, ICARDA, among others, carried on a series of faba bean collection missions in North Africa. During that period, several faba bean samples were retrieved from different regions, in particular from Tunisia. At the end of the 20th century, taking into account the evident loss of genetic diversity in the Mediterranean [24], specific attention was paid to the practice of in situ conservation of crops. This is the conservation of agricultural genetic resources on farms located in the same areas where local communities had developed them, with specific attention to neglected crops [25,26]. According to Duc et al. [23], in situ conservation of biodiversity may contribute to the development of the best-adapted materials for local agronomic practices and involve farmers in the selection process through participatory breeding [27]. Within this frame, NGBT started an ongoing program of genetic resource collection in different areas of Tunisia. The aim was to preserve Tunisian crop gene-pools from genetic erosion and characterize local germplasm, thanks to an integrated approach, including on-farm conservation of local germplasm and landraces.

In order to better plan an in situ conservation strategy for faba bean and to understand the possible loss of genetic diversity in Tunisian *Vicia faba* germplasm, the genetic structure of the samples collected in recent years was compared with those collected by ICARDA in the 1970s. This paper reports on the results of this comparison.

2. Materials and Methods

2.1. Plant Material

The plant material used in the present study consisted of a collection of 51 Tunisian local faba bean samples (Table 1). It included 29 samples collected during 2016–2018 by the NGBT and 22 faba bean accessions collected by ICARDA, starting from the seventies until the early nineties. The NGBT samples were collected in the governorates of Beja and Jendouba (Figure S1) characterized by annual average rainfall of 800 and 600 mm, respectively. The passport data and ethno-botanical information of the NGBT samples are available at NGBT. The samples of ICARDA (labelled ICAR) were derived from collection missions conducted from the seventies until the early nineties in North Tunisia (Beja, Bizerte, and Siliana).

Table 1. List of faba bean samples included in the study.

Id Name	Local Name	Governorate	Location	Longitude (E)	Latitude (N)	Seed Type
NGBT 1	Malti	Beja	El Hamra	9.011641	36.52219	Large
NGBT 3	Malti	"	"	9.011641	36.52219	Large
NGBT 4	Malti	"	"	9.011641	36.52219	Large
NGBT 5	Chemchali	"	"	9.011641	36.52219	Small
NGBT 8	Chemchali	"	"	9.011641	36.52219	Small
NGBT 9		"	"	9.012502	36.52136	Medium
NGBT 10	-	"	"	9.012502	36.52136	Small
NGBT 13	-	"	"	9.012502	36.52136	Small
NGBT 34	-	"	"	9.012502	36.52136	Small
NGBT 35	-	"	"	9.012502	36.52136	Small
NGBT 50	Chemchali	"	"	9.011641	36.52219	Small
NGBT 51	Chemchali	"	"	9.011641	36.52219	Small
NGBT 52	Chemchali	"	"	9.011641	36.52219	Small
NGBT 53	Chemchali	"	"	9.011641	36.52219	Small
NGBT 55	-	"	"	9.012502	36.52136	Medium
NGBT 66	-	"	"	9.012502	36.52136	Medium
NGBT 16	-	Jendouba	Oued Ghrib	8.412815	36.37286	Small
NGBT 18	-	"	"	8.412815	36.37286	Large
NGBT 21	Bachar	"	"	8.414203	36.37533	Small
NGBT 56	-	"	"	8.414203	36.37533	Large
NGBT 57	-	"	"	8.414203	36.37533	Large
NGBT 60	-	"	"	8.412815	36.37286	Small
NGBT 61	-	"	"	8.412815	36.37286	Large
NGBT 63	-	"	"	8.414203	36.37533	Large
NGBT 36	Malti	"	Fouazia	8.404811	36.40103	Large
NGBT 38	Malti	"	"	8.404811	36.40103	Large
NGBT 48	Malti	"	"	8.404811	36.40103	Large
NGBT 62	Malti	"	"	8.404811	36.40103	Large
NGBT 64	Malti	"	"	8.404811	36.40103	Large
ICAR 22	Local	n.a.	n.a.	n.a.	n.a.	Small
ICAR 23	Seville	n.a.	n.a.	n.a.	n.a.	Large
ICAR 24	Misri 32	Siliana	n.a.	9.616670	36.35000	Small
ICAR 25	Local	n.a.	n.a.	n.a.	n.a.	Large
ICAR 26	-	Bizerte	n.a.	n.a.	n.a.	Large
ICAR 27	-	Beja	n.a.	n.a.	n.a.	Medium
ICAR 28	-	"	n.a.	n.a.	n.a.	Large
ICAR 29	-	n.a.	n.a.	n.a.	n.a.	Large
ICAR 30	Local	Bizerte	n.a.	n.a.	n.a.	Small
ICAR 31	Local	n.a.	n.a.	n.a.	n.a.	Large
ICAR 32	Local	n.a.	n.a.	n.a.	n.a.	Small
ICAR 33	-	n.a.	n.a.	n.a.	n.a.	Small

Table 1. *Cont.*

Id Name	Local Name	Governorate	Location	Longitude (E)	Latitude (N)	Seed Type
ICAR 39	Local	Bizerte	n.a.	n.a.	n.a.	Medium
ICAR 40	Malti 24	"	n.a.	9.666670	37.05000	Medium
ICAR 41	Malti 25	"	n.a.	9.666680	37.06000	Small
ICAR 42	Misri 39	Beja	n.a.	9.583330	36.66670	Small
ICAR 43	Misri 41	"	n.a.	9.216670	36.71670	Small
ICAR 44	-	"	n.a.	n.a.	n.a.	Medium
ICAR 45	-	"	n.a.	n.a.	n.a.	Small
ICAR 46	Local	Bizerte	n.a.	n.a.	n.a.	Large
ICAR 47	Local	n.a.	n.a.	n.a.	n.a.	Small
ICAR 68	Local	n.a.	n.a.	n.a.	n.a.	Large

n.a. data not available.

2.2. Seed Phenotypic Traits

Five seeds of each sample were randomly selected to determine average seed size. The three axial dimensions of seed length (L), width (W), and thickness (T) were measured using a Vernier caliper (Gilson Tools, Japan) with accuracy of 0.05 mm. The geometric mean diameter (Dg) was calculated by using the equation reported by Mohsenin [28]:

$$Dg = (L \times W \times T)^{1/3} \tag{1}$$

The sphericity (φ) of faba bean seeds was calculated using the following formula:

$$\varphi = [(L \times W \times T)^{1/3}/L]*100 \tag{2}$$

2.3. DNA Extraction and SSR Assays

Total genomic DNA was extracted from fresh young leaves—five plants per sample—using the cetyltrimethyl ammonium bromide (CTAB) method, as described by Fulton et al. [29]. DNA concentration was determined using a NanoDrop™ ND-2000 (Thermo Scientific, MA, USA) and the quality was verified by separation on 0.8% agarose gel. Equal DNA quantities of the five plants of the sample were then pooled, and all DNA samples were diluted to a standard working concentration of 50 ng/μl by adding ultrapure water (Gibco, Invitrogen, USA).

A set of 11 simple sequence repeats (SSRs) markers, retrieved from the literature [30–32] were used for this study (Table S1). A preliminary assay was carried out in order to evaluate the robustness of PCR reaction and the reproducibility of the fragments. In particular, different faba samples randomly chosen were analysed considering technical replicates. The PCR conditions for each SSR markers were set up at best conditions, considering an annealing temperature ranging from 45 to 60 °C. The fragments produced were separated on 2.0% agarose gel containing Nancy-520 DNA Gel Stain (Sigma Genosys, St. Louis, MO, USA), and visualized under UV light. The amplification reactions were performed in a final volume of 20 μl, containing the template DNA (50 ng), the 5′end of each forward primer with the M13 (21 bp) tail, the reverse primer (M13) labeled with fluorescent dye (FAM, VIC, PET, or NED). PCR reactions were performed in a thermal cycler (Bio-Rad Laboratories, Hercules, CA, USA) as follows: an initial denaturing step at 95 °C for 3 min, followed by 36 cycles of 94 °C for 20 s, 56 °C for 50 s, 72°C for 1 min, and a final extension step at 72 °C for 7 min. The amplification products were detected by automatic capillary sequencer ABI PRISM 3100 Genetic Analyzer (Applied Biosystems, Waltham, MA, USA), and the fragments were analyzed with GeneMapper genotyping software version 5.0 (Thermo Fisher Scientific, Waltham, MA, USA). The internal molecular weight standard was GeneScanTM 600 LIZ dye Size Standard (Thermo Fisher Scientific, Waltham, MA, USA).

2.4. Data Analysis

Hierarchical ascending classification (HAC) clustering analysis based on dissimilarity matrix of morphometric seed data (L, W, T, Dg, and φ) was performed to evaluate the relationship among the faba samples using XLSTAT statistical software ver. 2016.2 (Addinsoft Inc, New York, USA).

The genetic indices, number of different alleles (*Na*), Shannon's information index (*I*), observed Heterozygosity (*Ho*), expected Heterozygosity (*He*), Fixation Index (*F*), and private alleles were calculated using GenAlEx version 6.5 [33]. The allelic data were used to obtain a similarity matrix, from which a dendrogram was constructed using the UPGMA algorithm with MEGA ver. 4 [34]. The molecular data were processed using STRUCTURE ver. 2.3.4 [35]. The number of sub-populations (K) was estimated by 10 independent runs for each K (from 1 to 10), applying the admixture model, 500000 Markov Chain Monte Carlo (MCMC) repetitions, and a 100000 burn-in period. The means of the log-likelihood estimates for each K were calculated. The true K was determined with the Evanno test [36] using STRUCTURE HARVESTER [37]. Analysis of molecular variance (AMOVA) was used to partition the genetic variation into inter- and intra-gene pool diversities in faba using GenAlEx program ver. 6.5, with 1000 permutations.

3. Results

3.1. Variation in Seeds Phenotypic Traits

Morphometric seed traits (L, W, T, Dg, and φ) were measured in the samples belonging to both the NGBT and ICARDA faba bean collection. The average mean of the three principal axial dimensions (L, W, and T) and Dg of the NGBT and ICAR groups are shown in Figure 1. No statistically significant differences for these morphometric seed traits were detected between the two groups. The values of faba bean sphericity (φ) were calculated by using the geometric mean diameter and length data. No significant difference was found for these traits when comparing the NGBT and ICAR groups.

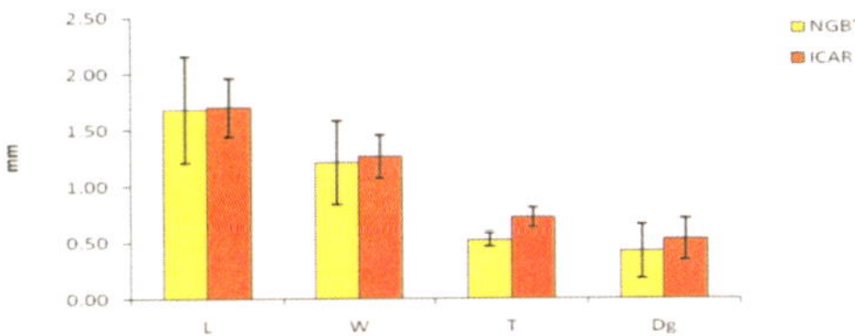

Figure 1. Length (L), width (W), thickness (T), and geometric mean diameter (Dg) averages of samples collected by the National Gene Bank of Tunisia (NGBT) and Agricultural Research in the Dry Areas (ICAR). Lines represent the standard deviation.

The values of morphometric seed traits of each samples were used to calculate a dissimilarity matrix based on Euclidean distances. A dendrogram was obtained starting from the matrix, in which three main clusters are evidenced, corresponding to three seed types—large, medium (called also equina), and small (Figure 2). Each cluster included both NGBT and ICAR samples with no reference to the area of origin.

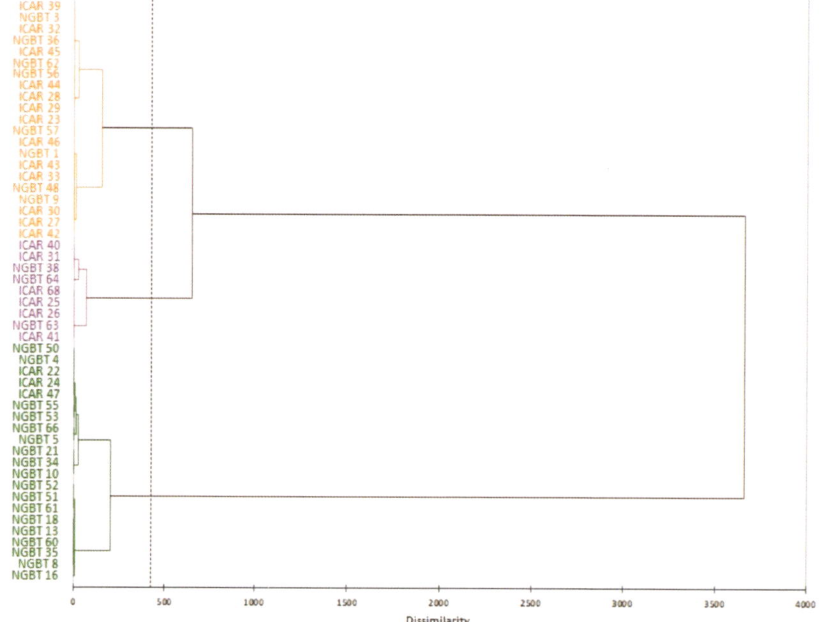

Figure 2. Dendrogram based on dissimilarity matrix calculated from morphometric seed traits in the faba bean collection split in samples collected by the National Gene Bank of Tunisia (NGBT) and Agricultural Research in the Dry Areas (ICAR). The colors orange, purple, and green were used to distinguish between the medium, large, and small seed type clusters, respectively.

3.2. Molecular Variation of Faba Bean Collection

In order to evaluate the genetic diversity of 51 faba bean samples, a set of 11 SSR markers was used (Table 1 and Table S1). A total number of 94 alleles were identified, ranging from 3 (loci M22 and M46) to 22 (locus VFG41), with an average of 10.2 alleles per locus (Table S2).

AMOVA did not show a molecular diversity among the 29 NGBT faba beans when these were grouped according to the collection sites (El Hamra, Oued Ghrib and Fouazia), suggesting that the NGBT faba bean samples belonged to one genetically cohesive population (Table S3).

According to these results, we investigated the genetic diversity among samples collected in recent years by NGBT and ICAR during the 1970s, in the same Tunisian areas. Similar values for the number of different alleles, Shannon's information index, Heterozygosity expected, Heterozygosity observed, Diversity index, and Fixation index were found between the two groups, although the ICAR samples appeared slightly more fixed (Table 2).

Table 2. Number of different alleles (*Na*), Shannon's information index (*I*), Heterozygosity observed (*Ho*), Heterozygosity expected (*He*), Fixation Index (*F*), and private alleles of faba bean collection split in samples recently collected by the National Gene Bank of Tunisia (NGBT) and Agricultural Research in the Dry Areas (ICARDA).

	N° Samples	Na	I	Ho	He	F	Private Alleles
NGBT	29	7	1.428	0.621	0.658	0.027	23
ICARDA	22	6	1.392	0.546	0.657	0.145	12
Whole Collection	51	7	1.410	0.584	0.657	0.086	

AMOVA indicated that the NGBT and ICAR groups have no statistical difference, at the molecular level (Table 3), as most of the diversity clearly appeared within groups, with only a limited variation occurring among them.

Table 3. Analysis of molecular variance (AMOVA) of faba bean collection split in samples collected by the National Gene Bank of Tunisia (NGBT) and Agricultural Research in the Dry Areas (ICARDA).

Source of Variation	df	SS	MS	Est. Var.	*p*-Values
Among groups	1	3.650	3.65	0.01	n.s
Within groups	49	161.94	3.31	3.3	
Total	50	165.59		3.31	

df = degree of freedom, SS = sum of squares, MS = mean squares, Est. Var. = estimate of variance, *p*-value.

To define the genetic relationships among the faba bean samples, the similarity matrix obtained was also used to produce a UPGMA dendrogram (Figure 3). The clustering showed that the NGBT samples were admixed with ICAR ones, thus suggesting that the two groups could be considered a unique meta-population. Total faba bean collection was also evaluated with Bayesian clustering modeling, performed using SSR allelic data generated according to 11 SSR markers. As the clustering model presumes the underlying existence of K clusters, an Evanno test [36] was performed that yielded K=3 as the highest log-likelihood (Figure S2). Nevertheless, each sample analyzed showed to belong to all three clusters identified, and none of them predominantly pertained to a specific cluster (Figure 4). These results seemingly indicate that the Tunisian faba collection was structured in three subpopulations that do not correspond to NGTB and ICAR samples' distinction.

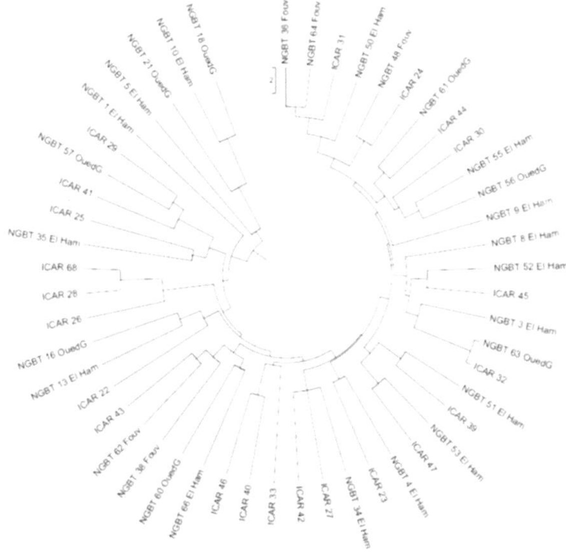

Figure 3. Dendrogram of faba bean collection split in samples collected by the National Gene Bank of Tunisia (NGBT) and Agricultural Research in the Dry Areas (ICAR) resulting from the UPGMA cluster analysis based on similarity matrix obtained from 11 SSR allelic data.

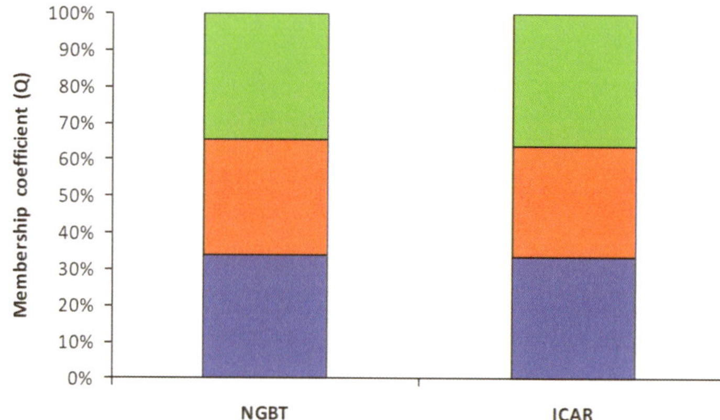

Figure 4. Membership coefficient (Q) mean of the faba bean collection split in samples collected by the National Gene Bank of Tunisia (NGBT) and Agricultural Research in the Dry Areas (ICARDA). The different colors indicate the three subpopulations detected using a Bayesian approach (blue: subpopulation 1; red: subpopulation 2, and green: subpopulation 3).

4. Discussion

Faba bean is an important crop for sustainable agriculture in marginal areas and advanced agricultural systems, as it plays an important role in soil fertility and nitrogen fixation, being able to grow in diverse climatic and soil conditions. Although faba bean is less consumed in western countries as human food, it is considered one of the main sources of cheap protein and energy for many people in Africa, parts of Asia, and Latin America, where many people cannot afford to buy meat [38]. Seed size and shape are characters of polygenic control [39,40] and have undergone strong selection measures by farmers during evolution of the crop. Seed size is considered a key trait in the study of the historical evolution of this crop based on archaeological remains and findings [2]. This selection pressure can still be found in the habits of farmers, who manually select seeds to be sown in the next season [41], a habit that was followed in many Mediterranean regions until modern times and lost only on introduction of improved varieties. This manual selection procedure resulted in the formation of peculiar landraces, such as "Larga di Leonforte" [42]. However, this study noted that Tunisian farmers did not practice seed selection, and have not done so for the last 50 years.

The analysis of morphometric seed traits and molecular markers carried out in this work did not distinguish patterns in the distribution of morphological and genetic variations. All the samples collected by the NGBT teams appeared to belong to a single, genetically cohesive population. Similar results were observed for the faba bean samples obtained by ICARDA and collected from the same areas. When all faba samples were analyzed together, there were no differences between the NGBT and ICARDA groups. In fact, our molecular analyses and seed morphometric variation study demonstrated that the two groups belonged to a single meta population.

Seed size is one of the most important morphological traits responsible for yield, and a major target for breeding. Several studies have led to the mapping of QTLs/genes for seed weight/size in soybean [43], chickpea [44], and lentil [45]. In faba bean, the genetic control of these traits is still unclear, although consensus linkage maps have been produced [43,46].

Our data confirmed that morphometric seed traits were not associated to the markers used. This implies that any selection of lines based on seed traits still retains a variable level of genetic diversity potentially associated to other traits, such as adaptation or resistance, an issue to be taken into account in any faba bean-breeding program based on phenotypic data.

The absence of genetic differentiation in different collection sites might depend on many factors. The absence of human selection pressure does not force crop adaptation in a specific direction.

The outbreeding habit of faba bean is a second factor; in fact, while the advanced breeding varieties favor inbreeding in search of higher stability, as requested by UPOV standards, the landraces are generally quite allogamous [47,48]. Recent studies have also demonstrated that the level of allogamy might depend on the species of pollinating insects [49]. A further, effective mechanism could be derived by the spontaneous seed exchange practice occurring among farmers. In informal seed markets, seed exchange by farmers favors the establishment of a landrace in a given environment with uniform agro-climatic characteristics. The farmers cultivating these landraces are actually the relics of a once wider cultivation area. All these factors act synergically to produce the observed genetic differentiation patterns.

The Evanno test yielded three subpopulations within the faba bean collection. In addition, each sample had a coefficient membership lower than 0.50, thus denying that it predominantly belonged to one of these subpopulations. This further supports the fact that the two meta-populations, NGBT and ICARDA, are subsamples of a unique population collected in the same area at different times.

The biological significance of the three subpopulations detected with STRUCTURE is unclear, and might depend on the presence of genetic signals that do not parallel clear-cut characteristics. Nevertheless, several authors have reported that hierarchical analyses, such as those based on STRUCTURE or similar software, relay on strict assumptions that might not completely apply to the case studies, thus resulting in incorrect evaluation of the population structure [50].

In conclusion, unintended conservation of ancient faba bean germplasm in Tunisian farms is witnessed, because the farmers cultivating faba bean landraces do not follow seed selection, as well as owing to concurring factors such as the high level of cross-pollination, and the consistent presence of pollinators (due to the limited use or absence of insecticides). This beneficial situation is an empirical application of best practices recommended by research institutions for on-farm conservation of plant genetic resources. In their quest to feed their families and their precious animals, the Tunisian farmers in the small villages mentioned above serendipitously stored and protected their faba bean genetic resources. These findings and implications should be further discussed in the broadest context possible. Future research directions should also be highlighted.

Supplementary Materials: The following are available online at http://www.mdpi.com/2073-4425/11/2/236/s1, **Figure S1:** NGBT sampling **Figure S2:** Delta K for differing numbers of subpopulations (K) estimated for faba bean collection with 11 markers; **Table S1:** List of SSR markers used for genetic analysis of faba bean collection; **Table S2:** Number of different alleles (*Na*), Shannon's information index (*I*), Heterozygosity observed (*Ho*), Heterozygosity expected (*He*), and Fixation Index (*F*) of 11 SSR markers used in faba bean collection; **Table S3:** Analysis of molecular variance (AMOVA) of 29 NGTB faba bean split according to collection sites (El Hamra, Oued Ghrib, and Fouazia).

Author Contributions: Conceptualization, E.B., K.K., D.P., M.M.F.-S., and G.M.; Formal analysis, M.M.M., D.D., and G.M.; Funding acquisition, C.M. and D.P.; Investigation, E.B., K.K., W.S., and M.M.F.-S.; Methodology, D.P., M.M.F.-S., and G.M.; Resources, E.B. and K.K.; Supervision, D.P., M.M.F.-S., and G.M.; Writing—original draft, E.B., K.K., D.P., M.M.F.-S., and G.M.; Writing—review & editing, D.P., M.M.F.-S., and G.M. All authors have read and agreed to the published version of the manuscript.

Funding: This work was carried out within the framework of the project "Ressources phytogénétiques tunisiennes mieux conservées et valorisées", funded by the Italian Cooperation to the Tunisian Ministry of Local Affairs and Environment.

Acknowledgments: The authors wish to thank Amine Hmid, Anis Khlij, and Alberto Dragotta from the Mediterranean Agronomic Institute of Bari, Italy (CIHEAM-IAM); Olfa Saddoud and Mbarek Ben Naceur from the National Gene Bank of Tunisia for project management and administration. Special thanks are also given to Mr. Mostafa Khemiri and Mr. Ali Ayari, agronomic engineer-extensionists and head of CTV Fernana and CTV Amdoun, respectively, for facilitating contacts with farmers and help during the collection missions. We give special thanks to Nicoletta Rapanà for excellent technical assistance.

Conflicts of Interest: The authors declare that they have no competing interests for this research.

References

1. Leht, M.; Jaaska, V. Cladistic and phenetic analysis of relationships in *Vicia subgenus Vicia* (Fabaceae) by morphology and isozymes. *Plant Syst. Evol.* **2002**, *232*, 237–260. [CrossRef]

2. Caracuta, V.; Barzilai, O.; Khalaily, H.; Milevski, I.; Paz, Y.; Vardi, J.; Regev, L.; Boaretto, E. The onset of faba bean farming in the Southern Levant. *Sci. Rep.* **2015**, *5*, 14370. [CrossRef] [PubMed]
3. Tanno, K.; Willcox, G. The origins of *Cicer arietinum* L. and *Vicia faba* L.: Early finds from Tell el Kerkh, north-west Syria, late 10th millennium B.P. *Veg. Hist. Archaeobot.* **2006**, *15*, 197–204.
4. Caracuta, V.; Weinstein-Evron, M.; Kaufman, D.; Yeshurun, R.; Silvent, J.; Boaretto, E. 14,000-year-old seeds indicate the Levantine origin of the lost progenitor of faba bean. *Sci. Rep.* **2016**, *6*, 37399. [CrossRef]
5. Kosterin, O.E. The lost ancestor of the broad bean (*Vicia faba* L.) and the origin of plant cultivation in the Near East. *Vavilov J Genet. PL. BR.* **2014**, *18*, 831–840. [CrossRef]
6. Hole, F. A Reassessment of the Neolithic Revolution. *Paléorien* **1984**, *10*, 49–60. [CrossRef]
7. Cubero, J.I. On the evolution of *Vicia faba* L. *Theor. Appl. Genet.* **1974**, *45*, 47–51. [CrossRef]
8. Valderrábano, M.; Gil, T.; Heywood, V.; de Montmollin, B. *Conserving Wild Plants in the South and East Mediterranean Region*; IUCN: Gland, Switzerland; Málaga, Spain, 2018; p. 146.
9. Anonymous. *Direction Générale de la Production Agricole (DGPA)*; Ministry of Agriculture Fisheries and Hydraulic Resources of Tunisia: El Ain, Tunisia, 2018.
10. Rebaa, F.; Abid, G.; Aouida, M.; Abdelkarim, S.; Aroua, I.; Muhovski, Y.; Baudoin, J.P.; M'hamdi, M.; Sassi, K.; Jebara, M. Genetic variability in Tunisian populations of faba bean (*Vicia faba* L. var. major) assessed by morphological and SSR markers. *Physiol. Mol. Biol. Plants* **2017**, *23*, 397–409. [CrossRef]
11. Kharrat, M.; Ouchari, H. Faba bean status and prospects in Tunisia. *Grain Legumes* **2011**, *56*, 11–12.
12. Hamza, N. La valorization et l'utilisation des ressources génétiques Locals, moyens de leur conservation durable en Tunisie. *Ann. de l'INRAT 2ème Numéro Spécial Centen.* **2015**, *88*, 26–32.
13. Singh, R.K.; Bohra, N.; Sharm, L. Valorizing faba bean for animal feed supplements via biotechnological approach: Opinion. *Biocatal. Agric. Biotechnol.* **2019**, *17*, 366–368. [CrossRef]
14. De Giovanni, C.; Pavan, S.; Taranto, F.; Di Rienzo, V.; Miazzi, M.M.; Marcotrigiano, A.R.; Mangini, G.; Montemurro, C.; Ricciardi, L.; Lotti, C. Genetic variation of a global germplasm collection of chickpea (*Cicer arietinum* L.) including Italian accessions at risk of genetic erosion. *Physiol. Mol. Biol. Plants* **2016**. [CrossRef]
15. Oliveira, H.R.; Tomàs, D.; Silva, M.; Viegas, W.; Veloso, M.M. Genetic diversity and population structure in *Vicia faba* L. landraces and wild related species assessed by nuclear SSRr. *PLoS ONE* **2016**, *11*, e015480. [CrossRef]
16. Liu, Z.; Li, H.; Wen, Z.; Fan, X.; Li, Y.; Guan, R.; Guo, Y.; Wang, S.; Wang, D.; Qiu, L. Comparison of genetic diversity between Chinese and American soybean (*Glycine max* (L.)) accessions revealed by high-density SNPs. *Front. Plant Sci.* **2017**, *8*, 2014. [CrossRef]
17. Pavan, S.; Bardaro, N.; Fanelli, V.; Marcotrigiano, A.R.; Mangini, G.; Taranto, F.; Catalano, D.; Montemurro, C.; De Giovanni, C.; Lotti, C.; et al. Genotyping by sequencing of cultivated lentil (*Lens culinaris* Medik.) highlights population structure in the Mediterranean gene pool associated with geographic patterns and phenotypic variables. *Front. Genet.* **2019**, *10*, 872. [CrossRef]
18. Ouji, A.; Suso, M.J.; Rouaissi, M.; Abdellaoui, R.; El Gazzah, M. Genetic diversity of nine faba bean (*Vicia faba* L.) populations revealed by isozyme markers. *Genes Genom.* **2011**, *33*, 31–38. [CrossRef]
19. Ouji, A.; El Bok, S.; Syed, N.H.; Abdellaoui, R.; Rouaissi, M.; Flavell, A.J.; El Gazzah, M. Genetic diversity of faba bean (*Vicia faba* L.) populations revealed by sequence specific amplified polymorphism (SSAP) markers. *Afr. J. Biotechnol.* **2012**, *11*, 2162–2168.
20. Yahia, Y.; Hannachi, H.; Monforte, A.J.; Cockram, J.; Loumerem, M.; Zarouri, B.; Ferchichi, A. Genetic diversity in *Vicia faba* L. populations cultivated in Tunisia revealed by simple sequence repeat analysis. *Plant Genet. Res.* **2014**, 278–285. [CrossRef]
21. Yahia, Y.; Guetat, A.; Walid, E.; Ferchichi, A.; Yahia, H.; Loumerem, M. Analysis of agromorphological diversity of southern Tunisia faba bean (*Vicia faba* L.) germplasm. *Afr. J. Biotechnol.* **2012**, *11*, 11913–11924.
22. Dempsey, G.J. *In Situ Conservation of Crops and Their Relatives: A Review of Current Status and Prospects for Wheat and Maize*; Paper 96-08; CIMMYT NRG: Veracruz, Mexico, 1996.
23. Damania, A.B. History, achievements, and status of genetic resources conservation. *Agron. J.* **2008**, *100*, S-27–S-39. [CrossRef]
24. Maxted, N.; Bennett, S.J. Conservation, diversity and use of Mediterranean legumes. In *Plant Genetic Resources of Legumes in the Mediterranean*; Current Plant Science and Biotechnology in Agriculture; Maxted, N., Bennett, S.J., Eds.; Springer: Dordrecht, The Netherlands, 2001; Volume 39.

25. Altieri, M.A.; Merrick, L. In situ conservation of crop genetic resources through maintenance of traditional farming systems. *Econ. Bot.* **1987**, *41*, 41–86. [CrossRef]
26. Khoury, C.; Laliberté, B.; Guarino, L. Trends in ex situ conservation of plant genetic resources: A review of global crop and regional conservation strategies. *Genet. Resour. Crop Evol.* **2010**, *5*, 57–625. [CrossRef]
27. Enjalbert, J.; Dawson, J.C.; Paillard, S.; Rhoné, B.; Rousselle, Y.; Thomas, M.; Goldringer, I. Dynamic management of crop diversity: From an experimental approach to on-farm conservation. *C. R. Biol.* **2011**, *334*, 458–468. [CrossRef]
28. Mohsenin, N.N. *Physical Properties of Plant and Animal Materials*; Gordon and Breach Science Publishers: New York, NY, USA, 1970.
29. Fulton, T.; Chunwongse, J.; Tanksley, S. Microprep protocol for extraction of DNA from tomato and other herbaceous plants. *Plant Mol. Biol. Rep.* **1995**, *13*, 207–209. [CrossRef]
30. Ma, Y.; Yang, T.; Guan, J.; Wang, S.; Wang, H.; Sun, X.; Zong, X. Development and characterization of 21 EST-derived microsatellite markers in Vicia faba (fava bean). *Am. J. Bot.* **2011**, *98*, e22–e24. [CrossRef]
31. Gong, Y.M.; Xu, S.C.; Mao, W.H.; Hu, Q.Z.; Zhang, G.W.; Ding, J.; Li, Z.Y. Generation and characterization of 11 novel est derived microsatellites from *Vicia faba* (Fabaceae). *Am. J. Bot.* **2010**, *97*, 69–71. [CrossRef]
32. Zeid, M.M.; Mitchell, S.; Link, W.; Carter, W.; Nawar, A.; Fulton, T.; Kresovich, S. Simple sequence repeats (SSRs) in faba bean: New loci from Orobanche-resistant cultivar 'Giza 402'. *Plant Breed.* **2009**, *128*, 149–155. [CrossRef]
33. Peakall, R.; Smouse, P.E. GenAlEx 6.5: Genetic analysis in Excel. Population genetic software for teaching and research an update. *Bioinformatics* **2012**, *28*, 2537–2539. [CrossRef]
34. Tamura, K.; Dudley, J.; Nei, M.; Kumar, S. MEGA4: Molecular Evolutionary Genetics Analysis (MEGA) software version 4.0. *Mol. Biol. Evol.* **2007**, *24*, 1596–1599. [CrossRef]
35. Pritchard, J.K.; Stephens, M.; Donnelly, P. Inference of population structure using multilocus genotype data. *Genetics* **2000**, *155*, 945–959.
36. Evanno, G.; Regnaut, S.; Goudet, J. Detecting the number of clusters of individuals using the software structure: A simulation study. *Mol. Ecol.* **2005**, *14*, 2611–2620. [CrossRef] [PubMed]
37. Earl, D.A.; von Holdt, B.M. Structure harvester: A website and program for visualizing structure output and implementing the Evanno method. *Conserv. Genet. Resour.* **2012**, *4*, 359–361. [CrossRef]
38. Alghamdi, S.S.; Migdadi, H.M.; Ammar, M.H.; Paull, J.G.; Siddique, K.H.M. Faba bean genomics: Current status and future prospects. *Euphytica* **2012**, *186*, 609–624. [CrossRef]
39. Satovic, Z.; Avila, C.M.; Cruz-Izquierdo, S.; Díaz-Ruíz, R.; García-Ruíz, G.M.; Palomino, C.; Gutiérrez, N.; Vitale, S.; Ocaña-Moral, S.; Gutiérrez, M.V.; et al. A reference consensus genetic map for molecular markers and economically important traits in faba bean (*Vicia faba* L.). *BMC Genom.* **2013**, *14*, 932. [CrossRef]
40. Torres, A.M.; Avila, C.M.; Gutierrez, N.; Palomino, C.; Moreno, M.T.; Cubero, J.I. Marker-assisted selection in faba bean (*Vicia faba* L.). *Field crops Res.* **2010**, *115*, 243–252. [CrossRef]
41. Negri, V. Landraces in central Italy: Where and why they are conserved and perspectives for their on-farm conservation. *Genet. Resour. Crop Evol.* **2003**, *50*, 871–885. [CrossRef]
42. Gresta, F.; Avola, G.; Albertini, E.; Raggi, L.; Abbate, V. A study of variability in the Sicilian faba bean landrace 'Larga di Leonforte'. *Genet. Resour. Crop Evol.* **2010**, *57*, 523–531. [CrossRef]
43. Sun, Y.N.; Pan, J.B.; Shi, X.L.; Du, X.Y.; Wu, Q.; Qi, Z.M.; Jiang, H.W.; Xin, D.W.; Liu, C.Y.; Hu, G.H.; et al. Multi-environment mapping and meta-analysis of 100-seed weight in soybean. *Mol. Biol. Rep.* **2012**, *39*, 9435–9443. [CrossRef]
44. Verma, S.; Gupta, S.; Bandhiwal, N.; Kumar, T.; Bharadwaj, C.; Bhatia, S. High-density linkage map construction and mapping of seed trait QTLs in chickpea (*Cicer arietinum* L.) using Genotyping-by-Sequencing (GBS). *Sci. Rep.* **2015**, *5*, 17512. [CrossRef]
45. Verma, P.; Goyal, R.; Chahota, R.K.; Sharma, T.R.; Abdin, M.Z.; Bhatia, S. Construction of a genetic linkage map and identification of QTLs for seed weight and seed size traits in lentil, *Lens culinaris* Medik. *PLoS ONE* **2015**, *10*, e0139666. [CrossRef]
46. Webb, A.; Cottage, A.; Wood, T. A SNP-based consensus genetic map for synteny based trait targeting in faba bean (*Vicia faba* L.). *Plant Biotechnol. J.* **2016**, *14*, 177–185.
47. Suso, M.J.; Pierre, J.; Moreno, M.T.; Esnault, R.; Le Guen, J. Variation in outcrossing levels in faba bean cultivars: Role of ecological factors. *J. Agric. Sci.* **2001**, *136*, 399–405. [CrossRef]
48. Link, W.; Edered, W.; Metz, P.; Buiel, H.; Melchinger, A.E. Genotypic and environmental variation for degree of cross-fertilization in faba bean. *Crop Sci.* **1994**, *34*, 960–964. [CrossRef]

49. Marzinzig, B.; Brünjes, L.; Biagioni, S.; Behling, H.; Link, W.; Westphal, C. Bee pollinators of faba bean (*Vicia faba* L.) differ in their foraging behaviour and pollination efficiency. *Agric. Ecosyst. Environ.* **2018**, *264*, 24–33. [CrossRef]
50. Lawson, D.J.; Van Dorp, L.; Falush, D. A tutorial on how not to over-interpret STRUCTURE and ADMIXTURE bar plots. *Nat. Commun.* **2018**, *9*, 3258. [CrossRef] [PubMed]

MDPI

St. Alban-Anlage 66

4052 Basel

Switzerland

Tel. +41 61 683 77 34

Fax +41 61 302 89 18

www.mdpi.com

Genes Editorial Office

E-mail: genes@mdpi.com

www.mdpi.com/journal/genes

CPSIA information can be obtained
at www.ICGtesting.com
Printed in the USA
LVRC080915250621
690828LV00007BA/337

9 783036 508542